Advances in Experimental Medicine and Biology

Volume 1264

Advances in Experimental Medicine and Biology provides a platform for scientific contributions in the main disciplines of the biomedicine and the life sciences. This series publishes thematic volumes on contemporary research in the areas of microbiology, immunology, neurosciences, biochemistry, biomedical engineering, genetics, physiology, and cancer research. Covering emerging topics and techniques in basic and clinical science, it brings together clinicians and researchers from various fields.

Advances in Experimental Medicine and Biology has been publishing exceptional works in the field for over 40 years, and is indexed in SCOPUS, Medline (PubMed), Journal Citation Reports/Science Edition, Science Citation Index Expanded (SciSearch, Web of Science), EMBASE, BIOSIS, Reaxys, EMBiology, the Chemical Abstracts Service (CAS), and Pathway Studio.

2019 Impact Factor: 2.450 5 Year Impact Factor: 2.324.

More information about this series at http://www.springer.com/series/5584

Eric Murillo-Rodriguez •
S. R. Pandi-Perumal • Jaime M. Monti
Editors

Cannabinoids and Neuropsychiatric Disorders

Springer

Editors
Eric Murillo-Rodriguez
División Ciencias de la Salud
Univ Anahuac Mayab
Merida, Yucatán, Mexico

S. R. Pandi-Perumal
Somnogen Canada Inc.
Toronto, ON, Canada

Jaime M. Monti
Clinics Hospital
Montevideo, Uruguay

ISSN 0065-2598 ISSN 2214-8019 (electronic)
Advances in Experimental Medicine and Biology
ISBN 978-3-030-57371-3 ISBN 978-3-030-57369-0 (eBook)
https://doi.org/10.1007/978-3-030-57369-0

This Springer imprint is published by the registered company Springer Nature Switzerland AG.
The registered company address is: Gewerbestrasse 11, 6330 Cham, Switzerland

Cannabis is the most versatile herbal remedy, and the most useful plant on Earth. No other single plant contains as wide a range of medically active herbal constituents.

Dr. Ethan Russo, Cannabinoid Research Institute

The illegality of cannabis is outrageous, an impediment to full utilization of a drug which helps produce the serenity and insight, sensitivity and fellowship so desperately needed in this increasingly mad and dangerous world.

Carl Sagan

This volume is dedicated to our respective families

Foreword

For most researchers, and certainly for the general population, "cannabis" relates to the plant and its constituents alone. However, since the mid-1980s and early 1990s, research has expanded our knowledge. Today, the cannabinoid field of science covers the cannabinoid receptors, the endocannabinoids (particularly anandamide and 2-AG), their synthetic and degradation pathways, and endogenous anandamide-like compounds, which are fatty acid amides with amino acids or ethanolamines. All these entities are parts of a major new physiological system—the endocannabinoid one. Most probably, the field will expand further.

Plant Cannabinoids While many dozens of plant cannabinoids are known today, most research and acquired knowledge are on Δ^9-tetrahydrocannabinol (THC) and cannabidiol (CBD). CBD was isolated in the 1930s in the labs of Adams in the USA and Todd in the UK, but its structure was elucidated many years later—in 1963; THC was isolated in its pure form, and its structure was elucidated only in 1964. These chemical advances were made many decades after the isolation of morphine and cocaine—the two other major illicit plant constituents. The reason for this discrepancy seems to be the technical difficulty in isolating the plant cannabinoids in their pure form, due to the stupendous mixture of this family of compounds produced by the plant. Modern methods for separation and purification not available previously were needed.

In addition to THC and CBD, there are indications that cannabigerol (CBG) and possibly cannabichromene (CBC) are likewise of medicinal interest. Very little is known about the rest of the plant cannabinoids, except on the cannabinoid acids, which are the precursors of the neutral cannabinoids. These acids are not stable, which seems to be the main reason why their biological properties were not well investigated. However, recently CBD acid was stabilized (by esterification to its methyl ester). It seems to parallel CBD in its actions. We already know that it is a potent anti-nociceptive, antiemetic, and anxiolytic compound. Shall we see it in the market, like CBD, in all kinds of industrial-prepared foods and beauty lotions? I hope not.

Most "medical cannabis" sold today is in the form of mixtures in which the amounts of specific cannabinoids vary. Can we guess where we shall be with such mixtures or pure cannabinoids in about a decade from now? Given the huge market today, the mixtures, as well as pure CBD, will probably be still around. However, we can expect to have better-defined mixtures, as well as semi-synthetic CBD and CBG derivatives, as drugs in many areas. Numerous pharmaceutical companies have cannabinoid programs. In addition to the pain and anxiety mentioned above, we shall probably see synthetic and semi-synthetic cannabinoids in additional areas of psychiatry and neurology as well as, presumably, in gastroenterology and immunology.

Endogenous Cannabinoids The discovery of a receptor in the 1980s led to the isolation of endogenous cannabinoids (endocannabinoids) in the 1990s. Two of these, anandamide and 2-AG, have been the topic of thousands of publications. We have learned much about their chemistry, including the syntheses and degradation of these molecules in the animal body, as well as their bioactivities. The endocannabinoid system has turned out to be a central one in animal physiology. Indeed in a recent review, it was stated that "...modulating endocannabinoid activity may have therapeutic potential in almost all diseases affecting humans." Even the dopaminergic or cholinergic systems have not been so described.

What are the research pathways ahead of us in this area?

A. Will the endocannabinoids be investigated in humans? More than 25 years after their discovery human studies are almost unavailable!
B. Shall we see additional endocannabinoids, which have not been isolated so far? They may differ in their activity from anandamide and 2-AG.
C. Has the activity of endocannabinoids been looked into in all animal biochemical systems?
D. Do we know enough about the role of the endocannabinoids in our emotions and personality?
E. Can we expect to see endocannabinoid derivatives as drugs?

Anandamide-like Endogenous Molecules The biosynthesis of anandamide is based on fatty acid (arachidonic acid) and amino acid derivatives (an ethanolamine). The animal body has numerous fatty acids and amino acids, and indeed, it uses the established biosynthetic pathway of anandamide for the synthesis of many additional, chemically related molecules, most of which do not bind to the cannabinoid receptors. Over the last two decades, several groups have investigated these anandamide-like endogenous molecules. A few examples of such compounds (tested only in mice and rats so far) are as follows:

Arachidonoyl serine is neuroprotective after brain trauma. It causes vasodilation, thus allowing better blood flow into damaged areas.

Oleoyl serine acts on osteoblasts and prevents bone loss in osteoporosis by increasing bone formation and restraining bone resorption.

Oleoyl glycine has powerful anti-nicotine addiction properties. It blocks the establishment of nicotine place preference—a test for addiction formation—and reduces withdrawal responses in nicotine-dependent mice. In morphine-dependent rats, it was also found to reduce withdrawal responses but did not affect morphine addiction, thus demonstrating selectivity.

These are just a few examples. Many other anandamide-like compounds are present in the animal body and act in numerous biological processes. Indeed, it has been speculated that the huge number of such compounds—the concentration levels of which may differ from person to person—may be involved in the personality differences.

I would like to end with a look at the future of cannabinoid drugs—as seen from afar. At present, most patients who use cannabinoid-based drugs are prescribed "medical marijuana"—a term that from a medical point of view is unacceptable. "Medical marijuana" reaching the public has to be better defined as regards constituents, whose levels in many cases are not even mentioned. The level of constituents in cannabis depends to a large degree not only on the genetics of the plant but also on the conditions under which it was grown. Hence, today consumption of "medical marijuana" is to a large extent a medical gamble. As mentioned above, I believe that in most countries, within the next few years, strict regulations will be enacted, so that patients will always be able to get the same material as regards constituents.

A second point—many of the drugs we use today are derivatives of natural products. Thus, we have not prescribed cortisone (an important hormone), but derivatives of cortisone. Such derivatives are better suited to be used as drugs than natural constituents are. It seems reasonable to expect that within a decade pharmaceutical companies will develop derivatives of CBD and THC, and possibly CBG, which will be used as novel drugs. We may also have synthetic drugs, unrelated to the plant cannabinoids, which bind to the cannabinoid receptor, particularly to the CB2 receptor, whose activation does not lead to marijuana-like activity.

In summary, I assume that within a decade we shall have both new cannabinoid drugs and well-defined extracts, used in parallel. Let us hope so.

Hebrew University, Medical Faculty,
Pharmacy School, Institute for Drug
Research, Jerusalem, Israel

Raphael Mechoulam

Preface

The editors are pleased to present the first edition of *Cannabinoids and Neuropsychiatric Disorders*, which has been included in the prestigious *Advances in Experimental Medicine and Biology* series (volume 1264). As editors, we are very happy about this decision as our volume fits perfectly in this landmark biomedicine and the life sciences series.

The plant *Cannabis sativa* has been used both recreationally and medicinally for thousands of years; it was only in 1964 that chemists YehielGaoni and Raphael Mechoulam at the Hebrew University of Jerusalem identified and isolated the psychoactive components in cannabis, Δ 9-tetrahydrocannabinol (Δ 38 9-THC; Gaoni and Mechoulam 1964).

To give an overview on this subject, we have included nine chapters. The topics covered include the constituents of *Cannabis sativa*, the molecular mechanism of cannabis and its neuropharmacological effects, emerging roles of cannabinoids and synthetic cannabinoids in clinical medicine, and exploring the use of cannabis in neuropsychiatric disorders.

We are privileged to have compiled this volume. During the course of our assignment, we learned much in the process of editing this important volume. We sincerely hope that the readers will find this volume uniquely valuable as a research and clinical resource. We sincerely hope that our volume will be useful to researchers and practicing clinicians.

Merida, Mexico — Eric Murillo-Rodriguez
Toronto, Canada — S. R. Pandi-Perumal
Montevideo, Uruguay — Jaime M. Monti

Reference

Gaoni Y, Mechoulam R (1964) Isolation, structure, and partial synthesis of an active constituent of hashish. J Am Chem Soc 86(8):1646–1647. https://doi.org/10.1021/ja01062a046

Acknowledgments

Cannabinoids and Neuropsychiatric Disorders provides scientific and medical information on cannabis to all healthcare workers interested in basic, translational, and clinical medicine. It is our pleasure to acknowledge the contributions of those who were instrumental in the production of this book.

Our sincere appreciation goes to Prof.Raphael Mechoulam at the Hebrew University of Jerusalem who identified and isolated the psychoactive components in cannabis, Delta-9-tetrahydrocannabinol (Δ9-THC), who agreed to write the foreword. We wish to express our appreciation for his contribution.

We would like to express our deep appreciation to all the contributors for their scholarly contributions that facilitated the development of this volume. These authors have done a superb job of producing authoritative chapters that synthesize vast amounts of scientific and clinical data to create informative chapters. The expertise of contributors to *Cannabinoids and Neuropsychiatric Disorders* reflects the broad diversity and knowledge concerning cannabis research, which has continued to grow over the last several decades. These authors represent the cutting edge of basic and applied research and provide the most recent information regarding how such knowledge can be utilized in clinical settings. Their informed opinions and insights have significantly contributed to our scientific understanding of cannabinoids and have provided important interpretations regarding future research directions.

The highly talented people of Springer USA made this project an especially pleasurable one. We were delighted to have the professional and highly enthusiastic support of Dr. Beatrice Menz, Senior Editor, Springer Nature, Switzerland AG. Without her continuous and unstinting support, this volume would not have been possible.

It was a pleasure to work with the entire production team of Springer. Their guidance, technical expertise, and commitment to excellence were invaluable. We wish to acknowledge the help of Amrei Strehl, Senior Editor, Springer Vienna, Austria; Coral Zhou, Project Coordinator, Springer Nature, Beijing City, China; Mr. Daniel Ignatius Jagadisan, Project Coordinator (Books), Springer Nature, India; and Mahalakshmi Rajendran of Spi Global, Chennai, India.

Finally, and most importantly, we want to thank our spouses and families for their support and understanding during the development of this book.

Eric Murillo-Rodriguez
S. R. Pandi-Perumal
Jaime M. Monti

Contents

Constituents of *Cannabis Sativa*

1

Erin M. Rock and Linda A. Parker

Abstract

The *Cannabis sativa* plant has been used medicinally and recreationally for thousands of years, but recently only relatively some of its constituents have been identified. There are more than 550 chemical compounds in cannabis, with more than 100 phytocannabinoids being identified, including Δ^9-tetrahydrocannabinol (THC) and cannabidiol (CBD). These phytocannabinoids work by binding to the cannabinoid receptors, as well as other receptor systems. Also within cannabis are the aromatic terpenes, more than 100 of which have been identified. Cannabis and its constituents have been indicated as therapeutic compounds in numerous medical conditions, such as pain, anxiety, epilepsy, nausea and vomiting, and post-traumatic stress disorder. This chapter provides an overview of some of the biological effects of a number of the cannabinoids and terpenes, as well as discussing their known mechanisms of action and evidence of potential therapeutic effects.

E. M. Rock · L. A. Parker (✉)
Department of Psychology and Collaborative Neuroscience Program, University of Guelph, Guelph, ON, Canada
e-mail: parkerl@uoguelph.ca

Keywords

Cannabis sativa · Phytocannabinoid · Terpene · Δ^9-tetrahydrocannabinol · Cannabidiol · Cannabinoid receptors

1.1 Introduction

Although *Cannabis sativa* has been used both recreationally and medicinally for thousands of years, it was only in 1964 that chemists Yehiel Gaoni and Raphael Mechoulam at the Hebrew University of Jerusalem identified and isolated the psychoactive component in cannabis, Δ^9-tetrahydrocannabinol (Δ^9-THC; Gaoni and Mechoulam 1964). This discovery of the psychoactive component allowed for scientific investigations of how the plant produced its psychotropic effects. It was not until about 20 years later when Allyn Howlett's group at the St. Louis University Medical School, discovered the target for Δ^9-THC, the cannabinoid 1 (CB_1) receptor (Devane et al. 1988; Howlett et al. 1986). Shortly thereafter, Mechoulam's group discovered the endogenous cannabinoids (eCBs) anandamide (AEA; Devane et al. 1992) and 2-arachidonyl glycerol (2-AG; Mechoulam et al. 1995; Sugiura et al. 1995), which also act on CB_1 receptors. Additionally, a second cannabinoid receptor was discovered, CB_2, which was found primarily in the peripheral immune system (Munro et al. 1993). The original identification of the

E. Murillo-Rodriguez et al. (eds.), *Cannabinoids and Neuropsychiatric Disorders*, Advances in Experimental Medicine and Biology 1264, https://doi.org/10.1007/978-3-030-57369-0_1

psychoactive component of cannabis by Mechoulam's group led to the discovery of a whole new system that is crucially involved in regulatory functions of health and disease.

1.2 Constituents of *Cannabis sativa*

Over the last few decades, the total number of compounds identified from cannabis has risen. To date, 554 compounds in cannabis have been identified, including 113 phytocannabinoids (Ahmed et al. 2015; ElSohl and Gul 2014) and 120 terpenes (ElSohly and Slade 2005). The cannabinoids identified from cannabis are of 11 types (ElSohl and Gul 2014; Mechoulam 2005): Δ^9-THC, Δ^8-THC, cannabidiol (CBD), cannabigerol (CBG), cannabichromene (CBC), cannabinodiol (CBND), cannabielsoin (CBE), cannabicyclol (CBL), cannabinol (CBN), cannabitriol (CBT), and miscellaneous cannabinoids. The concentration of cannabinoids within the plant is dependent upon growing conditions such as moisture, temperature, soil nutrients, and UV radiation (see Pate 1994 for review). Over the past two decades, the Δ^9-THC content of recreational cannabis has risen dramatically, while the CBD content has remained stable or decreased to negligible levels. ElSohly et al. (2016) indicate that the Δ^9-THC content of recreational cannabis in the United States has risen from 4% in 1995 to 12% in 2014. In contrast, the cannabis supplied to researchers by the National Institute of Drug Abuse has typically contained less than 4%. This suggests that today's cannabis differs considerably from the cannabis that was available years ago, both in its effects on mental health and cognitive functions. The focus of this chapter is to provide a description of the biological effects of some of the identified cannabinoids and terpenes, and discuss some of their mechanisms of action and therapeutic effects.

1.2.1 Cannabinoids

The term cannabinoid usually refers to the chemical substances isolated from the cannabis plant, which possess the typical C_{21} terpenophenolic skeleton. In addition, this term encompasses their derivatives as well, with the term phytocannabinoid referring to those compounds originating from the plant. Phytocannabinoids show different affinities for the CB_1 and CB_2 receptors, with other molecular targets also being identified (for an excellent review see Morales et al. 2017).

1.2.1.1 THC-Type Cannabinoids

Δ^9-THC

Δ^9-THC and cannabinoids similar in structure to Δ^9-THC have been the most extensively studied cannabinoids. After its identification by Gaoni and Mechoulam, Δ^9-THC was then tested for cannabinoid activity in rhesus monkeys, dogs, gerbils, mice, and rats (Edery et al. 1971; Grunfeld and Edery 1969; Mechoulam et al. 1970), and only Δ^9-THC was found to produce the typical psychoactive effects of cannabis. Some of the effects produced by Δ^9-THC in these early animal studies were severe motor disturbances, redness of the mucous membrane that covers the eyeball, slow movements, decline of aggression, sleepy state, and decreased spontaneous locomotion. Subsequently, Roger Pertwee (1972) tested Δ^9-THC in the ring test, a quantitative in vivo assay for catalepsy (muscular rigidity and fixed posture), and concluded that indeed Δ^9-THC was producing catalepsy. Billy Martin's group (Martin et al. 1991) then incorporated the ring test and three other bioassays into the mouse tetrad assay, including catalepsy, hypokinesia (inactivity), hypothermia (reduced body temperature), and antinociception (pain relief). The mouse tetrad assay is commonly used to screen for psychotropic cannabinoids. Δ^9-THC is a partial agonist at the CB_1 and CB_2 receptors.

One of the most common uses of medical cannabis is to treat pain. Indeed, Δ^9-THC has been shown to reduce acute and chronic pain (for a review see Costa and Comelli 2014), especially neuropathic pain (Ware et al. 2010; Wilsey et al. 2008, 2013). Δ^9-THC also displays synergistic effects for most opioid drugs (e.g. Abrams et al. 2011; Cichewicz et al. 1999; Lynch and Clark 2003), suggesting opioid-sparing effects. In addition, Δ^9-THC also eliminates the nightmares of traumatic events and improves sleep (Babson et al. 2017; Pedersen and Sandberg 2013), particularly in war veterans suffering from post-traumatic stress disorder (e.g. Betthauser et al. 2015; Jetly et al. 2015). Δ^9-THC has also shown beneficial effects on Tourette's syndrome (tic reduction; Müller-Vahl et al. 2002, 2003), appetite stimulation in patients with advanced cancer (Nelson et al. 1994), and reduction of nausea and vomiting in chemotherapy patients (e.g. Chang et al. 1979; Frytak et al. 1979).

Animal models also suggest a therapeutic potential for Δ^9-THC in a number of conditions. Δ^9-THC reduced inflammation and in vitro motility disturbances in rat colitis (Jamontt et al. 2010). Rodent studies demonstrate a biphasic effect of Δ^9-THC on anxiety-related behaviors such that low doses reduce anxiety, while high doses produce anxiogenic effects (Hill and Gorzalka 2004). Δ^9-THC produces antidepressant (e.g. Bambico et al. 2012), antinausea (Parker et al. 2003), and antiemetic effects (Cluny et al. 2008; Parker et al. 2004) in animal models. Δ^9-THC also reduces neurological deficits in animal models of neurodegeneration (e.g. Louw et al. 2000), delays motor impairment, increases survival in a mouse model of Amyotrophic Lateral Sclerosis (Raman et al. 2004), and improves activity and hand-eye coordination in animal models of Parkinson's Disease (van Vliet et al. 2008). Clearly, Δ^9-THC has a number of therapeutic effects, likely with more medicinal potential that is yet to be discovered.

Δ^8-THC

Small quantities of Δ^8-THC (Hively et al. 1966) and Δ^8-THC acid (Hanuš and Krejčí 1975) have also been identified in cannabis. Δ^8-THC has been shown to be a partial agonist at the CB_1 and CB_2 receptors (Huffman et al. 1999; Razdan 1986). Δ^8-THC has been used in children undergoing chemotherapy to prevent vomiting, with few reported side effects (Abrahamov et al. 1995). Low doses of Δ^8-THC (0.001 mg/kg) have been shown to increase food consumption in mice, but produced an overall decrease in body weight, without typical cannabinoid side effects (Avraham et al. 2004). Furthermore, Δ^8-THC has also been shown to cause a decrease in body weight in female rats (without impacting food intake; Sjödén et al. 1973), suggesting it may be beneficial in weight loss.

Synthetic Δ^9-THC

Two synthetic analogs of Δ^9-THC have been approved by the US Food and Drug Administration in the form of capsules that may be prescribed for chemotherapy-induced nausea and vomiting: nabilone (Cesamet, Valeant Pharmaceuticals North America) and dronabinol (Marinol; Solvay Pharmaceuticals). Indeed, early clinical studies (Einhorn et al. 1981; Herman et al. 1977, 1979) demonstrated the efficacy, safety, and tolerability of nabilone in reducing nausea and vomiting in cancer patients. Nabilone reduced vomiting frequency and nausea severity in 77% of patients (Herman et al. 1977), demonstrating its efficacy as rescue or adjunct therapy for cancer patients. Dronabinol is currently being evaluated for its analgesic properties in patients with bone metastases from breast cancer (early phase I study; NCT03661892), and as an adjunct therapy to opiates in patients with chronic pain (NCT00153192).

Nabiximols (Sativex, GW Pharmaceuticals), the cannabis-based medicine containing approximately equal amounts of Δ^9-THC and the nonintoxicating cannabinoid cannabidiol (CBD), is administered as a sublingual spray, and is approved in Canada for the relief of Multiple Sclerosis (MS) or cancer pain and to reduce MS spasticity (Mechoulam et al. 2014).

Δ^9-THC-Acid (THCA)

Other identified THC-type compounds in the plant are not psychoactive but may have therapeutic potential. The carboxylic acidic precursor of THC, Δ^9-THC-acid (THCA; Mechoulam et al. 1969), is decarboxylated to Δ^9-THC by heating (smoking and baking), as well as storage, at room temperature. Indeed, storage at 4 °C resulted in instability after 1 month, with 91% still detectable when THCA was stored in methanol, and 68% still detectable when stored in chloroform (Smith and Vaughan 1977). Also, even when stored at −18 °C, THCA was still lost (Smith and Vaughan 1977). The stability of THCA is improved in olive oil, (with 78% of THCA detectable after 10 days at 25°), over that of ethanol (with only 33% detectable; Citti et al. 2016). Interestingly, THCA produced no psychoactive effects when administered to rhesus monkeys at doses up to 5 mg/kg (intravenously, i.v.), to mice at doses up to 20 mg/kg (intraperitoneally, i.p.), and to dogs at doses up to 7 mg/kg (Grunfeld and Edery 1969).

Clinical interest in THCA is growing due to its apparent lack of psychoactivity (Grunfeld and Edery 1969; Edery et al. 1972), which may be because of its reported low-binding affinity at CB_1. The affinity studies of THCA at the CB_1 receptor are mixed, with reports of equivalent (Rosenthaler et al. 2014) or weaker (Verhoeckx et al. 2006) binding in comparison to Δ^9-THC, or lack of affinity for the CB_1 receptor (Ahmed et al. 2008; Husni et al. 2014). It is suggested that this disparity in binding affinity may be due to the inherent sample contamination of THCA's decarboxylation into Δ^9-THC (Edery et al. 1972; McPartland et al. 2017).

An excellent review by Moreno-Sanz (2016) has discussed THCA's molecular targets, which include phospholipids' metabolism, prostaglandins' metabolism, transient receptor potential (TRP) channel signaling, anandamide, and 2-arachidonoylglycerol signaling. Rock et al. (2013a) reported that THCA is 10 times more potent than Δ^9-THC in reducing nausea and vomiting in animal models, an effect that was blocked by the CB_1 receptor antagonist/inverse agonist SR141716. However, THCA did not induce the classic CB_1 receptor-mediated effects such as hypothermia or reduced motor activity (Rock et al. 2013a) and only THCA (not Δ^9-THC) was detected in the plasma of rats treated with THCA, suggesting that it may be acting at a peripheral rather than at the central site of action. These findings suggest that THCA may be a more desirable therapeutic treatment than Δ^9-THC for nausea and vomiting due to its increased potency and lack of psychoactive properties.

Tetrahydrocannabivarin

Tetrahydrocannabivarin (THCV), identified in the 1970s (Gill 1971; Merkus 1971), was initially thought to share Δ^9-THC's catalepsy effects in mice and to produce mild psychoactive effects in humans (Hollister 1974). These effects have since been shown to be dose-dependent, with such effects only occurring at very high doses, while at low doses, THCV acts as a neutral antagonist at the CB_1 receptor (Pertwee 2005; Pertwee et al. 2007). Indeed, THCV reduces food intake and body weight at low doses (like the CB_1 receptor antagonist/inverse agonist SR141716; Riedel et al. 2009). THCV is also a partial agonist at the CB_2 receptor (Bolognini et al. 2010). Interestingly, in animal models, unlike SR141716, THCV does not produce nausea (Rock et al. 2013b) or anxiety-like behavior (O'Brien et al. 2013), and at a high dose (10 mg.kg, i.p.) actually reduces nausea (Rock et al. 2013b). As anxiety and nausea were two side effects produced by the inverse agonism of the CB_1 receptor with SR141716, these results suggest that THCV may be a useful weight loss treatment, devoid of the negative side effects of SR141716.

1.2.1.2 Cannabidiol-Type Cannabinoids

Cannabidiol

The primary nonpsychoactive cannabinoid of cannabis (particularly in hemp) is cannabidiol (CBD), which was first isolated from Mexican marijuana by Adams et al. (1940). In 1963, Mechoulam and Shvo isolated CBD from Lebanese hashish and established its structure (Mechoulam and Shvo 1963). CBD lacks the psychotropic effects of Δ^9-THC and has great therapeutic potential. Unlike Δ^9-THC, CBD does not activate the CB_1 and CB_2 receptors, likely explaining CBD's lack of psychoactive effects. Instead, CBD acts through multiple mechanisms. At very low (nanomolar to micromolar) concentrations, CBD acts as an antagonist at the orphan G-protein-coupled receptor GPR55, and the transient receptor potential of the melastatin type-8 (TRPM8) channel (Pertwee 2008). CBD is also an agonist at the nuclear peroxisome proliferator-activated receptor-γ (PPAR-γ), and the transient receptor potential of vanilloid types 1 (TRPV1) and 2 (TRPV2) channels (Bisogno et al. 2001). Cannabidiol also acts as a noncompetitive CB_1 receptor antagonist, as well as an inverse agonist at the CB_2 receptor (Thomas et al. 2007) Furthermore, cannabidiol inhibits the degradation of the endogenous cannabinoid anandamide (Bisogno et al. 2001). Finally, Russo et al. (2005) were the first to suggest that CBD also acts as an agonist at a specific serotonin receptor, $5\text{-}HT_{1A}$.

CBD has anxiolytic, antipsychotic, and neuroprotective properties. There is also evidence suggesting its potential use in epilepsy, substance abuse and dependence, schizophrenia, social phobia, post-traumatic stress, depression, bipolar disorder, sleep disorders, and Parkinson's disease (for a recent review see Crippa et al. 2018).

Preclinical animal models suggest that CBD has beneficial effects such as reversing cognitive deficits in mouse models of Alzheimer's disease (Cheng et al. 2014), reducing nausea in rats and vomiting in shrews (Rock et al. 2012), attenuating Δ^9-THC's debilitating effect on cognition in rhesus monkeys (Jacobs et al. 2016), producing anxiolytic-like effects in rats (Guimarães et al. 1990), and producing antidepressant-like effects in mice in the forced swim test (Zanelati et al. 2010).

One of CBD's most promising therapeutic effects is its use as an anticonvulsant drug, especially for children with epileptic syndromes (for an excellent review see Fraguas-Sánchez and Torres-Suárez 2018). Indeed, for Dravet syndrome (early-onset encephalopathic epilepsy with a high mortality rate), a recent randomized, controlled trial showed that CBD reduced convulsive-seizure frequency among children and young adults with drug-resistant Dravet syndrome (Devinsky et al. 2017). Furthermore, results from an ongoing expanded-access program showed that CBD may be an effective long-term treatment option for treatment-resistant epilepsy (Szaflarski et al. 2018). In fact, an oral solution based on pure plant-derived CBD (Epidiolex®) (NCT02397863) has been recently approved in the United States for the treatment of childhood epileptic syndromes such as Dravet syndrome and Lennox–Gastaut syndrome in patients 2 years of age and older. Interestingly, although Mechoulam's group (Cunha et al. 1980) demonstrated the antiepileptic effects of CBD, it has taken quite some time to reach approval by the United States Food and Drug Administration.

CBD also has beneficial effects in a number of other conditions. In Parkinson's patients, CBD ameliorated motor symptoms and improved the quality of life (Chagas et al. 2014). CBD also decreases anxiety for public speaking in socially anxious individuals (Bergamaschi et al. 2011; Crippa et al. 2011). CBD reduces the detrimental effects of Δ^9-THC on cognition (Bhattacharyya et al. 2010). CBD (when added to antipsychotic medications) lowers positive psychotic scores in patients with schizophrenia (McGuire et al. 2018). Likely due to its multiple mechanisms of action, CBD seems to have great therapeutic potential without the adverse psychoactive effects associated with Δ^9-THC.

Cannabidiolic Acid

Cannabidiolic acid (CBDA) is the nonpsychoactive precursor of CBD that is present in the fresh cannabis plant (particularly in its industrial hemp forms). It slowly decarboxylates (that is, loses its acidic function) in response to heating (e.g. when marijuana is smoked). In 2018, the cannabinoid content was analyzed in 15 cannabis strains, with CBDA being detected in 13/15 of these strains, with the percentage ranging from 0.1–12.6%, whereas CBD was detected in only 4/15 strains with the percentage ranging from 0.1–11.4% (Baron et al. 2018). Recently, the analysis of 200 cannabis oils detected CBDA concentrations ranging from 0 to 6 mg/ml (Carcieri et al. 2018). A recent study evaluated the pharmacokinetics of oral cannabis, with CBDA having a much higher peak serum concentration than that of CBD, suggesting that much higher levels of CBDA than CBD are present after oral consumption (Pellesi et al. 2018). Clearly, more research is needed on CBDA. To date, no controlled clinical trials with CBDA have been published.

Recent rodent studies indicate that CBDA may be 100–1000 times more potent than CBD in reducing toxin-induced vomiting and nausea in animal models. It may be particularly effective in treating the side effect of anticipatory nausea (for which no selective treatment is currently available) in chemotherapy patients (Bolognini et al. 2013; Rock et al. 2014a, b). Interestingly, the doses of THC or CBDA that are ineffective alone, when given in combination, become particularly effective as a treatment for acute nausea and vomiting in animal models (Rock and Parker 2015; Rock et al. 2015, 2016). CBDA has also been shown to prevent stress-induced anxiogenic-like responding in rodents (an anxiolytic-like effect; Rock et al. 2017). In addition, CBDA (as well as very low doses of combined CBDA and THC) has anti-inflammatory effects and reduces enhanced pain sensation in an animal model of acute inflammation (Rock et al. 2018). Finally, CBDA also inhibits highly aggressive human breast cancer cell migration (Takeda et al. 2012). Taken together, these results suggest an important role for CBDA in cancer treatment, acting not only to reduce the symptoms of nausea and vomiting but also to reduce cancer cell migration (an important factor in cancer metastasis), as well as reducing stress-induced anxiety and pain.

Cannabidivarin

Cannabidivarin (CBDV) lacks psychoactive properties and is a very weak agonist at the CB_1 and CB_2 receptors (Hill et al. 2013; Rosenthaler et al. 2014), and the TRPV1, TRPV2, and TRPV3 cation channels (De Petrocellis et al. 2011, 2012). CBDV reduces behavioral alterations and brain atrophy in a mouse model of Rett syndrome, a rare neurodevelopmental disorder (Vigli et al. 2018). CBDV has been shown to have antiepileptic action (Amada et al. 2013; Hill et al. 2012, 2013), as well as antinausea potential in animal models (Rock et al. 2013b). In fact, an ongoing phase II double-blind, placebo-controlled trial is assessing the efficacy and safety of CBV for controlling focal seizures in adults (NCT02365610).

Cannabigerol

Cannabigerol is a nonpsychoactive phytocannabinoid (Izzo et al. 2009) with low affinity for the cannabinoid CB_1 and CB_2 receptors (Rosenthaler et al. 2014), and has also been shown to block the $5\text{-}HT_{1A}$ receptor (Cascio et al. 2010). In fact, CBG dose-dependently blocked the CBD-induced suppression of nausea in rats and vomiting in shrews, but on its own reduced nausea at a low dose (Rock et al. 2011). In addition, CBG also acts as a weak TRPV1 agonist and TRPV2 agonist, a potent TRPM8 antagonist, and a potent TRPA1 agonist (De Petrocellis et al. 2008, 2011). CBG has been shown to have anti-inflammatory and neuroprotective effects in neurodegenerative disease models (Borrelli et al. 2013; Gugliandolo et al. 2018; Valdeolivas et al. 2015), suggesting that it may be a potential treatment against neuroinflammation and oxidative stress. CBG has also been shown to increase food intake in rats (Brierley et al. 2016, 2017) and enhance the liking of sweet saccharin in the taste reactivity test (O'Brien et al. 2013). These data suggest that CBG may have potential as a treatment for cancer patients, possibly reducing nausea, stimulating appetite, and reducing inflammation.

1.2.2 Terpenes

It is the terpenes in cannabis that cause the plant's aroma and reported "flavor". Terpenes are the odorous compounds present in essential oils. More than 100 terpenes have been identified in *C. sativa* (Brenneisen 2007; Rothschild et al. 2005), but the most terpenes to be identified in a single plant sample is 40, although many more terpenes are simply unnamed (Merli et al. 1980). It is possible that there may be unnamed terpenes that are unique to *C. sativa*. The presence and distribution of terpenes vary in *C. sativa*, due to the process of obtaining the essential oil, environmental growing conditions, or plant maturity when harvested (Brenneisen 2007; Meier and Mediavilla 1998). Pre-clinical evidence indicates that terpenes may have therapeutic potential. D-limonene, β-myrcene, and α-pinene are some of the most common terpenes in *C. sativa*. The literature suggests that terpenes may also act synergistically with cannabinoids to produce beneficial effects (see Russo 2011 for review). Indeed, combinations of cannabinoids and terpenes could provide promising therapeutic tools, which may ultimately reveal why people attribute relief from certain symptoms to particular cannabis strains.

1.2.2.1 D-Limonene

Present in cannabis, and also common in lemons and other citrus fruits, D-limonene, is a terpenoid that has been studied very little in *C. sativa*. It has been shown to have potent anticancer, anxiolytic, and immunostimulating properties in humans (Komori et al. 1995). D-Limonene has also been shown to have antifungal and antibacterial properties (Uemura et al. 1997). More recently, D-limonene has been shown to have anxiolytic effects mediated by serotonin and dopamine in the prefrontal cortex and hippocampal region of mice (Yun 2014). Future in vivo research with this terpene may reveal further therapeutic potential.

1.2.2.2 β-Myrcene

β-myrcene is a prominent terpene in *C. sativa*. Myrcene has shown analgesic effects in mouse models (de Cássia da Silveira et al. 2017), anti-inflammatory activity (Russo 2011), antibiotic properties (McPartland and Russo 2001), and anxiolytic properties (Cleemput et al. 2009). These results suggest that β-myrcene may contribute to these classic therapeutic effects seen with whole-plant cannabis.

1.2.2.3 α-Pinene

α-pinene is present in cannabis as well as conifers, exerting anti-inflammatory effects (Kim et al. 2015), and inhibiting prostate cancer growth in a xenograft mouse model (Zhao et al. 2018). In addition, α-pinene has been shown to be an acetylcholinesterase inhibitor, suggesting it may modulate cognitive effects (Kennedy et al. 2011), which could counteract THC-induced memory deficits. A recent study suggests that it is, however, devoid of anticonvulsant action in a mouse model (Felipe et al. 2018).

1.2.2.4 β-Caryophyllene

The terpene β-caryophyllene, which is a major compound of *C. sativa* essential oil, is also a well-known active principle of black pepper. β-caryophyllene produces effects in preclinical models such as antidepressant-like effects in mice (Bahi et al. 2014; de Oliveira et al. 2018), decreased seizures in mice (Tchekalarova et al. 2011), alleviation of ischemic brain damage (Yang et al. 2017), reduction of peripheral neuropathy in mice (Segat et al. 2017), interference with motor paralysis and neuroinflammation in a mouse model of Multiple Sclerosis (Alberti et al. 2017), and reduction of anxiety-like behavior in mice (Bahi et al. 2014). β-caryophyllene also possesses anti-inflammatory and gastric cytoprotective properties (Singh and Sharma 2015). Interestingly, it has been shown to bind to the CB_2 receptor and could therefore actually be considered as a phytocannabinoid (Gertsch et al. 2008).

1.2.3 Conclusions

In the following chapters, the effects of cannabinoid constituents on various neuropsychiatric disorders will be described. As the cannabinoid field evolves, additional cannabinoid constituents may be identified and their therapeutic potential may be revealed. Furthermore, the beneficial effects of terpenes, alone and in combination with cannabinoids may be realized as more terpenes are identified and named. As more investigators access these compounds for scientific investigation, more of the beneficial effects of this plant may come to light.

References

Abrahamov A, Abrahamov A, Mechoulam R (1995) An efficient new cannabinoid antiemetic in pediatric oncology. Life Sci 56:2097–2102

Abrams DI, Couey P, Shade SB et al (2011) Cannabinoid-opioid interaction in chronic pain. Clin Pharmacol Ther 90:844–851. https://doi.org/10.1038/clpt.2011.188

Adams R, Hunt M, Clark JH (1940) Structure of cannabidiol, a product isolated from the marihuana extract of Minnesota wild hemp. I. J Am Chem Soc 62:196–200. https://doi.org/10.1021/ja01858a058

Ahmed SA, Ross SA, Slade D et al (2008) Cannabinoid ester constituents from high-potency Cannabis sativa. J Nat Prod 71:536–542. https://doi.org/10.1021/np070454a

Ahmed SA, Ross SA, Slade D et al (2015) Minor oxygenated cannabinoids from high potency Cannabis sativa. Phytochemistry 117:194–199. https://doi.org/10.1016/j.phytochem.2015.04.007

Alberti TB, Barbosa WL, Vieira JL et al (2017) (−)-β-Caryophyllene, a CB2 receptor-selective phytocannabinoid, suppresses motor paralysis and neuroinflammation in a murine model of multiple sclerosis. Int J Mol Sci 18:E691. https://doi.org/10.3390/ijms18040691

Amada N, Yamasaki Y, Williams CM et al (2013) Cannabidivarin (CBDV) suppresses pentylenetetrazole (PTZ)-induced increases in epilepsy-related gene expression. Peer J 1:e214. https://doi.org/10.7717/peerj.214

Avraham Y, Ben-Shushan D, Breuer A et al (2004) Very low doses of Δ8-THC increase food consumption and alter neurotransmitter levels following weight loss. Pharmacol Biochem Behav 77:675–684. https://doi.org/10.1016/j.pbb.2004.01.015

Babson KA, Sottile J, Morabito D (2017) Cannabis, cannabinoids, and sleep: a review of the literature. Curr Psychiatry Rep 19:23. https://doi.org/10.1007/s11920-017-0775-9

Bahi A, Al Mansouri S, Al Memari E et al (2014) β-Caryophyllene, a CB2 receptor agonist produces multiple behavioral changes relevant to anxiety and depression in mice. Physiol Behav 135:119–124. https://doi.org/10.1016/j.physbeh.2014.06.003

Bambico FR, Hattan PR, Garant JP et al (2012) Effect of delta-9-tetrahydrocannabinol on behavioral despair and on pre- and postsynaptic serotonergic transmission. Prog Neuro-Psychopharmacol Biol Psychiatry 38:88–96. https://doi.org/10.1016/j.pnpbp.2012.02.006

Baron EP, Lucas P, Eades J et al (2018) Patterns of medicinal cannabis use, strain analysis, and substitution effect among patients with migraine, headache, arthritis, and chronic pain in a medicinal cannabis cohort. J Headache Pain 19:37. https://doi.org/10.1186/s10194-018-0862-2

Bergamaschi MM, Queiroz RH, Chagas MH et al (2011) Cannabidiol reduces the anxiety induced by simulated public speaking in treatment-naive social phobia patients. Neuropsychopharmacology 36:1219–1226. https://doi.org/10.1038/npp.2011.6

Betthauser K, Pilz J, Vollmer LE (2015) Use and effects of cannabinoids in military veterans with posttraumatic stress disorder. Am J Health Syst Pharm 72:1279–1284. https://doi.org/10.2146/ajhp140523

Bhattacharyya S, Morrison PD, Fusar-Poli P et al (2010) Opposite effects of delta-9-tetrahydrocannabinol and cannabidiol on human brain function and psychopathology. Neuropsychopharmacology 35:764–774. https://doi.org/10.1038/npp.2009.184

Bisogno T, Hanus L, De Petrocellis L (2001) Molecular targets for cannabidiol and its synthetic analogues: effect on vanilloid VR1 receptors and on the cellular uptake and enzymatic hydrolysis of anandamide. Br J Pharmacol 134:845–852. https://doi.org/10.1038/sj.bjp.0704327

Bolognini D, Costa B, Maione S et al (2010) The plant cannabinoid Δ9-tetrahydrocannabivarin can decrease signs of inflammation and inflammatory pain in mice. Br J Pharmacol 160:677–687. https://doi.org/10.1111/j.1476-5381.2010.00756.x

Bolognini D, Rock EM, Cluny NL et al (2013) Cannabidiolic acid prevents vomiting in Suncus murinus and nausea-induced behaviour in rats by enhancing 5-HT1A receptor activation. Br J Pharmacol 168:1456–1470. https://doi.org/10.1111/bph.12043

Borrelli F, Fasolino I, Romano B et al (2013) Beneficial effect of the non-psychotropic plant cannabinoid cannabigerol on experimental inflammatory bowel disease. Biochem Pharmacol 85:1306–1316. https://doi.org/10.1016/j.bcp.2013.01.017

Brenneisen R (2007) Chemistry and analysis of phytocannabinoids and other cannabis constituents. In: ElSohly M (ed) Marijuana and the cannabinoids forensic science and medicine. Humana Press, New York, pp 17–49

Brierley DI, Samuels J, Duncan M et al (2016) Cannabigerol is a novel, well-tolerated appetite stimulant in pre-satiated rats. Psychopharmacology 233:3603–3613. https://doi.org/10.1007/s00213-016-4397-4

Brierley DI, Samuels J, Duncan M et al (2017) A cannabigerol-rich Cannabis sativa extract, devoid of 9-tetrahydrocannabinol, elicits hyperphagia in rats. Behav Pharmacol 28:280–284. https://doi.org/10.1097/FBP.0000000000000285

Carcieri C, Tomasello C, Simiele M et al (2018) Cannabinoids concentration variability in cannabis olive oil galenic preparations. J Pharm Pharmacol 70:143–149. https://doi.org/10.1111/jphp.12845

Cascio MG, Gauson LA, Stevenson LA et al (2010) Evidence that the plant cannabinoid cannabigerol is a highly potent α2-adrenoceptor agonist and moderately potent 5HT1A receptor antagonist. Br J Pharmacol 159:129–141. https://doi.org/10.1111/j.1476-5381.2009.00515.x

Chagas MH, Zuardi AW, Tumas V et al (2014) Effects of cannabidiol in the treatment of patients with Parkinson's disease: an exploratory double-blind trial. J Psychopharmacol 28:1088–1098. https://doi.org/10.1177/0269881114550355

Chang AE, Shiling DJ, Stillman RC et al (1979) Delta-9-tetrahydrocannabinol as an antiemetic in cancer patients receiving high-dose methotrexate. A prospective, randomized evaluation. Ann Intern Med 91:819–824

Cheng D, Low JK, Logge W et al (2014) Chronic cannabidiol treatment improves social and object recognition in double transgenic APPswe/PS1E9 mice. Psychopharmacology 231:3009–3017. https://doi.org/10.1007/s00213-014-3478-5

Cichewicz DL, Martin ZL, Smith FL et al (1999) Enhancement mu opioid antinociception by oral delta9-tetrahydrocannabinol: dose-response analysis and receptor identification. J Pharmacol Exp Ther 289:859–867

Citti C, Ciccarella G, Braghiroli D et al (2016) Medicinal cannabis: principal cannabinoids concentration and their stability evaluated by a high performance liquid chromatography coupled to diode array and quadrupole time of flight mass spectrometry method. J Pharm Biomed Anal 128:201–209. https://doi.org/10.1016/j.jpba.2016.05.033

Cleemput MV, Cattoor K, Bosscher KD et al (2009) Hop (Humulus lupulus)-derived bitter acids as multipotent bioactive compounds. J Nat Prod 72:1220–1230. https://doi.org/10.1021/np800740m

Cluny NL, Naylor RJ, Whittle BA et al (2008) The effects of cannabidiol and tetrahydrocannabinol on motion-induced emesis in Suncus murinus. Basic Clin Pharmacol Toxicol 103:150–156. https://doi.org/10.1111/j.1742-7843.2008.00253.x

Costa B, Comelli F (2014) Pain. In: Pertwee RG (ed) Handbook of cannabis. Oxford University Press, United Kingdom, pp 473–486

Crippa JA, Derenusson GN, Ferrari TB et al (2011) Neural basis of anxiolytic effects of cannabidiol (CBD) in generalized social anxiety disorder: a preliminary report. J Psychopharmacol 25:121–130. https://doi.org/10.1177/0269881110379283

Crippa JA, Guimarães FS, Campos AC et al (2018) Translational investigation of the therapeutic potential of cannabidiol (CBD): toward a new age. Front Immunol 9:2009. https://doi.org/10.3389/fimmu.2018.02009

Cunha JM, Carlini EA, Pereira AE et al (1980) Chronic administration of cannabidiol to healthy volunteers and epileptic patients. Pharmacology 21:175–185

de Cássia da Silveira E Sá R, Lima TC, da Nóbrega FR et al (2017) Analgesic-like activity of essential oil constituents: an update. Int J Mol Sci 18:E2392. https://doi.org/10.3390/ijms18122392

de Oliveira DR, da Silva DM, Florentino IF et al (2018) Monoamine involvement in the antidepressant-like effect of β-caryophyllene. CNS Neurol Disord Drug Targets 17:309–320. https://doi.org/10.2174/1871527317666180420150249

De Petrocellis L, Ligresti A, Moriello AS et al (2011) Effects of cannabinoids and cannabinoid-enriched cannabis extracts on TRP channels and endocannabinoid metabolic enzymes. Br J Pharmacol 163:1479–1494. https://doi.org/10.1111/j.1476-5381.2010.01166.x

De Petrocellis L, Orlando P, Moriello AS et al (2012) Cannabinoid actions at TRPV channels: effects on TRPV3 and TRPV4 and their potential relevance to gastrointestinal inflammation. Acta Physiol 204:255–266. https://doi.org/10.1111/j.1748-1716.2011.02338.x

De Petrocellis L, Vellani V, Schiano-Moriello A et al (2008) Plant-derived cannabinoids modulate the activity of transient receptor potential channels of ankyrin type-1 and melastatin type-8. J Pharmacol Exp Ther 325:1007–1015. https://doi.org/10.1124/jpet.107.134809

Devane WA, Dysarz FA, Johnson MR et al (1988) Determination and characterization of a cannabinoid receptor in rat brain. Mol Pharmacol 34:605–613

Devane WA, Hanus L, Breuer A et al (1992) Isolation and structure of a brain constituent that binds to the cannabinoid receptor. Science 258:1946–1949

Devinsky O, Cross JH, Laux L et al (2017) Trial of cannabidiol for drug-resistant seizures in the Dravet syndrome. N Engl J Med 376:2011–2020. https://doi.org/10.1056/NEJMoa1611618

Edery H, Grunfeld Y, Ben-Zvi Z et al (1971) Structural requirements for cannabinoid activity. Ann N Y Acad Sci 191:40–53

Edery H, Grunfeld Y, Porath G et al (1972) Structure-activity relationships in the tetrahydrocannabinol series. Modifications on the aromatic ring and it the side-chain. Arzneimittelforschung 22:1995–2003

Einhorn LH, Nagy C, Furnas B et al (1981) Nabilone: an effective antiemetic in patients receiving cancer chemotherapy. J Clin Pharmacol 21:64–69

ElSohl MA, Gul W (2014) Constituents of Cannabis sativa. In: Pertwee RG (ed) Handbook of cannabis. Oxford University Press, UK, pp 3–22

ElSohly MA, Mehmedic Z, Foster S et al (2016) Changes in cannabis potency over the last 2 decades (1995–2014): analysis of current data in the United States. Biol Psychiatry 79:613–619. https://doi.org/10.1016/j.biopsych.2016.01.004

ElSohly MA, Slade D (2005) Chemical constituents of marijuana: the complex mixture of natural cannabinoids. Life Sci 78:539–548. https://doi.org/10.1016/j.lfs.2005.09.011

Felipe CFB, Albuquerque AMS, de Pontes JLX et al (2018) Comparative study of alpha- and beta-pinene effect on PTZ-induced convulsions in mice. Fundam Clin Pharmacol 33(2):181–190. https://doi.org/10.1111/fcp.12416

Fraguas-Sánchez AI, Torres-Suárez AI (2018) Medical use of cannabinoids. Drugs 78:1665–1703. https://doi.org/10.1007/s40265-018-0996-1

Frytak S, Moertel CG, O'Fallon JR et al (1979) Delta-9-tetrahydrocannabinol as an antiemetic for patients receiving cancer chemotherapy. A comparison with prochlorperazine and a placebo. Ann Intern Med 91:825–830

Gaoni Y, Mechoulam R (1964) Isolation, structure, and partial synthesis of an active constituent of hashish. J Am Chem Soc 86:1646–1647

Gertsch J, Leonti M, Raduner S et al (2008) Beta-caryophyllene is a dietary cannabinoid. Proc Natl Acad Sci U S A 105:9099–9104. https://doi.org/10.1073/pnas.0803601105

Gill EW (1971) Propyl homologue of tetrahydrocannabinol: its isolation from cannabis, properties, and synthesis. J Chem Soc, Section C: Organic 0:579–582

Grunfeld Y, Edery H (1969) Psychopharmacological activity of the active constituents of hashish and some related cannabinoids. Psychopharmacologia 14:200–210

Gugliandolo A, Pollastro F, Grassi G et al (2018) In vitro model of neuroinflammation: efficacy of cannabigerol, a non-psychoactive cannabinoid. Int J Mol Sci 19: E1992. https://doi.org/10.3390/ijms19071992

Guimarães FS, Chiaretti TM, Graeff FG et al (1990) Anti-anxiety effect of cannabidiol in the elevated plus-maze. Psychopharmacology 100:558–559. https://doi.org/10.1007/BF02244012

Hanuš L, Krejčí Z (1975) Isolation of two new cannabinoid acids from Cannabis sativa L. of Czechoslovak origin. Acta Univ Olomuc Fac Med 74:161–166

Herman TS, Einhorn LH, Jones SE et al (1979) Superiority of nabilone over prochlorperazine as an antiemetic in patients receiving cancer chemotherapy. N Engl J Med 300:1295–1297

Herman TS, Jones SE, Dean J et al (1977) Nabilone: a potent antiemetic cannabinol with minimal euphoria. Biomedicine 27:331–334

Hill TD, Cascio MG, Romano B et al (2013) Cannabidivarin-rich cannabis extracts are anticonvulsant in mouse and rat via a CB1 receptor-independent mechanism. Br J Pharmacol 170:679–692. https://doi.org/10.1111/bph.12321

Hill MN, Gorzalka BB (2004) Enhancement of anxiety-like responsiveness to the cannabinoid CB(1) receptor agonist HU-210 following chronic stress. Eur J Pharmacol 499:291–295. https://doi.org/10.1016/j.ejphar.2004.06.069

Hill AJ, Mercier MS, Hill TD et al (2012) Cannabidivarin is anticonvulsant in mouse and rat. Br J Pharmacol 167:1629–1642. https://doi.org/10.1111/j.1476-5381.2012.02207.x

Hively RL, Mosher WA, Hoffmann FW (1966) Isolation of trans-Δ6-tetrahydrocannabinol from marijuana. J Am Chem Soc 88:1832–1833

Hollister LE (1974) Structure-activity relationships in man of cannabis constituents, and homologs and metabolites of delta9-tetrahydrocannabinol. Pharmacology 11:3–11

Howlett AC, Qualy JM, Khachatrian LL (1986) Involvement of Gi in the inhibition of adenylate cyclase by cannabimimetic drugs. Mol Pharmacol 29:307–313

Huffman JW, Liddle J, Yu S et al (1999) 3-(1′,1-′-Dimethylbutyl)-1-deoxy-Δ8-THC and related compounds: synthesis of selective ligands for the CB2 receptor. Bioorg Med Chem 7:2905–2914

Husni AS, McCurdy CR, Radwan MM et al (2014) Evaluation of phytocannabinoids from high potency *Cannabis sativa* using *in vitro* bioassays to determine structure-activity relationships for cannabinoid receptor 1 and cannabinoid receptor 2. Med Chem Res 23:4295–4300. https://10.1007/s00044-014-0972-6

Izzo AA, Borrelli F, Capasso R (2009) Non-psychotropic plant cannabinoids: new therapeutic opportunities from an ancient herb. Trends Pharmacol Sci 30:515–527. https://10.1016/j.tips.2009.07.00

Jacobs DS, Kohut SJ, Jiang S et al (2016) Acute and chronic effects of cannabidiol on Δ9-tetrahydrocannabinol (Δ9-THC)-induced disruption in stop signal task performance. Exp Clin Psychopharmacol 24:320–330. https://10.1037/pha0000081

Jamontt JM, Molleman A, Pertwee RG et al (2010) The effects of Delta-tetrahydrocannabinol and cannabidiol alone and in combination on damage, inflammation and in vitro motility disturbances in rat colitis. Br J Pharmacol 160:712–723. https://10.1111/j.1476-5381.2010.00791.x

Jetly R, Heber A, Fraser G et al (2015) The efficacy of nabilone, a synthetic cannabinoid, in the treatment of PTSD-associated nightmares: a preliminary randomized, double-blind, placebo-controlled cross-over design study. Psychoneuroendocrinology 51:585–588. https://10.1016/j.psyneuen.2014.11.002

Kennedy DO, Dodd FL, Robertson BC et al (2011) Monoterpenoid extract of sage (salvia lavandulaefolia) with cholinesterase inhibiting properties improves cognitive performance and mood in healthy adults. J Psychopharmacol 25:1088–1100. https://10.1177/0269881110385594

Kim DS, Lee HJ, Jeon YD et al (2015) Alpha-pinene exhibits anti-inflammatory activity through the suppression of MAPKs and the NF-κB pathway in mouse peritoneal macrophages. Am J Chin Med 43:731–742. https://10.1142/S0192415X15500457

Komori T, Fujiwara R, Tanida M et al (1995) Effects of citrus fragrance on immune function and depressive states. Neuroimmunology 2:174–180. https://10.1159/000096889

Louw DF, Yang FW, Sutherland GR (2000) The effect of delta-9-tetrahydrocannabinol on forebrain ischemia in rat. Brain Res 857:183–187. https://doi.org/10.1016/S0006-8993(99)02422-1

Lynch ME, Clark AJ (2003) Cannabis reduces opioid dose in the treatment of chronic non-cancer pain. J Pain Symptom Manag 25:496–498. https://doi.org/10.1016/S0885-3924(03)00142-8

Martin BR, Compton DR, Thomas BF et al (1991) Behavioral, biochemical, and molecular modeling evaluations of cannabinoid analogs. Pharmacol Biochem Behav 40:471–478

McGuire P, Robson P, Cubala WJ et al (2018) Cannabidiol (CBD) as an adjunctive therapy in schizophrenia: a multicenter randomized controlled trial. Am J Psychiatry 175:225–231. https://doi.org/10.1176/appi.ajp.2017.17030325

McPartland JM, MacDonald C, Young M et al (2017) Affinity and efficacy studies of tetrahydrocannabinolic acid a at cannabinoid receptor types one and two. Cannabis Cannabinoid Res 2:87–95. https://10.1089/can.2016.0032

McPartland JM, Russo EB (2001) Cannabis and cannabis extracts: greater than the sum of their parts? J Cann Therap 1:103–132. https://doi.org/10.1300/J175v01n03_08

Mechoulam R (2005) Plant cannabinoids: a neglected pharmacological treasure trove. Br J Pharmacol 146:913–915. https://doi.org/10.1038/sj.bjp.0706415

Mechoulam R, Ben-Shabat S, Hanus L et al (1995) Identification of an endogenous 2-monoglyceride, present in canine gut, that binds to cannabinoid receptors. Biochem Pharmacol 50:83–90. https://doi.org/10.1016/0006-2952(95)00109-D

Mechoulam R, Ben-Zvi Z, Yagnitinsky B et al (1969) A new tetrahydrocannabinolic acid. Tetrahedron Lett 28:2339–2341

Mechoulam R, Hanus LO, Pertwee R et al (2014) Early phytocannabinoid chemistry to endocannabinoids and beyond. Nat Rev Neurosci 15:757–764. https://doi.org/10.1038/nrn3811

Mechoulam R, Shani A, Edery H et al (1970) Chemical basis of hashish activity. Science 169:611–612

Mechoulam R, Shvo Y (1963) Hashish. I. the structure of cannabidiol. Tetrahedron 19:2073–2078

Meier C, Mediavilla V (1998) Factors influencing the yield and the quality of hemp (Cannabis sativa L.) essential oil. J Int Hemp Assoc 5:16–20

Merkus FWHM (1971) Cannabivarin and tetrahydrocannabivarin, two new constituents of hashish. Nature 232:579–580

Merli F, Novotny M, Lee ML (1980) Fractionation and gas chromatographic analysis of aza-arenes in complex mixtures. J Chromatogr 199:371–378. https://doi.org/10.1016/S0021-9673(01)91388-3

Morales P, Hurst DP, Reggio PH (2017) Molecular targets of the phytocannabinoids: a complex picture. Prog Chem Org Nat Prod 103:103–131. https://doi.org/10.1007/978-3-319-45541-9_4

Moreno-Sanz G (2016) Can you pass the acid test? Critical review and novel therapeutic perspectives of Δ9-tetrahydrocannabinolic acid a. Cannabis Cannabinoid Res 1:124–130. https://doi.org/10.1089/can.2016.0008

Müller-Vahl KR, Schneider U, Koblenz A et al (2002) Treatment of Tourette's syndrome with delta 9-tetrahydrocannabinol (THC): a randomized crossover trial. Pharmacopsychiatry 3:57–61. https://doi.org/10.1055/s-2002-25028

Müller-Vahl KR, Schneider U, Prevedel H et al (2003) Delta 9-tetrahydrocannabinol (THC) is effective in the treatment of tics in Tourette syndrome: a 6- week randomized trial. J Clin Psychiatry 64:459–465

Munro S, Thomas KL, Abu-Shaar M (1993) Molecular characterization of a peripheral receptor for cannabinoids. Nature 365:61–65. https://doi.org/10.1038/365061a0

Nelson K, Walsh D, Deeter P et al (1994) A phase II study of delta-9-tetrahydrocannabinol for appetite stimulation in cancer-associated anorexia. J Palliat Care 10:14–18

O'Brien LD, Wills KL, Segsworth B et al (2013) Effect of chronic exposure to rimonabant and phytocannabinoids on anxiety-like behavior and saccharin palatability. Pharmacol Biochem Behav 103:597–602. https://doi.org/10.1016/j.pbb.2012.10.008

Parker LA, Kwiatkowska M, Burton P et al (2004) Effect of cannabinoids on lithium-induced vomiting in the Suncus murinus (house musk shrew). Psychopharmacology 171:156–161. https://doi.org/10.1007/s00213-003-1571-2

Parker LA, Mechoulam R, Schlievert C et al (2003) Effects of cannabinoids on lithium-induced conditioned rejection reactions in a rat model of nausea. Psychopharmacology 166:156–162. https://doi.org/10.1007/s00213-002-1329-2

Pate DW (1994) Chemical ecology of cannabis. J Int Hemp Association 2(29):32–37

Pedersen W, Sandberg S (2013) The medicalisation of revolt: a sociological analysis of medical cannabis users. Sociol Health Illn 35:17–32. https://doi.org/10.1111/j.1467-9566.2012.01476.x

Pellesi L, Licata M, Verri P et al (2018) Pharmacokinetics and tolerability of oral cannabis preparations in patients with medication overuse headache (MOH)—a pilot study. Eur J Clin Pharmacol 74:1427–1436. https://doi.org/10.1007/s00228-018-2516-3

Pertwee RG (1972) The ring test: a quantitative method for assessing the 'cataleptic' effect of cannabis in mice. Br J Pharmacol 46:753–763

Pertwee RG (2005) Inverse agonism and neutral antagonism at cannabinoid CB1 receptors. Life Sci 76:1307–1324. https://doi.org/10.1016/j.lfs.2004.10.025

Pertwee RG (2008) The diverse CB1 and CB2 receptor pharmacology of three plant cannabinoids: Δ9-tetrahydrocannabinol, cannabidiol and Δ9-tetrahydrocannabivarin. Br J Pharmacol 153:199–215. https://doi.org/10.1038/sj.bjp.0707442

Pertwee RG, Thomas A, Stevenson LA et al (2007) The psychoactive plant cannabinoid, delta9-tetrahydrocannabinol, is antagonized by delta8- and delta9- tetrahydrocannabivarin in mice in vivo. Br J Pharmacol 150:586–594. https://doi.org/10.1038/sj.bjp.0707124

Raman C, McAllister SD, Rizvi G (2004) Amyotrophic lateral sclerosis: delayed disease progression in mice by treatment with a cannabinoid. Amyotroph Lateral Scler Other Motor Neuron Disord 5:33–39. https://doi.org/10.1080/14660820310016813

Razdan RK (1986) Structure-activity relationships in cannabinoids. Pharmacol Rev 38:75–149
Riedel G, Fadda P, McKillop-Smith S (2009) Synthetic and plant-derived cannabinoid receptor antagonists show hypophagic properties in fasted and non-fasted mice. Br J Pharmacol 156:1154–1166. https://doi.org/10.1111/j.1476-5381.2008.00107.x
Rock EM, Bolognini D, Limebeer CL et al (2012) Cannabidiol, a non-psychotropic component of cannabis, attenuates vomiting and nausea-like behaviour via indirect agonism of 5-HT(1A) somatodendritic autoreceptors in the dorsal raphe nucleus. Br J Pharmacol 165:2620–2634. https://doi.org/10.1111/j.1476-5381.2011.01621.x
Rock EM, Connolly C, Limebeer CL et al (2016) Effect of combined oral doses of Δ(9)-tetrahydrocannabinol (THC) and cannabidiolic acid (CBDA) on acute and anticipatory nausea in rat models. Psychopharmacology 233:3353–3360. https://doi.org/10.1007/s00213-016-4378-7
Rock EM, Goodwin JM, Limebeer CL et al (2011) Interaction between non-psychotropic cannabinoids in marihuana: effect of cannabigerol (CBG) on the anti-nausea or anti-emetic effects of cannabidiol (CBD) in rats and shrews. Psychopharmacology 215:505–512. https://doi.org/10.1007/s00213-010-2157-4
Rock EM, Kopstick RL, Limebeer CL et al (2013a) Tetrahydrocannabinolic acid reduces nausea-induced conditioned gaping in rats and vomiting in Suncus murinus. Br J Pharmacol 170:641–648. https://doi.org/10.1111/bph.12316
Rock EM, Limebeer CL, Navaratnam R (2014a) A comparison of cannabidiolic acid with other treatments for anticipatory nausea using a rat model of contextually elicited conditioned gaping. Psychopharmacology 231:3207–3215. https://doi.org/10.1007/s00213-014-3498-1
Rock EM, Limebeer CL, Parker LA (2014b) Anticipatory nausea in animal models: a review of potential novel therapeutic treatments. Exp Brain Res 232:2511–2534. https://doi.org/10.1007/s00221-014-3942-9
Rock EM, Limebeer CL, Parker LA (2015) Effect of combined doses of Δ(9)-tetrahydrocannabinol (THC) and cannabidiolic acid (CBDA) on acute and anticipatory nausea using rat (Sprague-Dawley) models of conditioned gaping. Psychopharmacology 232:4445–4454. https://doi.org/10.1007/s00213-015-4080-1
Rock EM, Limebeer CL, Parker LA (2018) Effect of cannabidiolic acid and Δ9-tetrahydrocannabinol on carrageenan-induced hyperalgesia and edema in a rodent model of inflammatory pain. Psychopharmacology 235:3259–3271. https://doi.org/10.1007/s00213-018-5034-1
Rock EM, Limebeer CL, Petrie GN et al (2017) Effect of prior foot shock stress and Δ9-tetrahydrocannabinol, cannabidiolic acid, and cannabidiol on anxiety-like responding in the light-dark emergence test in rats. Psychopharmacology 234:2207–2217. https://doi.org/10.1007/s00213-017-4626-5
Rock EM, Parker LA (2015) Synergy between cannabidiol, cannabidiolic acid, and Δ^9-tetrahydrocannabinol in the regulation of emesis in the Suncus murinus (house musk shrew). Behav Neurosci 129:368–370. https://doi.org/10.1037/bne0000057
Rock EM, Sticht MA, Duncan M et al (2013b) Evaluation of the potential of the phytocannabinoids, cannabidivarin (CBDV) and Δ(9) -tetrahydrocannabivarin (THCV), to produce CB1 receptor inverse agonism symptoms of nausea in rats. Br J Pharmacol 170:671–678. https://doi.org/10.1111/bph.12322
Rosenthaler S, Pöhn B, Kolmanz C et al (2014) Differences in receptor binding affinity of several phytocannabinoids do not explain their effects on neural cell cultures. Neurotoxicol Teratol 46:49–56. https://doi.org/10.1016/j.ntt.2014.09.003
Rothschild M, Bergstrom G, Wangberg S (2005) Cannabis sativa: volatile compounds from pollen and entire male and female plants of two variants, northern lights and Hawaian Indica. Bot J Linn Soc 147:387–397. https://doi.org/1111/j.1095-8339.2005.00417.x
Russo EB (2011) Taming THC: potential cannabis synergy and phytocannabinoid-terpenoid entourage effects. Br J Pharmacol 163:1344–1364. https://doi.org/10.1111/j.1476-5381.2011.01238.x
Russo EB, Burnett A, Hall B et al (2005) Agonistic properties of cannabidiol at 5-HT1a receptors. Neurochem Res 30:1037–1043. https://doi.org/10.1007/s11064-005-6978-1
Segat GC, Manjavachi MN, Matias DO et al (2017) Antiallodynic effect of β-caryophyllene on paclitaxel-induced peripheral neuropathy in mice. Neuropharmacology 125:207–219. https://doi.org/10.1016/j.neuropharm.2017.07.015
Singh B, Sharma R (2015) Plant terpenes: defense responses, phylogenetic analysis, regulation and clinical applications. 3 Biotech 5:129–151. https://doi.org/10.1007/s13205-014-0220-2
Sjödén PO, Järbe TU, Henriksson BG (1973) Influence of tetrahydrocannabinols (delta8-THC and delta9-THC) on body weight, food, and water intake in rats. Pharmacol Biochem Behav 1:395–399
Smith RN, Vaughan CG (1977) The decomposition of acidic and neutral cannabinoids in organic solvents. J Pharm Pharmacol 29:286–290
Sugiura T, Kondo S, Sukagawa A et al (1995) 2-Arachidonoylglycerol: a possible endogenous cannabinoid receptor ligand in brain. Biochem Biophys Res Commun 215:89–97. https://doi.org/10.1006/bbrc.1995.2437
Szaflarski JP, Bebin EM, Comi AM et al (2018) Long-term safety and treatment effects of cannabidiol in children and adults with treatment-resistant epilepsies: expanded access program results. Epilepsia 59:1540–1548. https://doi.org/10.1111/epi.14477

Takeda S, Okajima S, Miyoshi H et al (2012) Cannabidiolic acid, a major cannabinoid in fiber-type cannabis, is an inhibitor of MDA-MB-231 breast cancer cell migration. Toxicol Lett 214:314–319. https://doi.org/10.1016/j.toxlet.2012.08.029

Tchekalarova J, da Conceição MK, Gomes Júnior AL et al (2011) Pharmacological characterization of the cannabinoid receptor 2 agonist, β-caryophyllene on seizure models in mice. Seizure 57:22–26. https://doi.org/10.1016/j.seizure.2018.03.009

Thomas GL, Baillie AM, Phillips RK et al (2007) Cannabidiol displays unexpectedly high potency as an antagonist of CB1 and CB2 receptor agonists in vitro. Br J Pharmacol 150:613–623. https://doi.org/10.1038/sj.bjp.0707133

Uemura M, Hata GI, Toda T et al (1997) Effectiveness of eucalyptol and d-limonene as gutta-percha solvents. J Endod 23:739–741. https://doi.org/10.1016/S0099-2399(97)80346-9

Valdeolivas S, Navarrete C, Cantarero I (2015) Neuroprotective properties of cannabigerol in Huntington's disease: studies in r6/2 mice and 3-nitropropionate-lesioned mice. Neurotherapeutics 12:185–199. https://doi.org/10.1007/s13311-014-0304-z

van Vliet SA, Vanwersch RA, Jongsma MJ et al (2008) Therapeutic effects of delta9-THC and modafinil in a marmoset Parkinson model. Eur Neuropsychopharmacol 18:383–389. https://doi.org/10.1016/j.euroneuro.2007.11.003

Verhoeckx KC, Korthout HA, van Meeteren-Kreikamp AP (2006) Unheated Cannabis sativa extracts and its major compound THC-acid have potential immuno-modulating properties not mediated by CB1 and CB2 receptor coupled pathways. Int Immunopharmacol 6:656–665. https://doi.org/10.1016/j.intimp.2005.10.002

Vigli D, Cosentino L, Raggi C et al (2018) Chronic treatment with the phytocannabinoid Cannabidivarin (CBDV) rescues behavioural alterations and brain atrophy in a mouse model of Rett syndrome. Neuropharmacology 140:121–129. https://doi.org/10.1016/j.neuropharm.2018.07.029

Ware MA, Wang T, Shapiro S et al (2010) Smoked cannabis for chronic neuropathic pain: a randomized controlled trial. CMAJ 182:E694–E701. https://doi.org/10.1503/cmaj.091414

Wilsey B, Marcotte T, Deutsch R et al (2013) Low-dose vaporized cannabis significantly improves neuropathic pain. J Pain 14:136–148. https://doi.org/10.1016/j.jpain.2012.10.009

Wilsey B, Marcotte T, Tsodikov A et al (2008) A randomized, placebo-controlled, crossover trial of cannabis cigarettes in neuropathic pain. J Pain 9:506–521. https://doi.org/10.1016/j.jpain.2007.12.010

Yang M, Lv Y, Tian X et al (2017) Neuroprotective effect of β-caryophyllene on cerebral ischemia-reperfusion injury via regulation of necroptotic neuronal death and inflammation: in vivo and in vitro. Front Neurosci 11:583. https://doi.org/10.3389/fnins.2017.00583

Yun J (2014) Limonene inhibits methamphetamine-induced locomotor activity via regulation of 5-HT neuronal function and dopamine release. Phytomedicine 21:883–887. https://doi.org/10.1016/j.phymed.2013.12.004

Zanelati TV, Biojone C, Moreira FA et al (2010) Antidepressant-like effects of cannabidiol in mice: possible involvement of 5-HT1A receptors. Br J Pharmacol 159:122–128. https://doi.org/10.1111/j.1476-5381.2009.00521.x

Zhao Y, Chen R, Wang Y et al (2018) α-Pinene inhibits human prostate cancer growth in a mouse xenograft model. Chemotherapy 63:1–7. https://doi.org/10.1159/000479863

Neuromolecular Mechanisms of *Cannabis* Action

2

Yousra Adel and Stephen P. H. Alexander

Abstract

Most of our current understanding of the neuromolecular mechanisms of *Cannabis* action focusses on two plant cannabinoids, THC and CBD. THC acts primarily through presynaptic CB cannabinoid receptors to regulate neurotransmitter release in the brain, spinal cord and peripheral nerves. CBD action, on the other hand, is probably mediated through multiple molecular targets.

Keywords

Δ^9-Tetrahydrocannabinol · cannabidiol · cannabinoid receptors

Abbreviations

2AG	2-Arachidonoylglycerol
2-AGE	2-Arachidonoylglycerol ether
CB_1	Cannabinoid receptor type 1
CB_2	Cannabinoid receptor type 2
CBC	Cannabichromene
CBD	Cannabidiol
CBDA	Cannabidiolic acid
CBDV	Cannabidivarin
CBG	Cannabigerol
COX	Cyclooxygenase
DAGL	Diacylglycerol lipase
FAAH	Fatty acid amide hydrolase
GPCR	G protein-coupled receptors
MAGL	Monoacylglycerol lipase
PLC	Phospholipase C
PPAR	Peroxisome proliferator-activated receptor
THC	Δ^9-Tetrahydrocannabinol
THCA-A	Δ^9-Tetrahydrocannabinolic acid
THCV	Δ^9-Tetrahydrocannabivarin
TRPV	Transient receptor potential vanilloid family

Y. Adel · S. P. H. Alexander (✉)
University of Nottingham. Faculty of Medicine & Health Sciences, Nottingham, UK
e-mail: steve.alexander@nottingham.ac.uk

2.1 Introduction and Scope of this Chapter

Cannabis, like many natural products, is a complex and variable mix of metabolites, some of which are common across many plant species, such as terpenoids and flavonoids, and some of which appear to be unique to *Cannabis*. These phytocannabinoids constitute a group of C21 or C22 terpeno-phenolic constituents, with the principal constituents being acids, including Δ^9-tetrahydrocannabinolic acid, cannabidiolic acid, cannabinolic acid, cannabinodiolic acid, cannabigerolic acid and cannabichromenic acid, for review, see Andre et al. (2016). Intriguingly, the bioactivity of the acids has drawn little attention. By contrast, the decarboxylated products of the acids have enjoyed the vast majority of

E. Murillo-Rodriguez et al. (eds.), *Cannabinoids and Neuropsychiatric Disorders*, Advances in Experimental Medicine and Biology 1264, https://doi.org/10.1007/978-3-030-57369-0_2

attention from both scientific and non-scientific audiences. A particular focus has been on THC, Δ^9-tetrahydrocannabinol, which is the predominant psychoactive cannabinoid and the primary reason for the nonmedicinal consumption of *Cannabis*. This compound was first isolated from *Cannabis* preparations over 50 years ago (Gaoni and Mechoulam 1964). The second most investigated cannabinoid is CBD, cannabidiol, which lacks the psychoactivity of THC. Our understanding of the bioactivity of the remainder of the phytocannabinoids falls off a knowledge cliff.

Accordingly, this chapter reviews the targets and neural functions of the cannabinoids. We will describe the receptor targets of THC, which are well established. We will consider the evidence for the molecular targets of CBD, which are less well-established. For the wider family of phytocannabinoids, we will review the evidence for their molecular and cellular functions.

2.1.1 Neuromolecular Targets of THC: The Cannabinoid Receptors

In 1990, the orphan G protein-coupled receptor (GPCR) SKR6 cloned from rat brain was reported to respond with similar potency to THC and its lower abundance isomer Δ^8-THC, but not CBD or cannabinol (CBN), in recombinant expression (Matsuda et al. 1990). Shortly thereafter, the human orthologue was cloned from the brainstem and testes, and identified as a cannabinoid receptor with over 97% identity to the rat protein (Gerard et al. 1991). The following year saw the first endogenous cannabinoid being identified in extracts from pig brain; this arachidonic acid conjugate was termed anandamide by Will Devane and Raphi Mechoulam (Devane et al. 1992). Three years after the cloning of the first cannabinoid receptor, a second, quite distinct GPCR was cloned from the HL60 human promyelocytic leukaemia cell line (Munro et al. 1993). This was initially described as a peripheral receptor for cannabinoids and bound THC and CBN with similar affinities, anandamide with lower affinity and CBD with much lower affinity. In 1995, a second endocannabinoid was identified; this was also an arachidonic acid conjugate, 2-arachidonoylglycerol, 2AG (Mechoulam et al. 1995; Sugiura et al. 1995).

The Nomenclature and Standards Committee of the Union of Basic and Clinical Pharmacology (NC-IUPHAR) currently recognises just two cannabinoid receptors, termed as CB_1 and CB_2 (Howlett et al. 2002; Pertwee et al. 2010), corresponding to the 'central' and 'peripheral' receptors, respectively. The two GPCRs share 44% amino acid sequence homology, although this correspondence increases to 68% for the ligand-binding domains of the transmembrane regions. They belong to the rhodopsin or family A of GPCR, which signal through pertussis toxin-sensitive $G_{i/o}$ proteins and function by activating the mitogen-activated protein kinase (MAPK) family and inhibiting adenylyl cyclase (Howlett et al. 2002; Pertwee et al. 2010).

2.1.1.1 CB_1 Cannabinoid Receptor Characterisation: Protein, Distribution, Signalling and Pharmacology

The CB_1 receptor, coded in humans by the *CNR1* gene (Pertwee et al. 2010), is of relatively long length for the rhodopsin family, 472 amino acids, having an N-terminus over 110-amino acid-long (Gerard et al. 1991). The N-terminus contains two asparagine residues, Asn^{77} and Asn^{83}, which are putative locations for glycosylation, a feature of most, if not all, GPCR. For the rat receptor, glycosylation increases the apparent molecular size from 53 to 64 kDa (Song and Howlett 1995). A similar phenomenon has been reported for the CB_1 receptor in human brain preparations (De Jesus et al. 2006). Towards the C-terminus, Cys^{415} has been described to be palmitoylated (Oddi et al. 2012), a common but not universal post-translational modification for GPCR. Palmitoylation was reported to anchor the receptor in lipid rafts of the plasma membrane and to assist in coupling to G proteins (Oddi et al. 2012). Two N-terminal splice variants of the CB_1 cannabinoid receptor have been described that differ in length and possess different ligand-binding properties (Ryberg et al. 2005) and are expressed at significantly lower levels in various tissues (Ryberg et al. 2005; Shire et al. 1995) when compared to the full-length receptor. The physiological and pharmacological effects of these genetic variants are yet to be fully elucidated.

The first CB_1 receptor antagonist identified was rimonabant (Rinaldi-Carmona et al. 1995), which gained approval for the treatment of metabolic disorder/obesity for a brief period in Europe (Di Marzo and Despres 2009). The antagonist structure was modified to produce AM6538, which was recently crystallised with the CB_1 receptor (Hua et al. 2016). A further structurally-unrelated antagonist, taranabant, was co-crystallised in a contemporaneous study (Shao et al. 2016). Two additional crystal structures of the CB_1 cannabinoid receptor have been reported where agonists were involved. AM841 and AM11542 are structural analogues of THC (Hua et al. 2017). In an attempt to gain further understanding of the signalling mechanisms of the CB_1 receptor, the structure of a CB_1-Gi signalling complex bound to a high affinity, high efficacy agonist, MDMB-Fubinaca (Krishna Kumar et al. 2019). This characterisation of the distinct structures of the CB_1 receptor by various classes of compounds will likely aid future drug discovery centred on the cannabinoid receptors (Krishna Kumar et al. 2019).

As identified above, the CB_1 cannabinoid receptor is $G_{i/o}$-coupled, which is associated with the inhibition of cyclic AMP production (Howlett et al. 2002). In neuronal cells, however, alternative signalling pathways are thought to predominate. Thus, the CB_1 receptor couples to the potassium channel opening leading to cellular hyperpolarisation, while inhibiting voltage-gated calcium channels (Mackie et al. 1995). In recombinant expression and in particular cells in culture, the CB_1 receptor also enhances intracellular calcium release via the G protein-dependent (apparently $G\beta\gamma$ subunit) stimulation of phospholipase C-β (PLC-β) leading to inositol-1,4,5-trisphosphate generation (Lauckner et al. 2005).

Investigations utilising immunocytochemistry, quantitative autoradiography, and in situ hybridisation (Howlett et al. 2002) revealed that the CB_1 receptors are expressed abundantly at the terminals of central and peripheral neurons, where they inhibit the release of multiple neurotransmitters (Pertwee et al. 2010). There is high expression specifically in the cerebellum, olfactory bulb, neocortex, basal ganglia, brain stem and hippocampus (Herkenham et al. 1991). Peripherally, the CB_1 receptor is expressed in the testes, vascular endothelium, spleen, in the enteric nervous system of the gastrointestinal tract (Izzo and Sharkey 2010), adipocytes and retina.

The physiological role of CB_1 receptors in the CNS is best understood in the Schaffer collateral commissural pathway. Modelling of the operation of endocannabinoids at a glutamatergic synapse highlighted the post-junctional location of ligand-gated ion channels and G_q-coupled GPCR activated by high-frequency stimulation-evoked release of high levels of presynaptic glutamate. These receptors evoke generation of diacylglycerol (and inositol 1,4,5-trisphosphate), which is metabolised by perisynaptic diacylglycerol lipase to produce 2AG. By as yet unidentified mechanisms, 2AG leaves the postjunctional neuron to activate presynaptic CB_1 receptors, leading to the inhibition of neuronal excitability (Zachariou et al. 2013). This adaptation of synaptic efficiency is termed short-term depression or depolarisation-evoked suppression of excitation. It is attractive to hypothesise that this phenomenon is related to the observation that *Cannabis* and THC elicit impairments of short-term memory in vivo (Melges et al. 1970).

Endocannabinoids also play a role in a related phenomenon termed as depolarisation-evoked suppression of inhibition. This phenomenon plays out in much the same way as depolarisation-evoked suppression of excitation, except that the 2AG-generated post-junctionally acts on CB_1 receptors on a GABAergic nerve terminal. This leads to a reduction of GABA release, which thereby elicits disinhibition of synaptic efficacy, and enhanced neurotransmission through the affected pathway. There appears to be a predominance of CB_1 receptors on GABAergic nerve terminals compared to glutamatergic nerve terminals (Marsicano and Lutz 1999), which makes it likely that the administration of *Cannabis* or THC leads to changes in GABA signalling pathways.

Prolonged exposure to THC has been found to produce an array of effects in humans, including analgesia, dysphoria, dependence and tolerance (Mechoulam and Parker 2013) and the majority of these effects were prevented following pretreatment with the CB_1 selective blocker rimonabant (Kendall and Yudowski 2016). In animal models, long-term treatment with THC resulted in a region-dependent reduction in the CB_1 receptor radioligand binding (Romero et al. 1997). In the hippocampus, long-term potentiation (a corollary of learning and memory) exhibited a long-lasting inhibition with repeated THC administration, persisting for up to 14 days after treatment was halted (Hoffman et al. 2007).

2.1.1.2 CB2 Cannabinoid Receptor Characterisation: Protein, Distribution, Signalling and Pharmacology

The CB_2 receptor is coded by the *CNR2* gene (Pertwee et al. 2010) located on chromosome 1p36 and is composed of 360 amino acids in humans (Munro et al. 1993). Compared to the CB_1 cannabinoid receptor, the CB_2 receptor has a much shorter N terminus, with a base molecular size of ~42 kDa, which was increased to ~55 kDa by glycosylation (Zhang et al. 2007). Analogous to the CB_1 cannabinoid receptor, a cysteine residue is expressed in the C-terminus close to the seventh transmembrane domain, Cys^{320}. As yet, there are no published data on the possibility that Cys320 is palmitoylated. In 2009, a second splice variant of the CB_2 receptor was recognised based on a human neuroblastoma cDNA library (Liu et al. 2009). The two CB_2 receptor isoforms were found to display tissue-specific expression patterns. The previously identified CB_2 receptor isoform was primarily identified in the spleen and the immune system, while the novel isoform was recognised abundantly in the testes and brain regions of the reward system.

Very recently, a crystal structure of the CB_2 receptor was published and described (Li et al. 2019). In this report, a ligand derived from rimonabant was used, which was modified to enhance CB_2 receptor affinity, AM10257. An interesting feature of the crystal structure is that the antagonist-bound conformation of the CB_2 receptor has more similarity to the agonist-bound conformation of the CB_1 receptor than the antagonist-bound version (Li et al. 2019). It will be interesting to see the impact that this divergence in structure will have on future drug design.

Both the CB_1 and CB_2 receptors share some common pharmacology (activation by the same endocannabinoids and THC) and both couple to the same family of G proteins; however, they differ profoundly in their signalling profiles. As opposed to the CB_2 receptor, the CB_1 receptor has been reported to couple in addition to Gs and to stimulate adenylate cyclase activity (Glass and Felder 1997). In early studies, comparing the two receptors in recombinant expression in the same host cells, there was a further distinction between the two receptors. When expressed in anterior pituitary cells, both receptors coupled to the inhibition of cAMP production, while only the CB_1 receptor coupled to the inhibition of voltage-gated calcium channels and opening of potassium channels (Felder et al. 1995).

Investigations using in situ hybridisation, northern blot and receptor autoradiography (Howlett et al. 2002; Pertwee 1997) revealed that the CB_2 receptor was predominantly expressed in the macrophages, spleen (Munro et al. 1993), tonsils (Carayon et al. 1998) and immune cells. The precise immune cells that abundantly express CB_2 include monocytes, B cells, polymorphonuclear neutrophils, natural killer cells, CD4+ and CD8+ T cells and when stimulated, they regulate immune cell migration and cytokine release (Galiegue et al. 1995; Schatz et al. 1997). Using quantitative PCR, the CB_2 receptor was also detected in the monocytes and macrophages of the spleen and certain leukocyte populations, precisely the eosinophils (Galiegue et al. 1995). CB_2 receptor expression was also evaluated in other human organs and it was determined that the receptor was absent from the majority of non-immune organs with the exception of the uterus, pancreas and lungs, which exhibited low levels of mRNA (Turcotte et al.

2016). The CB_2 receptor was detected in the reproductive tissues of both sexes (Battista et al. 2012; Grimaldi et al. 2009) and was suggested to perform a function in affecting the fertility of both males and females. As mentioned above, the CB_2 receptor is often referred to as the peripheral cannabinoid receptor, given its abundant peripheral presence (Howlett et al. 2002), compared to its limited CNS expression (Gong et al. 2006). Nevertheless, recent investigations have detected the expression of the CB_2 receptor in the CNS, by the microglia (Atwood and Mackie 2010) following neuroinflammation, degeneration (Ashton and Glass 2007), in neuropathic pain (Zhang et al. 2003) in multiple sclerosis and amyotrophic lateral sclerosis (Yiangou et al. 2006). The expression level of the CB_2 receptor was variable and determined by the state of the cell; i.e. microglia do not express CB_2 in healthy human brain (Stella 2004). The degree of CB_2 receptor expression in the neurons and their physiological role remains to be fully elucidated.

Presumably because of the immune location of CB_2 receptors, there are fewer investigations of the potential for tolerance with repeated administration in humans. In preclinical models, however, CB_2 receptor-selective agonists failed to exhibit tolerance in a chronic pain model (Romero-Sandoval et al. 2008). Intriguingly, pregnancy seemed to influence CB receptor expression in B lymphocytes, such that CB_2 receptors were down-regulated, while CB_1 expression was increased (Wolfson et al. 2016).

2.1.2 Neuromolecular Targets of THC: Beyond the Cannabinoid Receptors

Other than the well-identified and investigated CB_1 and CB_2 receptors, other GPCRs, ion channel and nuclear receptors have been described to be stimulated by cannabinoid ligands (Pertwee et al. 2010). There are three GPCRs, which have been described as cannabinoid foster children, rather than orphan receptors; these are GPR18, GPR55 and GPR119 (Irving et al. 2017). Although there is little sequence homology with CB_1 or CB_2 receptors (given the low homology between the CB_1 and CB_2 receptors, this may not have a deeper implication), there is some homology between the putative endogenous ligands thought to activate them. Thus, GPR119 is activated by monounsaturated fatty acid analogues of anandamide and 2AG, N-oleoylethanolamine and 2-oleoylglycerol (Overton et al. 2006), but does not appear to respond to plant-derived or synthetic cannabinoids. GPR18 and GPR55, however, have been suggested to be targets for these agents. GPR18 is proposed to be activated endogenously by an oxidised version of anandamide, N-arachidonoylglycine (Kohno et al. 2006), while GPR55 is proposed to be activated endogenously by a conjugated version of 2AG, lysophosphatidylinositol (Oka et al. 2009). GPR18 has also been reported to be activated in vitro by THC (McHugh et al. 2012), as has GPR55 (Lauckner et al. 2008). However, whether these receptors are targets for THC in vivo has not yet been determined, and the role of these receptors in neural pathways is also unclear.

The endocannabinoid anandamide (and other endogenous analogues) has also been observed to activate the TRPV1 receptor (Alharthi et al. 2018; Zygmunt et al. 1999), which is a target for the hot component of spicy food derived from chilli peppers, capsaicin (Voets et al. 2004). The TRPV1 receptor is expressed at high levels in primary afferent neurones, where it functions as a broad integrator of nociceptive signalling and is the best investigated of a large family of cation-gating ion channels, the transient receptor potential family. THC appears not to activate the TRPV1 (De Petrocellis et al. 2011), but has been observed to activate other family members: TRPA1 (De Petrocellis et al. 2008), TRPC1 (Rao and Kaminski 2006), TRPM8 (De Petrocellis et al. 2008), TRPV2 (De Petrocellis et al. 2011), and TRPV3 (De Petrocellis et al. 2012). The contribution of these interactions to the action of THC in vivo is not clear.

Both anandamide and THC have been reported to act as positive allosteric modulators of the ligand-gated ion channel glycine receptor (Hejazi et al. 2006). There is evidence that this mechanism can contribute to the anti-nociceptive profile of these agents in vivo (Xiong et al. 2011). THC also inhibited human recombinant 5-HT_3 receptors, another ligand-gated ion channel, in vitro with high potency and efficacy (Barann et al. 2002). In contrast, THC also inhibited a third ligand-gated ion channel, α7 nicotinic acetylcholine receptors, but with a much lower potency and efficacy (Oz et al. 2004).

In terms of voltage-gated ion channels, THC was found to inhibit a number of human recombinant subtypes of voltage-gated calcium channels in vitro (Ross et al. 2008) and to inhibit an intrinsic sodium current in mouse neuroblastoma cells (Turkanis et al. 1991). Although it is attractive to hypothesise that these effects might contribute to the analgesic effect induced by THC in vivo, there is little evidence for this.

THC could be described as having an opportunistic nature in which it has a wide interactome. It has an unusual property in common with anandamide in which it is able to activate members of three of the four receptor superfamilies. Identified above are examples of GPCR and ligand-activated ion channels, which THC activates. It has also been described to activate members of the nuclear hormone receptor superfamily, particularly members of the peroxisome proliferator-activated receptors, PPARs, for review, see (O'Sullivan 2016). THC was an agonist in a reporter gene assay of PPARγ (O'Sullivan et al. 2005), although there are contrasting reports of THC action at PPARα (Sun et al. 2007; Takeda et al. 2014). PPARβ has been less thoroughly investigated in terms of responses to cannabinoids (O'Sullivan 2016).

2.1.3 Neuromolecular Targets of CBD

In contrast to THC, our knowledge of the neuromolecular mechanisms of cannabidiol is limited, which is a frustration. On the one hand, there are a number of putative targets through which CBD has been proposed to act, but none of these have features that are totally convincing as mechanisms of the in vivo action of CBD.

CBD action at the conventional cannabinoid receptors is contradictory. The majority of studies have failed to show occupancy of CB_1 or CB_2 receptors by CBD (Matsuda et al. 1990; Munro et al. 1993). However, CBD has been suggested to be a negative allosteric modulator of the CB_1 receptor in recombinant expression (Laprairie et al. 2015). There are reports that CBD can reduce the side effects caused by the administration of THC (Mechoulam and Hanus 2002) and it is attractive to hypothesise that the negative allosteric modulation of the CB_1 receptor might be the neuromolecular mechanism for this observation. CBD seems to have an interaction with the endocannabinoid system in which acute administration increased brain levels of a variety of N-acylethanolamines, including anandamide, with little impact on 2AG levels (Leishman et al. 2018). CBD has been reported to inhibit the anandamide hydrolysis enzyme fatty acid amide hydrolase activity in vitro with a range of mostly lower potencies (Bisogno et al. 2001; De Petrocellis et al. 2011; Watanabe et al. 1996) and it would be attractive to suggest this as an in vivo mechanism of action. However, the pattern of metabolite accumulation evoked by CBD and a selective inhibitor of fatty acid amide hydrolase differ, suggesting a distinct impact (Leishman et al. 2018).

CBD has been reported to interact with serotonergic signalling through multiple routes (Ledgerwood et al. 2011). It directly activated 5-HT_{1A} receptors (Russo et al. 2005), an effect implicated in anxiolytic effects (Campos and Guimaraes 2008), neuroprotection following hypoxia/ischemia (Mishima et al. 2005; Pazos et al. 2013), inhibition of nausea and vomiting behaviours (Rock et al. 2012) and inhibition of morphine-evoked reward (Katsidoni et al. 2013) of CBD in vivo. CBD also evoked an allosteric inhibition of 5-HT_{3A} receptors (Yang et al. 2010) in vitro, which may be mediated through accelerating the rate of receptor desensitisation (Xiong et al. 2011). However, this does not seem to be a major route for CBD effects in vivo.

As with THC, CBD has been reported to enhance glycine receptor function as a positive allosteric modulator (Ahrens et al. 2009), which appeared to be mediated through a transmembrane domain-located serine residue (Foadi et al. 2010). The analgesic effects of CBD in animal models have been suggested to be mediated through glycine receptors (Lu et al. 2018; Xiong et al. 2012).

As mentioned above for THC, CBD has a broad interactome. It also has been reported to modulate the function of the transient receptor potential family to stimulate TRPA1, inhibit TRPM8 (De Petrocellis et al. 2008), stimulate TRPV1 (Costa et al. 2004), TRPV2 (Qin et al. 2008), TRPV3 and TRPV4 (De Petrocellis et al. 2012). Other investigations reported that CBD decreases neuronal hyperactivity in epilepsy by causing activation followed by rapid desensitisation of TRPV1 and TRPV2 (Iannotti et al. 2014).

CBD also regulates calcium homeostasis in the hippocampal neurons as well as blocking the low-voltage T-type calcium channels, which are prominent modulators of neuronal excitability which specifically controlled partial or generalised seizures (Jones et al. 2010).

CBD was also reported to enhance adenosine signalling by inhibiting its uptake, which has been associated with its anti-inflammatory, neuroprotective and immunosuppressive roles (Carrier et al. 2006). Additionally, this mechanism for elevating extracellular adenosine leading to an indirect activation of adenosine receptors was implicated in CBD effects in vivo on pain modulation (Maione et al. 2011), and hypoxia/ischemia-induced brain damage (Castillo et al. 2010) and ventricular arrhythmias (Gonca and Darici 2015).

2.1.4 Neuromolecular Targets of Other Cannabinoids

A number of the other cannabinoids from the Cannabis plant have been described to have effects both in vitro and in vivo. However, there is a limited capacity to correlate the two profiles.

2.1.4.1 Δ^9-Tetrahydrocannabidivarin

Δ^9-Tetrahydrocannabidivarin (THCV) is a close structural analogue of THC, which also binds to CB receptors. Initially, THCV was suggested to act as an antagonist at both CB_1 and CB_2 receptors (Thomas et al. 2005), although later it was observed to act as a partial agonist at human recombinant CB_2 receptors (Bolognini et al. 2010). CB_2 activation appeared to translate to in vivo models of inflammation (Bolognini et al. 2010) and Parkinson's disease (Garcia et al. 2011). THCV was also described to potentiate 5-HT_{1A} receptor function in vitro and in an in vivo model of psychosis (Cascio et al. 2015). THCV also activated TRPA1, TRPV1, TRPV2 and blocked TRPM8 channels in recombinant expression (De Petrocellis et al. 2011). In vivo, THCV was able to reverse the effects of THC in a model of visceral pain (Booker et al. 2009).

2.1.4.2 Cannabinol

Cannabinol accumulates over time by the natural oxidation of THC. It was not thought to bind the CB_1 cannabinoid receptor (Matsuda et al. 1990), but a later report described agonist action at both CB_1 and CB_2 receptors (Rhee et al. 1997). Cannabinol was less potent than THC at the CB_1 receptor, but more potent than THC at the CB_2 receptor. In vivo, cannabinol evoked a reduction in pain behaviours in a model of visceral pain in a manner sensitive to a CB_1 receptor antagonist (Booker et al. 2009) and also evoked a CB_1 antagonist-sensitive increase in feeding behaviours (Farrimond et al. 2012).

Cannabinol activated TRPA1, inhibited TRPM8, was ineffective at TRPV1, inhibited TRPV2 (De Petrocellis et al. 2011) and showed limited agonist activity at TRPV3 and TRPV4 (De Petrocellis et al. 2012).

2.1.4.3 Δ^9-Tetrahydrocannabinolic Acid

Δ^9-Tetrahydrocannabinolic acid, THCA-A, is the naturally-occurring precursor of THC, which is abundant in the *Cannabis* plant. In binding studies, THCA-A was much weaker than THC at CB_1 or CB_2 cannabinoid receptors (McPartland et al. 2017). THCA-A was less potent than THC as an agonist at TRPA1, TRPV2, and TRPV3, as inactive at TRPV1, more active than THC at TRPV4 and equipotent as a TRPM8 antagonist (De Petrocellis et al. 2011; De Petrocellis et al. 2012). THCA was recently described as a potent PPARγ agonist in vitro, with beneficial effects in a seizure model in vivo, which could be reversed by a PPARγ antagonist (Nadal et al. 2017).

2.1.4.4 Cannabidivarin

Cannabidivarin is a close structural analogue of CBD, which displayed low-potency binding to CB_1 receptors (Hill et al. 2013), but showed sub-micromolar affinity at human recombinant CB_2 receptors (Rosenthaler et al. 2014). CBDV was a potent agonist at TRPA1, TRPV1, TRPV2, TRPV3 and TRPV4 and a potent antagonist at TRPM8 (De Petrocellis et al. 2011; De Petrocellis et al. 2012). In vivo, CBDV showed an anticonvulsant action through an unestablished mechanism (Amada et al. 2013; Hill et al. 2012; Hill et al. 2013). In addition, CBDV delayed memory deficits in mutant mice, again through an unidentified mechanism (Zamberletti et al. 2019).

2.1.4.5 Cannabidiolic Acid

Cannabidiolic acid, CBDA, is the naturally-occurring precursor of CBD. CBDA has been suggested to inhibit COX-2 (Takeda et al. 2008). CBDA was a low-potency agonist at TRPA1, less potent at TRPV1 and TRPV4, inactive at TRPV2 and TRPV3 and a low-potency antagonist at TRPM8 (De Petrocellis et al. 2011; De Petrocellis et al. 2012). CBDA antinociceptive behavioural effects were blocked by a TRPV1 antagonist in vivo (Rock et al. 2018). CBDA evoked an inhibition of nausea and vomiting behaviours in vivo; effects of which were reduced by 5-HT_{1A} receptor blockade (Bolognini et al. 2013). In vitro, CBDA appeared to act as a positive allosteric modulator of 5-HT_{1A} receptors (Bolognini et al. 2013).

2.1.4.6 Cannabigerol

Cannabigerol, CBG, has a distinct chemical structure from the other *Cannabis*-derived metabolites, with lower affinity at CB_1 and CB_2 receptors than THC (Rosenthaler et al. 2014). In contrast, it showed much higher affinity as an agonist at α_2-adrenoceptors and antagonist at 5-HT_{1A} receptors (Cascio et al. 2010). CBG was a potent agonist at TRPA1, less potent at TRPV1, TRPV2, TRPV3 and TRPV4 and a potent antagonist at TRPM8 (De Petrocellis et al. 2011; De Petrocellis et al. 2012). CBG was also able to activate both PPARα and PPARγ in vitro (D'Aniello et al. 2019). In vivo, CBG stimulated appetite in a manner yet to be explained (Brierley et al. 2016, 2017) and blocked the anti-nausea effect of CBD (Rock et al. 2011). CBG reduced colon cancer progression in vivo in a manner consistent with TRPM8 blockade (Borrelli et al. 2014).

2.1.4.7 Cannabichromene

A further chemical class of abundant *Cannabis* metabolite is cannabichromene, CBC, which exhibits lower potency at CB_1 and CB_2 receptors than THC (Rosenthaler et al. 2014). It was a very potent TRPA1 agonist with much lower potency at TRPV1, TRPV2 and TRPM8 and intermediate TRPV3 and TRPV4 potency (De Petrocellis et al. 2011; De Petrocellis et al. 2012). CBC activation of ERK activity in adult neural stem cells could be blocked by an A_1 adenosine receptor antagonist (Shinjyo and Di Marzo 2013). In vivo, CBC appeared to have anti-inflammatory properties, which was suggested to be mediated via TRPA1 channels (Romano et al. 2013).

2.1.5 Concluding Remarks

In this chapter, we have considered the evidence for bioactivity of the major cannabinoid metabolites from the *Cannabis* plant. It is clear that, although we have been aware of the predominant molecular mechanisms of action of THC for decades, there is much less knowledge of neuromolecular mechanisms for the remainder of the cannabinoids. A further point worth making is that many of these cannabinoids appear to have contradictory effects at the molecular targets which have been identified, particularly members of the TRP receptor family. Interpreting the impact of complex mixtures of these cannabinoids in vivo is consequently extremely difficult, complicated further by variation in pharmacokinetic profiles of these agents, an issue that has been researched only in limited detail.

References

Ahrens J, Leuwer M, Demir R, Krampfl K, de la Roche J, Foadi N, Karst M, Haeseler G (2009) Positive allosteric modulatory effects of ajulemic acid at strychnine-sensitive glycine alpha1- and alpha1beta-receptors. Naunyn Schmiedeberg's Arch Pharmacol 379:371–378

Alharthi N, Christensen P, Hourani W, Ortori C, Barrett DA, Bennett AJ, Chapman V, Alexander SPH (2018) N-3 polyunsaturated *N*-acylethanolamines are CB_2 cannabinoid receptor-preferring endocannabinoids. Biochim Biophys Acta Mol Cell Biol Lipids 1863:1433–1440

Amada N, Yamasaki Y, Williams CM, Whalley BJ (2013) Cannabidivarin (CBDV) suppresses pentylenetetrazole (PTZ)-induced increases in epilepsy-related gene expression. PeerJ 1:e214

Andre CM, Hausman JF, Guerriero G (2016) *Cannabis sativa*: the Plant of the Thousand and one Molecules. Front Plant Sci 7:19

Ashton JC, Glass M (2007) The cannabinoid CB2 receptor as a target for inflammation-dependent neurodegeneration. Curr Neuropharmacol 5:73–80

Atwood BK, Mackie K (2010) CB2: a cannabinoid receptor with an identity crisis. Br J Pharmacol 160:467–479

Barann M, Molderings G, Bruss M, Bonisch H, Urban BW, Gothert M (2002) Direct inhibition by cannabinoids of human 5-HT3A receptors: probable involvement of an allosteric modulatory site. Br J Pharmacol 137:589–596

Battista N, Meccariello R, Cobellis G, Fasano S, Di Tommaso M, Pirazzi V, Konje JC, Pierantoni R, Maccarrone M (2012) The role of endocannabinoids in gonadal function and fertility along the evolutionary axis. Mol Cell Endocrinol 355:1–14

Bisogno T, Hanus L, De Petrocellis L, Tchilibon S, Ponde DE, Brandi I, Moriello AS, Davis JB, Mechoulam R, Di Marzo V (2001) Molecular targets for cannabidiol and its synthetic analogues: effect on vanilloid VR1 receptors and on the cellular uptake and enzymatic hydrolysis of anandamide. Br J Pharmacol 134:845–852

Bolognini D, Costa B, Maione S, Comelli F, Marini P, Di Marzo V, Parolaro D, Ross RA, Gauson LA, Cascio MG, Pertwee RG (2010) The plant cannabinoid Δ^9-tetrahydrocannabivarin can decrease signs of inflammation and inflammatory pain in mice. Br J Pharmacol 160:677–687

Bolognini D, Rock EM, Cluny NL, Cascio MG, Limebeer CL, Duncan M, Stott CG, Javid FA, Parker LA, Pertwee RG (2013) Cannabidiolic acid prevents vomiting in *Suncus murinus* and nausea-induced behaviour in rats by enhancing $5\text{-}HT_{1A}$ receptor activation. Br J Pharmacol 168:1456–1470

Booker L, Naidu PS, Razdan RK, Mahadevan A, Lichtman AH (2009) Evaluation of prevalent phytocannabinoids in the acetic acid model of visceral nociception. Drug Alcohol Depend 105:42–47

Borrelli F, Pagano E, Romano B, Panzera S, Maiello F, Coppola D, De Petrocellis L, Buono L, Orlando P, Izzo AA (2014) Colon carcinogenesis is inhibited by the TRPM8 antagonist cannabigerol, a cannabis-derived non-psychotropic cannabinoid. Carcinogenesis 35:2787–2797

Brierley DI, Samuels J, Duncan M, Whalley BJ, Williams CM (2016) Cannabigerol is a novel, well-tolerated appetite stimulant in pre-satiated rats. Psychopharmacology 233:3603–3613

Brierley DI, Samuels J, Duncan M, Whalley BJ, Williams CM (2017) A cannabigerol-rich *Cannabis sativa* extract, devoid of Δ^9-tetrahydrocannabinol, elicits hyperphagia in rats. Behav Pharmacol 28:280–284

Campos AC, Guimaraes FS (2008) Involvement of 5HT1A receptors in the anxiolytic-like effects of cannabidiol injected into the dorsolateral periaqueductal gray of rats. Psychopharmacology 199:223–230

Carayon P, Marchand J, Dussossoy D, Derocq JM, Jbilo O, Bord A, Bouaboula M, Galiegue S, Mondiere P, Penarier G, Fur GL, Defrance T, Casellas P (1998) Modulation and functional involvement of CB2 peripheral cannabinoid receptors during B-cell differentiation. Blood 92:3605–3615

Carrier EJ, Auchampach JA, Hillard CJ (2006) Inhibition of an equilibrative nucleoside transporter by cannabidiol: a mechanism of cannabinoid immunosuppression. Proc Natl Acad Sci U S A 103:7895–7900

Cascio MG, Gauson LA, Stevenson LA, Ross RA, Pertwee RG (2010) Evidence that the plant cannabinoid cannabigerol is a highly potent α_2-adrenoceptor agonist and moderately potent $5HT_{1A}$ receptor antagonist. Br J Pharmacol 159:129–141

Cascio MG, Zamberletti E, Marini P, Parolaro D, Pertwee RG (2015) The phytocannabinoid, Δ^9-tetrahydrocannabivarin, can act through $5\text{-}HT_{1A}$ receptors to produce antipsychotic effects. Br J Pharmacol 172:1305–1318

Castillo A, Tolon MR, Fernandez-Ruiz J, Romero J, Martinez-Orgado J (2010) The neuroprotective effect of cannabidiol in an in vitro model of newborn hypoxic-ischemic brain damage in mice is mediated by CB_2 and adenosine receptors. Neurobiol Dis 37:434–440

Costa B, Giagnoni G, Franke C, Trovato AE, Colleoni M (2004) Vanilloid TRPV1 receptor mediates the antihyperalgesic effect of the nonpsychoactive cannabinoid, cannabidiol, in a rat model of acute inflammation. Br J Pharmacol 143:247–250

D'Aniello E, Fellous T, Iannotti FA, Gentile A, Allara M, Balestrieri F, Gray R, Amodeo P, Vitale RM, Di Marzo V (2019) Identification and characterization of phytocannabinoids as novel dual PPARα/γ agonists by a computational and in vitro experimental approach. Biochim Biophys Acta Gen Subj 1863:586–597

De Jesus ML, Salles J, Meana JJ, Callado LF (2006) Characterization of CB1 cannabinoid receptor immunoreactivity in postmortem human brain homogenates. Neuroscience 140:635–643

De Petrocellis L, Ligresti A, Moriello AS, Allara M, Bisogno T, Petrosino S, Stott CG, Di Marzo V (2011) Effects of cannabinoids and cannabinoid-enriched *Cannabis* extracts on TRP channels and endocannabinoid metabolic enzymes. Br J Pharmacol 163:1479–1494

De Petrocellis L, Orlando P, Moriello AS, Aviello G, Stott C, Izzo AA, Di Marzo V (2012) Cannabinoid actions at TRPV channels: effects on TRPV3 and TRPV4 and their potential relevance to gastrointestinal inflammation. Acta Physiol (Oxf) 204:255–266

De Petrocellis L, Vellani V, Schiano-Moriello A, Marini P, Magherini PC, Orlando P, Di Marzo V (2008) Plant-derived cannabinoids modulate the activity of transient receptor potential channels of ankyrin type-1 and melastatin type-8. J Pharmacol Exp Ther 325:1007–1015

Devane WA, Hanus L, Breuer A, Pertwee RG, Stevenson LA, Griffin G, Gibson D, Mandelbaum A, Etinger A, Mechoulam R (1992) Isolation and structure of a brain constituent that binds to the cannabinoid receptor. Science 258:1946–1949

Di Marzo V, Despres JP (2009) CB1 antagonists for obesity--what lessons have we learned from rimonabant? Nat Rev Endocrinol 5:633–638

Farrimond JA, Whalley BJ, Williams CM (2012) Cannabinol and cannabidiol exert opposing effects on rat feeding patterns. Psychopharmacology 223:117–129

Felder CC, Joyce KE, Briley EM, Mansouri J, Mackie K, Blond O, Lai Y, Ma AL, Mitchell RL (1995) Comparison of the pharmacology and signal transduction of the human cannabinoid CB1 and CB2 receptors. Mol Pharmacol 48:443–450

Foadi N, Leuwer M, Demir R, Dengler R, Buchholz V, de la Roche J, Karst M, Haeseler G, Ahrens J (2010) Lack of positive allosteric modulation of mutated α_1S267I glycine receptors by cannabinoids. Naunyn Schmiedeberg's Arch Pharmacol 381:477–482

Galiegue S, Mary S, Marchand J, Dussossoy D, Carriere D, Carayon P, Bouaboula M, Shire D, Le Fur G, Casellas P (1995) Expression of central and peripheral cannabinoid receptors in human immune tissues and leukocyte subpopulations. Eur J Biochem 232:54–61

Gaoni Y, Mechoulam R (1964) Isolation, structure and partial synthesis of active constituent of hashish. J Am Chem Soc 86:1646–1647

Garcia C, Palomo-Garo C, Garcia-Arencibia M, Ramos J, Pertwee R, Fernandez-Ruiz J (2011) Symptom-relieving and neuroprotective effects of the phytocannabinoid Δ^9-THCV in animal models of Parkinson's disease. Br J Pharmacol 163:1495–1506

Gerard CM, Mollereau C, Vassart G, Parmentier M (1991) Molecular cloning of a human cannabinoid receptor which is also expressed in testis. Biochem J 279 (Pt 1):129–134

Glass M, Felder CC (1997) Concurrent stimulation of cannabinoid CB1 and dopamine D2 receptors augments cAMP accumulation in striatal neurons: evidence for a G_s linkage to the CB1 receptor. J Neurosci 17:5327–5333

Gonca E, Darici F (2015) The effect of cannabidiol on ischemia/reperfusion-induced ventricular arrhythmias: the role of adenosine A1 receptors. J Cardiovasc Pharmacol Ther 20:76–83

Gong JP, Onaivi ES, Ishiguro H, Liu QR, Tagliaferro PA, Brusco A, Uhl GR (2006) Cannabinoid CB2 receptors: immunohistochemical localization in rat brain. Brain Res 1071:10–23

Grimaldi P, Orlando P, Di Siena S, Lolicato F, Petrosino S, Bisogno T, Geremia R, De Petrocellis L, Di Marzo V (2009) The endocannabinoid system and pivotal role of the CB2 receptor in mouse spermatogenesis. Proc Natl Acad Sci U S A 106:11131–11136

Hejazi N, Zhou C, Oz M, Sun H, Ye JH, Zhang L (2006) Δ^9-tetrahydrocannabinol and endogenous cannabinoid anandamide directly potentiate the function of glycine receptors. Mol Pharmacol 69:991–997

Herkenham M, Lynn AB, Johnson MR, Melvin LS, de Costa BR, Rice KC (1991) Characterization and localization of cannabinoid receptors in rat brain: a quantitative in vitro autoradiographic study. J Neurosci 11:563–583

Hill TD, Cascio MG, Romano B, Duncan M, Pertwee RG, Williams CM, Whalley BJ, Hill AJ (2013) Cannabidivarin-rich cannabis extracts are anticonvulsant in mouse and rat via a CB_1 receptor-independent mechanism. Br J Pharmacol 170:679–692

Hill AJ, Mercier MS, Hill TD, Glyn SE, Jones NA, Yamasaki Y, Futamura T, Duncan M, Stott CG, Stephens GJ, Williams CM, Whalley BJ (2012) Cannabidivarin is anticonvulsant in mouse and rat. Br J Pharmacol 167:1629–1642

Hoffman AF, Oz M, Yang R, Lichtman AH, Lupica CR (2007) Opposing actions of chronic Δ^9-tetrahydrocannabinol and cannabinoid antagonists on hippocampal long-term potentiation. Learn Mem 14:63–74

Howlett AC, Barth F, Bonner TI, Cabral G, Casellas P, Devane WA, Felder CC, Herkenham M, Mackie K, Martin BR, Mechoulam R, Pertwee RG (2002) International Union of Pharmacology. XXVII. Classification of cannabinoid receptors. Pharmacol Rev 54:161–202

Hua T, Vemuri K, Nikas SP, Laprairie RB, Wu Y, Qu L, Pu M, Korde A, Jiang S, Ho JH, Han GW, Ding K, Li X, Liu H, Hanson MA, Zhao S, Bohn LM, Makriyannis A, Stevens RC, Liu ZJ (2017) Crystal structures of agonist-bound human cannabinoid receptor CB_1. Nature 547:468–471

Hua T, Vemuri K, Pu M, Qu L, Han GW, Wu Y, Zhao S, Shui W, Li S, Korde A, Laprairie RB, Stahl EL, Ho JH,

Zvonok N, Zhou H, Kufareva I, Wu B, Zhao Q, Hanson MA, Bohn LM, Makriyannis A, Stevens RC, Liu ZJ (2016) Crystal structure of the human cannabinoid receptor CB1. Cell 167(750–762):e714

Iannotti FA, Hill CL, Leo A, Alhusaini A, Soubrane C, Mazzarella E, Russo E, Whalley BJ, Di Marzo V, Stephens GJ (2014) Nonpsychotropic plant cannabinoids, cannabidivarin (CBDV) and cannabidiol (CBD), activate and desensitize transient receptor potential vanilloid 1 (TRPV1) channels in vitro: potential for the treatment of neuronal hyperexcitability. ACS Chem Neurosci 5:1131–1141

Irving A, Abdulrazzaq G, Chan SLF, Penman J, Harvey J, Alexander SPH (2017) Cannabinoid receptor-related orphan G protein-coupled receptors. Adv Pharmacol 80:223–247

Izzo AA, Sharkey KA (2010) Cannabinoids and the gut: new developments and emerging concepts. Pharmacol Ther 126:21–38

Jones NA, Hill AJ, Smith I, Bevan SA, Williams CM, Whalley BJ, Stephens GJ (2010) Cannabidiol displays antiepileptiform and antiseizure properties in vitro and in vivo. J Pharmacol Exp Ther 332:569–577

Katsidoni V, Anagnostou I, Panagis G (2013) Cannabidiol inhibits the reward-facilitating effect of morphine: involvement of 5-HT1A receptors in the dorsal raphe nucleus. Addict Biol 18:286–296

Kendall DA, Yudowski GA (2016) Cannabinoid receptors in the central nervous system: their signaling and roles in disease. Front Cell Neurosci 10:294

Kohno M, Hasegawa H, Inoue A, Muraoka M, Miyazaki T, Oka K, Yasukawa M (2006) Identification of *N*-arachidonylglycine as the endogenous ligand for orphan G-protein-coupled receptor GPR18. Biochem Biophys Res Commun 347:827–832

Krishna Kumar K, Shalev-Benami M, Robertson MJ, Hu H, Banister SD, Hollingsworth SA, Latorraca NR, Kato HE, Hilger D, Maeda S, Weis WI, Farrens DL, Dror RO, Malhotra SV, Kobilka BK, Skiniotis G (2019) Structure of a signaling cannabinoid receptor 1-G protein complex. Cell 176(448–458):e412

Laprairie RB, Bagher AM, Kelly ME, Denovan-Wright EM (2015) Cannabidiol is a negative allosteric modulator of the cannabinoid CB_1 receptor. Br J Pharmacol 172:4790–4805

Lauckner JE, Hille B, Mackie K (2005) The cannabinoid agonist WIN55,212-2 increases intracellular calcium via CB_1 receptor coupling to $G_{q/11}$ G proteins. Proc Natl Acad Sci U S A 102:19144–19149

Lauckner JE, Jensen JB, Chen HY, Lu HC, Hille B, Mackie K (2008) GPR55 is a cannabinoid receptor that increases intracellular calcium and inhibits M current. Proc Natl Acad Sci U S A 105:2699–2704

Ledgerwood CJ, Greenwood SM, Brett RR, Pratt JA, Bushell TJ (2011) Cannabidiol inhibits synaptic transmission in rat hippocampal cultures and slices via multiple receptor pathways. Br J Pharmacol 162:286–294

Leishman E, Manchanda M, Thelen R, Miller S, Mackie K, Bradshaw HB (2018) Cannabidiol's Upregulation of N-acyl Ethanolamines in the central nervous system requires N-acyl Phosphatidyl ethanolamine-specific phospholipase D. Cannabis Cannabinoid Res 3:228–241

Li X, Hua T, Vemuri K, Ho JH, Wu Y, Wu L, Popov P, Benchama O, Zvonok N, Locke K, Qu L, Han GW, Iyer MR, Cinar R, Coffey NJ, Wang J, Wu M, Katritch V, Zhao S, Kunos G, Bohn LM, Makriyannis A, Stevens RC, Liu ZJ (2019) Crystal structure of the human cannabinoid receptor CB2. Cell 176(459–467):e413

Liu QR, Pan CH, Hishimoto A, Li CY, Xi ZX, Llorente-Berzal A, Viveros MP, Ishiguro H, Arinami T, Onaivi ES, Uhl GR (2009) Species differences in cannabinoid receptor 2 (CNR2 gene): identification of novel human and rodent CB2 isoforms, differential tissue expression and regulation by cannabinoid receptor ligands. Genes Brain Behav 8:519–530

Lu J, Fan S, Zou G, Hou Y, Pan T, Guo W, Yao L, Du F, Homanics GE, Liu D, Zhang L, Xiong W (2018) Involvement of glycine receptor alpha1 subunits in cannabinoid-induced analgesia. Neuropharmacology 133:224–232

Mackie K, Lai Y, Westenbroek R, Mitchell R (1995) Cannabinoids activate an inwardly rectifying potassium conductance and inhibit Q-type calcium currents in AtT20 cells transfected with rat brain cannabinoid receptor. J Neurosci 15:6552–6561

Maione S, Piscitelli F, Gatta L, Vita D, De Petrocellis L, Palazzo E, de Novellis V, Di Marzo V (2011) Non-psychoactive cannabinoids modulate the descending pathway of antinociception in anaesthetized rats through several mechanisms of action. Br J Pharmacol 162:584–596

Marsicano G, Lutz B (1999) Expression of the cannabinoid receptor CB1 in distinct neuronal subpopulations in the adult mouse forebrain. Eur J Neurosci 11:4213–4225

Matsuda LA, Lolait SJ, Brownstein MJ, Young AC, Bonner TI (1990) Structure of a cannabinoid receptor and functional expression of the cloned cDNA. Nature 346:561–564

McHugh D, Page J, Dunn E, Bradshaw HB (2012) Δ^9-Tetrahydrocannabinol and N-arachidonyl glycine are full agonists at GPR18 receptors and induce migration in human endometrial HEC-1B cells. Br J Pharmacol 165:2414–2424

McPartland JM, MacDonald C, Young M, Grant PS, Furkert DP, Glass M (2017) Affinity and efficacy studies of Tetrahydrocannabinolic acid a at cannabinoid receptor types one and two. Cannabis Cannabinoid Res 2:87–95

Mechoulam R, Ben-Shabat S, Hanus L, Ligumsky M, Kaminski NE, Schatz AR, Gopher A, Almog S, Martin BR, Compton DR et al (1995) Identification of an endogenous 2-monoglyceride, present in canine gut,

that binds to cannabinoid receptors. Biochem Pharmacol 50:83–90
Mechoulam R, Hanus L (2002) Cannabidiol: an overview of some chemical and pharmacological aspects Part I: chemical aspects. Chem Phys Lipids 121:35–43
Mechoulam R, Parker LA (2013) The endocannabinoid system and the brain. Annu Rev Psychol 64:21–47
Melges FT, Tinklenberg JR, Hollister LE, Gillespie HK (1970) Marihuana and temporal disintegration. Science 168:1118–1120
Mishima K, Hayakawa K, Abe K, Ikeda T, Egashira N, Iwasaki K, Fujiwara M (2005) Cannabidiol prevents cerebral infarction via a serotonergic 5-hydroxytryptamine$_{1A}$ receptor-dependent mechanism. Stroke 36:1077–1082
Munro S, Thomas KL, Abu-Shaar M (1993) Molecular characterization of a peripheral receptor for cannabinoids. Nature 365:61–65
Nadal X, Del Rio C, Casano S, Palomares B, Ferreiro-Vera C, Navarrete C, Sanchez-Carnerero C, Cantarero I, Bellido ML, Meyer S, Morello G, Appendino G, Munoz E (2017) Tetrahydrocannabinolic acid is a potent PPARγ agonist with neuroprotective activity. Br J Pharmacol 174:4263–4276
Oddi S, Dainese E, Sandiford S, Fezza F, Lanuti M, Chiurchiu V, Totaro A, Catanzaro G, Barcaroli D, De Laurenzi V, Centonze D, Mukhopadhyay S, Selent J, Howlett AC, Maccarrone M (2012) Effects of palmitoylation of Cys^{415} in helix 8 of the CB_1 cannabinoid receptor on membrane localization and signalling. Br J Pharmacol 165:2635–2651
Oka S, Toshida T, Maruyama K, Nakajima K, Yamashita A, Sugiura T (2009) 2-Arachidonoyl-*sn*-glycero-3-phosphoinositol: a possible natural ligand for GPR55. J Biochem 145:13–20
O'Sullivan SE (2016) An update on PPAR activation by cannabinoids. Br J Pharmacol 173:1899–1910
O'Sullivan SE, Tarling EJ, Bennett AJ, Kendall DA, Randall MD (2005) Novel time-dependent vascular actions of Δ^9-tetrahydrocannabinol mediated by peroxisome proliferator-activated receptor gamma. Biochem Biophys Res Commun 337:824–831
Overton HA, Babbs AJ, Doel SM, Fyfe MC, Gardner LS, Griffin G, Jackson HC, Procter MJ, Rasamison CM, Tang-Christensen M, Widdowson PS, Williams GM, Reynet C (2006) Deorphanization of a G protein-coupled receptor for oleoylethanolamide and its use in the discovery of small-molecule hypophagic agents. Cell Metab 3:167–175
Oz M, Zhang L, Ravindran A, Morales M, Lupica CR (2004) Differential effects of endogenous and synthetic cannabinoids on alpha7-nicotinic acetylcholine receptor-mediated responses in Xenopus oocytes. J Pharmacol Exp Ther 310:1152–1160
Pazos MR, Mohammed N, Lafuente H, Santos M, Martinez-Pinilla E, Moreno E, Valdizan E, Romero J, Pazos A, Franco R, Hillard CJ, Alvarez FJ, Martinez-Orgado J (2013) Mechanisms of cannabidiol neuroprotection in hypoxic-ischemic newborn pigs: role of 5HT(1A) and CB2 receptors. Neuropharmacology 71:282–291
Pertwee RG (1997) Pharmacology of cannabinoid CB_1 and CB_2 receptors. Pharmacol Ther 74:129–180
Pertwee RG, Howlett AC, Abood ME, Alexander SPH, Di Marzo V, Elphick MR, Greasley PJ, Hansen HS, Kunos G, Mackie K, Mechoulam R, Ross RA (2010) International Union of Basic and Clinical Pharmacology. LXXIX. Cannabinoid receptors and their ligands: beyond CB_1 and CB_2. Pharmacol Rev 62:588–631
Qin N, Neeper MP, Liu Y, Hutchinson TL, Lubin ML, Flores CM (2008) TRPV2 is activated by cannabidiol and mediates CGRP release in cultured rat dorsal root ganglion neurons. J Neurosci 28:6231–6238
Rao GK, Kaminski NE (2006) Induction of intracellular calcium elevation by Delta9-tetrahydrocannabinol in T cells involves TRPC1 channels. J Leukoc Biol 79:202–213
Rhee MH, Vogel Z, Barg J, Bayewitch M, Levy R, Hanus L, Breuer A, Mechoulam R (1997) Cannabinol derivatives: binding to cannabinoid receptors and inhibition of adenylylcyclase. J Med Chem 40:3228–3233
Rinaldi-Carmona M, Barth F, Heaulme M, Alonso R, Shire D, Congy C, Soubrie P, Breliere JC, Le Fur G (1995) Biochemical and pharmacological characterisation of SR141716A, the first potent and selective brain cannabinoid receptor antagonist. Life Sci 56:1941–1947
Rock EM, Bolognini D, Limebeer CL, Cascio MG, Anavi-Goffer S, Fletcher PJ, Mechoulam R, Pertwee RG, Parker LA (2012) Cannabidiol, a non-psychotropic component of cannabis, attenuates vomiting and nausea-like behaviour via indirect agonism of 5-HT (1A) somatodendritic autoreceptors in the dorsal raphe nucleus. Br J Pharmacol 165:2620–2634
Rock EM, Goodwin JM, Limebeer CL, Breuer A, Pertwee RG, Mechoulam R, Parker LA (2011) Interaction between non-psychotropic cannabinoids in marihuana: effect of cannabigerol (CBG) on the anti-nausea or anti-emetic effects of cannabidiol (CBD) in rats and shrews. Psychopharmacology 215:505–512
Rock EM, Limebeer CL, Parker LA (2018) Effect of cannabidiolic acid and Δ^9-tetrahydrocannabinol on carrageenan-induced hyperalgesia and edema in a rodent model of inflammatory pain. Psychopharmacology 235:3259–3271
Romano B, Borrelli F, Fasolino I, Capasso R, Piscitelli F, Cascio M, Pertwee R, Coppola D, Vassallo L, Orlando P, Di Marzo V, Izzo A (2013) The cannabinoid TRPA1 agonist cannabichromene inhibits nitric oxide production in macrophages and ameliorates murine colitis. Br J Pharmacol 169:213–229
Romero J, Garcia-Palomero E, Castro JG, Garcia-Gil L, Ramos JA, Fernandez-Ruiz JJ (1997) Effects of chronic exposure to Δ^9-tetrahydrocannabinol on cannabinoid receptor binding and mRNA levels in several rat brain regions. Brain Res Mol Brain Res 46:100–108

Romero-Sandoval A, Nutile-McMenemy N, DeLeo JA (2008) Spinal microglial and perivascular cell cannabinoid receptor type 2 activation reduces behavioral hypersensitivity without tolerance after peripheral nerve injury. Anesthesiology 108:722–734

Rosenthaler S, Pohn B, Kolmanz C, Huu CN, Krewenka C, Huber A, Kranner B, Rausch WD, Moldzio R (2014) Differences in receptor binding affinity of several phytocannabinoids do not explain their effects on neural cell cultures. Neurotoxicol Teratol 46:49–56

Ross HR, Napier I, Connor M (2008) Inhibition of recombinant human T-type calcium channels by Δ^9-tetrahydrocannabinol and cannabidiol. J Biol Chem 283:16124–16134

Russo EB, Burnett A, Hall B, Parker KK (2005) Agonistic properties of cannabidiol at 5-HT1a receptors. Neurochem Res 30:1037–1043

Ryberg E, Vu HK, Larsson N, Groblewski T, Hjorth S, Elebring T, Sjogren S, Greasley PJ (2005) Identification and characterisation of a novel splice variant of the human CB1 receptor. FEBS Lett 579:259–264

Schatz AR, Lee M, Condie RB, Pulaski JT, Kaminski NE (1997) Cannabinoid receptors CB1 and CB2: a characterization of expression and adenylate cyclase modulation within the immune system. Toxicol Appl Pharmacol 142:278–287

Shao Z, Yin J, Chapman K, Grzemska M, Clark L, Wang J, Rosenbaum DM (2016) High-resolution crystal structure of the human CB1 cannabinoid receptor. Nature 540:602–606

Shinjyo N, Di Marzo V (2013) The effect of cannabichromene on adult neural stem/progenitor cells. Neurochem Int 63:432–437

Shire D, Carillon C, Kaghad M, Calandra B, Rinaldi-Carmona M, Le Fur G, Caput D, Ferrara P (1995) An amino-terminal variant of the central cannabinoid receptor resulting from alternative splicing. J Biol Chem 270:3726–3731

Song C, Howlett AC (1995) Rat brain cannabinoid receptors are N-linked glycosylated proteins. Life Sci 56:1983–1989

Stella N (2004) Cannabinoid signaling in glial cells. Glia 48:267–277

Sugiura T, Kondo S, Sukagawa A, Nakane S, Shinoda A, Itoh K, Yamashita A, Waku K (1995) 2-Arachidonoylglycerol: a possible endogenous cannabinoid receptor ligand in brain. Biochem Biophys Res Commun 215:89–97

Sun Y, Alexander SP, Garle MJ, Gibson CL, Hewitt K, Murphy SP, Kendall DA, Bennett AJ (2007) Cannabinoid activation of PPARα; a novel neuroprotective mechanism. Br J Pharmacol 152:734–743

Takeda S, Ikeda E, Su S, Harada M, Okazaki H, Yoshioka Y, Nishimura H, Ishii H, Kakizoe K, Taniguchi A, Tokuyasu M, Himeno T, Watanabe K, Omiecinski CJ, Aramaki H (2014) Delta(9)-THC modulation of fatty acid 2-hydroxylase (FA2H) gene expression: possible involvement of induced levels of PPARalpha in MDA-MB-231 breast cancer cells. Toxicology 326:18–24

Takeda S, Misawa K, Yamamoto I, Watanabe K (2008) Cannabidiolic acid as a selective cyclooxygenase-2-inhibitory component in cannabis. Drug Metab Dispos 36:1917–1921

Thomas A, Stevenson LA, Wease KN, Price MR, Baillie G, Ross RA, Pertwee RG (2005) Evidence that the plant cannabinoid Δ^9-tetrahydrocannabivarin is a cannabinoid CB_1 and CB_2 receptor antagonist. Br J Pharmacol 146:917–926

Turcotte C, Blanchet MR, Laviolette M, Flamand N (2016) The CB2 receptor and its role as a regulator of inflammation. Cell Mol Life Sci 73:4449–4470

Turkanis SA, Karler R, Partlow LM (1991) Differential effects of delta-9-tetrahydrocannabinol and its 11-hydroxy metabolite on sodium current in neuroblastoma cells. Brain Res 560:245–250

Voets T, Droogmans G, Wissenbach U, Janssens A, Flockerzi V, Nilius B (2004) The principle of temperature-dependent gating in cold- and heat-sensitive TRP channels. Nature 430:748–754

Watanabe K, Kayano Y, Matsunaga T, Yamamoto I, Yoshimura H (1996) Inhibition of anandamide amidase activity in mouse brain microsomes by cannabinoids. Biol Pharm Bull 19:1109–1111

Wolfson ML, Muzzio DO, Ehrhardt J, Franchi AM, Zygmunt M, Jensen F (2016) Expression analysis of cannabinoid receptors 1 and 2 in B cells during pregnancy and their role on cytokine production. J Reprod Immunol 116:23–27

Xiong W, Cheng K, Cui T, Godlewski G, Rice KC, Xu Y, Zhang L (2011) Cannabinoid potentiation of glycine receptors contributes to cannabis-induced analgesia. Nat Chem Biol 7:296–303

Xiong W, Cui T, Cheng K, Yang F, Chen SR, Willenbring D, Guan Y, Pan HL, Ren K, Xu Y, Zhang L (2012) Cannabinoids suppress inflammatory and neuropathic pain by targeting α3 glycine receptors. J Exp Med 209:1121–1134

Xiong W, Koo BN, Morton R, Zhang L (2011) Psychotropic and nonpsychotropic cannabis derivatives inhibit human $5\text{-}HT_{3A}$ receptors through a receptor desensitization-dependent mechanism. Neuroscience 184:28–37

Yang KH, Galadari S, Isaev D, Petroianu G, Shippenberg TS, Oz M (2010) The nonpsychoactive cannabinoid cannabidiol inhibits 5-hydroxytryptamine3A receptor-mediated currents in Xenopus laevis oocytes. J Pharmacol Exp Ther 333:547–554

Yiangou Y, Facer P, Durrenberger P, Chessell IP, Naylor A, Bountra C, Banati RR, Anand P (2006) COX-2, CB2 and P2X7-immunoreactivities are increased in activated microglial cells/macrophages of multiple sclerosis and amyotrophic lateral sclerosis spinal cord. BMC Neurol 6:12

Zachariou M, Alexander SP, Coombes S, Christodoulou C (2013) A biophysical model of endocannabinoid-mediated short term depression in hippocampal inhibition. PLoS One 8:e58926

Zamberletti E, Gabaglio M, Piscitelli F, Brodie JS, Woolley-Roberts M, Barbiero I, Tramarin M, Binelli G, Landsberger N, Kilstrup-Nielsen C, Rubino T, Di Marzo V, Parolaro D (2019) Cannabidivarin completely rescues cognitive deficits and delays neurological and motor defects in male Mecp2 mutant mice. J Psychopharmacol 33:894–907

Zhang J, Hoffert C, Vu HK, Groblewski T, Ahmad S, O'Donnell D (2003) Induction of CB2 receptor expression in the rat spinal cord of neuropathic but not inflammatory chronic pain models. Eur J Neurosci 17:2750–2754

Zhang R, Kim TK, Qiao ZH, Cai J, Pierce WM Jr, Song ZH (2007) Biochemical and mass spectrometric characterization of the human CB2 cannabinoid receptor expressed in *Pichia pastoris*--importance of correct processing of the N-terminus. Protein Expr Purif 55:225–235

Zygmunt PM, Petersson J, Andersson DA, Chuang H, Sorgard M, Di Marzo V, Julius D, Hogestatt ED (1999) Vanilloid receptors on sensory nerves mediate the vasodilator action of anandamide. Nature 400:452–457

Neuropharmacological Effects of the Main Phytocannabinoids: A Narrative Review

3

Rafael G. dos Santos, Jaime E. C. Hallak, and José Alexandre S. Crippa

Abstract

Cannabis can synthetize more than 400 compounds, including terpenes, flavonoids, and more than 100 phytocannabinoids. The main phytocannabinoids are Δ-9-tetrahydrocannabinol (THC) and cannabidiol (CBD). Cannabis-based products are used as medicines in several countries. In this text, we present an overview of the main neurochemical mechanisms of action of the phytocannabinoids, especially THC and CBD. We also reviewed the indications and adverse effects of the main cannabis-based medicinal products. THC acts as a partial agonist at cannabinoid 1/2 receptors ($CB_{1/2}$). It is responsible for the characteristic effects of cannabis, such as euphoria, relaxation, and changes in perceptions. THC can also produce dysphoria, anxiety, and psychotic symptoms. THC is used therapeutically in nausea and vomiting due to chemotherapy, as an appetite stimulant, and in chronic pain. CBD acts as a noncompetitive negative allosteric modulator of the CB_1 receptor, as an inverse agonist of the CB_2 receptor, and as an inhibitor of the reuptake of the endocannabinoid anandamide. Moreover, CBD also activates $5\text{-}HT_{1A}$ serotonergic receptors and vanilloid receptors. Its use in treatment-resistant epilepsy syndromes is approved in some countries. CBD does not produce the typical effects associated with THC and has anxiolytic and antipsychotic effects. Some of the most common adverse effects of CBD are diarrhea, somnolence, nausea, and transaminase elevations (with concomitant use of antiepileptics). The mechanisms of action involved in both the therapeutic and adverse effects of the phytocannabinoids are not fully understood, involving not only the endocannabinoid system. This "promiscuous" pharmacology could be responsible for their wide therapeutic spectrum.

Keywords

Cannabinoids · Phytocannabinoids · Endocannabinoids · Mechanisms of action · Neuropharmacology

3.1 Introduction

Cannabis is a botanical genus composed of three species (*C. sativa*, *C. indica*, and *C. ruderalis*) that are broadly differentiated by their genetic and chemical ability to produce more or less of the two main phytocannabinoids: Δ-9-tetrahydrocannabinol (Δ-9-THC or simply THC) and

R. G. dos Santos · J. E. C. Hallak · J. A. S. Crippa (✉)
Department of Neurosciences and Behavior, Ribeirão Preto Medical School, University of São Paulo, Ribeirão Preto, Brazil

National Institute of Science and Technology–Translational Medicine, Ribeirão Preto, São Paulo, Brazil
e-mail: jcrippa@fmrp.usp.br

E. Murillo-Rodriguez et al. (eds.), *Cannabinoids and Neuropsychiatric Disorders*, Advances in Experimental Medicine and Biology 1264, https://doi.org/10.1007/978-3-030-57369-0_3

cannabidiol (CBD). Thus, species rich in THC are used for recreational and medicinal properties, while species with low THC content and high CBD content are used to produce seed and fiber and are also used for medicinal purposes (Hillig 2005; Andre et al. 2016). Cannabis can synthetize more than 400 compounds, including terpenes, flavonoids, and more than 100 phytocannabinoids including THC, CBD, Δ-8-tetrahydrocannabinol (Δ-8-THC), Δ-9-tetrahydrocannabivarin (Δ-9-THCV), Δ-9-tetrahydrocannabinolic acid (Δ-9-THCA), cannabinol (CBN), cannabidivarin (CBDV), cannabigerol (CBG), cannabichromene (CBC), cannabidiolic acid (CBDA), etc.(Andre et al. 2016; Izzo et al. 2009; Ranieri et al. 2016). Thus, the pharmacology, psychoactivity, therapeutic or toxic effects of cannabis varieties and "strains" will depend on the synergetic effects of all these compounds (Andre et al. 2016; MacCallum and Russo 2018). Accumulating evidence shows that *skunk*-like (high-potency) cannabis, rich in THC, is associated with a higher frequency of adverse reactions compared to low-potency (low THC/high CBD content) cannabis (Di Forti et al. 2015; Volkow et al. 2016).

Cannabis-derived products are available in different forms (Table 3.1). Herbal or raw cannabis (from nonstandardized to standardized varieties with known content THC and CBD, e.g., Bedrocan®, Bedrobinol®, Bediol®, Bedica®, Bedrolite®) and cannabis extracts/oils (from homemade to standardized medications, e.g., Sativex®, Epidiolex®, Purodiol®, TIL-TC150) are currently authorized for medicinal purposes including chronic pain, sleep disorders, anxiety and mood disorders, Parkinson disease, epilepsy, etc. in 30 States of the United States (US) and in some countries such as Canada, the Netherlands, Italy, Germany, Israel, and Brazil (Abuhasira et al. 2018; Bramness et al. 2018). In some countries such as the US and Brazil, several of the available extracts (except for Sativex® and Epidiolex®) are not standardized and show wide variation in cannabinoid content (Vandrey et al. 2015; Crippa et al. 2016). Moreover, some medicinal indications for these products were not assessed and approved after randomized controlled trials (RCTs) (for example, Parkinson disease in some US States and in Brazil). Furthermore, in other contexts, such as in some European countries, cannabis-based products are used only in rare or specific diseases (such as palliative care) (Abuhasira et al. 2018; Bramness et al. 2018). Nevertheless, although the recreational use of cannabis is associated with several adverse effects such as cognitive impairment and psychiatric disorders (Di Forti et al. 2015; Volkow et al. 2016), observational, open-label, and RCTs suggest that medicinal cannabis and cannabis-based products (standardized and nonstandardized) could be effective for some indications such as chronic pain, epilepsy, cancer-associated pain, and nausea, and are generally well tolerated (Gruber et al. 2016; Yassin and Robinson 2017; Abuhasira et al. 2018; Bellnier et al. 2018; de Hoop et al. 2018; Gruber et al. 2018; Hausman-Kedem et al. 2018; McCoy et al. 2018; Mondello et al. 2018; Sarid et al. 2018). However, most of these studies only reported short treatment periods and short follow-up periods, thus possible long-term effects of these cannabis-based products are largely unknown and should be further investigated. In fact, that is also true for pure cannabinoids such as THC- and CBD-based products.

Until June 2018, neither the US Food and Drug Administration (FDA) nor the European Medicines Agency (EMA) had approved a drug product containing or derived directly from herbal cannabis. This scenario has recently changed when the FDA approved on June 25, 2018 the use of Epidiolex® (a purified oral cannabis extract rich in CBD; GW Pharmaceuticals Plc.) for treating seizures associated with Lennox-Gastaut syndrome or Dravet syndrome in patients with 2 years of age and older (Kaufman 2018; Rubin 2018). This is the first FDA-approved drug that contains a purified component derived directly from cannabis. Another cannabis-based medicinal product from GW Pharmaceuticals is Nabiximols (Sativex®), an oromucosal spray containing THC and CBD in a 1:1 THC:CBD ratio approved in 29 countries for the treatment of multiple sclerosis associated spasticity and neuropathic pain (Abuhasira et al. 2018; Bramness et al. 2018).

Table 3.1 Summary of the main cannabis-derived products[1]

Product	Active compounds	Administration routes
Herbal cannabis (*Cannabis sp.*, Bedrocan®, Bedrobinol®, Bediol®, Bedica®, Bedrolite®) and nonstandardized cannabis extracts/oils	Mainly THC and CBD, but also dozens of phytocannabinoids, terpenes, etc.	Smoked, vaporized, oral
Dronabinol (Marinol®, Syndros®)	Synthetic THC	Oral
Nabilone (Cesamet®, Canemes®)	Synthetic THC analog	Oral
Nabiximols (Sativex®)	Cannabis extract with THC and CBD in a 1:1 THC:CBD ratio, with minor quantities of other phytocannabinoids, terpenes, etc.	Oromucosal spray
TIL-TC150	Cannabis extractcontaining only purified CBD and THC in50:1 CBD:THC ratio	Oral
CBD (Epidiolex®)	CBD extract, with minor quantities of phytocannabinoids, terpenes, etc.	Oral
CBD	Purified or synthetic CBD	Oral

[1]Adapted from the following references: Koppel et al. 2014; Whiting et al. 2015; Abuhasira et al. 2018; Bramness et al. 2018; MacCallum and Russo 2018; McCoy et al. 2018
CBD cannabidiol, *THC* tetrahydrocannabinol

Epidiolex® is not currently approved by the EMA, and Sativex® is not currently approved by the FDA, but things might change in 2019 for both substances in both agencies. More recently, TIL-TC150 (Tilray Inc.), a cannabis extract with purified CBD and THC in a 50:1 CBD:THC that complies with GMP standards is being investigated in Canada for the treatment seizures in children with Dravet syndrome (McCoy et al. 2018).

Moreover, the synthetic versions of the main phytocannabinoids (THC and CBD) are currently approved medications (THC) or are under clinical investigation (CBD) in some countries. For instance, synthetic THC or Dronabinol (Marinol®, AbbVie Inc.; Syndros®, Insys Therapeutics Inc.), used in capsules or as an oralsolution, is approved since the 1980s for the treatment of anorexia associated with weight loss in patients with AIDS (Acquired Immunodeficiency Syndrome) and for nausea and vomiting associated with cancer chemotherapy by the FDA and by some European countries (Whiting et al. 2015; Abuhasira et al. 2018; Bramness et al. 2018). A synthetic analog of THC, Nabilone (Cesamet®, Meda Pharmaceuticals Inc.; Canemes®, AOPOrphan Pharmaceuticals AG), is also approved since the 1980s for the treatment of nausea and vomiting associated with cancer chemotherapy by the FDA and by some European countries (Whiting et al. 2015; Abuhasira et al. 2018; Bramness et al. 2018). A synthetic pharmaceutical-grade version of CBD (STI Pharmaceuticals) is currently being investigated as an anticancer drug (Kenyon et al. 2018), and other synthetic derivatives CBD and other phytocannabinoids are being investigated in basic studies (Ranieri et al. 2016; Morales et al. 2017).

In this text we will present an overview of the main neurochemical mechanisms of action of the above mentioned phytocannabinoids, especially THC and CBD. We focused on human studies including both healthy volunteers and clinical samples. Human data for the other phytocannabinoids are very limited or do not exit at all, so when human studies were not available we tried to fulfil this gap with preclinical data.

3.2 Neuromolecular mechanisms of action of the main phytocannabinoids

3.2.1 THC

THC is the main psychotropic ingredient of cannabis, being responsible for its euphoriant effects, but also for some of its therapeutic effects (analgesia, increased appetite, hypnotic, etc.). THC acts as a partial agonist at the cannabinoid receptors 1 and 2 (or simply CB_1 and CB_2), and this is thought to be the main mechanism of action of this phytocannabinoid (Izzo et al. 2009; Weinstein et al. 2016; Colizzi and Bhattacharyya 2017; Sagar and Gruber 2018; Schonhofen et al. 2018). The cannabinoid receptors, their ligands (the endocannabinoids anandamide (AEA) and 2-arachidonoylglycerol (2-AG)), and the enzymes responsible for the synthesis and degradation of the endocannabinoids (fatty acid amide hydrolase (FAAH) and monoacylglycerol lipase (MAGL)) form the endocannabinoid system (ECS) (Ranieri et al. 2016; Schonhofen et al. 2018). The CB_1 receptoris distributed throughout the brain, with particularly high densities in the amygdala, hippocampus, striatum, frontal/prefrontal cortex, and motor areas. These areas are implicated in emotion processing and cognitive effects, including anxiety/relaxation (amygdala), learning/memory (hippocampus), reward processing (striatum), euphoria/ "high" (frontal/prefrontal cortex), and altered balance (motor areas). The ECS, mediated mainly by the CB_1 receptor, is also involved in regulating striatal dopamine release and glutamatergic and GABAergic neurons (Weinstein et al. 2016; Sagar and Gruber 2018; Schonhofen et al. 2018). CB_2 receptors are expressed in both the brain and peripheral organs and are involved in homeostasis, pain, and inflammation. The ECS is also implicated in the growth, differentiation, positioning, and connectivity among neurons and in neuroplasticity, including neurogenesis (Ranieri et al. 2016; Weinstein et al. 2016; Sagar and Gruber 2018; Schonhofen et al. 2018).

THC can also activate other receptors in nano/micromolar concentrations, such as the peroxisome proliferator-activated receptor γ (PPARγ) and the transient receptor potential Ankyrin 1 (TRPA1), which could be involved in the neuroprotective/inflammatoryand analgesic effects of THC (Izzo et al. 2009).

3.2.2 CBD

CBD is the second most abundant phytocannabinoid and the major noneuphoriant phytocannabinoid. CBD has several therapeutic potentials, including anxiolytic, antipsychotic, antiepileptic, and neuroprotective effects, but the mechanisms of these multiple pharmacological effects are complex and poorly understood. For instance, CBD neither directly binds to nor activates $CB_{1/2}$ receptors, as THC does. Some of the multiple mechanisms of action of CBD described in preclinical studies are summarized in Table 3.2.

3.2.3 Δ-9-THCV

This compound is derived from the phytocannabinoid cannabigerovarin (CBGV), and usually exists in very low quantities in cannabis varieties. Δ-9-THCV acts as a CB_1 receptor antagonist at lower doses and as an agonist of the same receptor at higher doses, and acts as a CB_2 receptor partial agonist (Izzo et al. 2009; dos Santos et al. 2015; Hill et al. 2010; Englund et al. 2016; Ranieri et al. 2016). Preclinical studies suggest that Δ-9-THCVdecreases food intake in animals and has antiepileptic properties (Izzo et al. 2009; dos Santos et al. 2015; Hill et al. 2010; Ranieri et al. 2016). In a study with 10 cannabis users, volunteers received 10 mg of Δ-9-THCV(oral) or placebo for 5 days, followed by 1 mg of intravenous THC on the fifth day.Δ-9-THCV was well tolerated and did not induce subjective effects, but it inhibited the heart rate increases produced by THC and potentiated the memory impairment induced by this phytocannabinoid (Englund et al. 2016). A recent neuroimaging study replicated the absence of subjective effects of Δ-9-THCV and showed that this compound reduced functional connectivity between the amygdala and parts of the default mode network (precuneus and the posterior cingulate cortex) and increased connectivity between the amygdala and parts of the executive control network (dorsal anterior cingulate cortex and premotor area) (Rzepa et al. 2016). These effects seem to be the neural basis

Table 3.2 Main mechanisms of action of cannabidiol (CBD)[a]

Target	Action
$\alpha 1/1\beta/3$ glycine receptors	Agonist/Positive allosteric modulator
Adenosine reuptake	**Inhibitor**
$A_{1/2A}$adenosine receptors	**Modulator**
Anandamide reuptake	**Inhibitor**
Ca^{2+} (intracellular)	**Regulator**
Ca^{2+} channels (voltage-gated T-type)	**Inhibitor**
CB_1cannabinoid receptor	**Noncompetitive antagonist/Noncompetitive negative allosteric modulator**
CB_2cannabinoid receptor	**Inverse agonist**
COX activity	**Inhibitor**
DA_2 dopamine receptor	**Partial agonist**
δ-opioid receptor	Positive allosteric modulator
FAAH	**Inhibitor**
Glutamate release	**Inhibitor**
GPR55 receptor	**Antagonist**
Hydroperoxide-induced oxidative damage	Inhibitor
mTOR pathway	**Activator**
μ-opioid receptor	Ligand/Positive allosteric modulator
NO production	Inhibitor
PGE2 production	**Inhibitor**
PPAR-γ receptor	**Agonist**
Putative abnormal-CBD receptor	Antagonist
$\sigma 1$ receptor	Antagonist
Na^+ channels	**Inhibitor**
TRPA1channel	Agonist
TRPM8 channel	Antagonist
TRPV1–4channels	**Agonist**
TNFα	**Modulator**
Tryptophan degradation	Inhibitor
VDAC1	Modulator
$5\text{-}HT_{1A}$	**Agonist**
$5\text{-}HT_{2A}$	Partial agonist
$5HT_{3A}$	Antagonist
5- and 15-lipoxygenase	Inhibitor

[a]Adapted from the following references: Izzo et al. 2009; dos Santos et al. 2015; Gobira et al. 2015; Ranieri et al. 2016; Seeman 2016; Campos et al. 2017; Morales et al. 2017; Perucca 2017; Crippa et al. 2018; Rodríguez-Muñoz et al. 2018; Schonhofen et al. 2018

The above list of targets/actions is not exhaustive. Targets/actions marked in bold seem to be the most relevant to the anxiolytic, antipsychotic, antiepileptic, and neuroprotector effects of CBD

CBD cannabidiol, *COX* cyclooxygenase, *FAAH* fatty acid amide hydrolase, *GPR55* G protein-coupled receptor 55, *mTOR* mammalian target of rapamycin intracellular pathway, *NO* nitric oxide, *PGE2* prostaglandin type E2, *PPAR-γ* nuclear peroxisome proliferator-activated receptor γ, *TNFα* tumor necrosis factor α, *TRPA1* transient receptor potential of ankyrin type 1, *TRPM8* transient receptor potential of the melastatin type 8, *TRPV1–4* transient receptor potential of vanilloid types 1–4, *VDAC1* voltage-dependent anion-selective channel protein type 1, *$5\text{-}HT_{1A}$* serotonin receptor subtype 1A

underlaying the possible use of this phytocannabinoid in the treatment of obesity. Studies with bigger samples and both healthy and clinical populations are needed to better understand the pharmacology of this compound and its possible therapeutic benefits.

3.2.4 Δ-9-THCA

This phytocannabinoid acts as a transient receptor potential of ankyrin type 1 (TRPA1) agonist and as a transient receptor potential of the melastatin type8 (TRPM8) antagonist, and preclinical studies showed that this compound has antiproliferative, antispasmodic, and analgesic properties (Izzo et al. 2009; dos Santos et al. 2015).

3.2.5 Δ-8-THC

Δ-8-THC results from the isomerization of THC and is found in very small amounts in cannabis. The pharmacology of Δ-8-THC and THC is similar, as both phytocannabinoids induce psychoactive and antiemetic effects in humans by agonism at the CB_1 receptor, butΔ-8-THC is less active (Izzo et al. 2009). Moreover, Δ-8-THC showed antiepileptic effects in animals (Colasanti et al. 1982; dos Santos et al. 2015).

3.2.6 CBDV

CBDV is a CBD analog derived from CBGV. Recent preclinical studies showed that this phytocannabinoid has antiepileptic effects that seem to be independent of $CB_{1/2}$ receptors (Hill et al. 2012, 2013). Furthermore, CBDV inhibits anandamide uptake and the synthetic enzyme of 2-AG and activates transient receptor potential of vanilloidtypes 1–2 (TRPV1/2) and TRPA1 channels (Hill et al. 2012, 2013; Iannotti et al. 2014; dos Santos et al. 2015; Ranieri et al. 2016; Morales et al. 2017).CBDV(800 mg once daily over 5 days) was well tolerated in phase I and II trials, and it is being investigated to treat seizure disorders, Rett syndrome, and autism spectrum disorder (Bialer et al. 2018).

3.2.7 CBN

CBNis a minor constituent of cannabis that is formed by the oxidation of THC. It was the first phytocannabinoid to be obtained in pure form, in 1896. Like CBD and CBDV, CBN inhibits cellular uptake of anandamide. Moreover, it also seems to act as a $CB_{1/2}$ partial agonist, although less potent than that of THC (10% of its psychoactivity) (Izzo et al. 2009). Few studies have investigated the pharmacology of CBN, but there is evidence that it has antiepileptic properties (Consroe and Wolkin 1977; dos Santos et al. 2015).

3.2.8 CBG

CBG is the precursor of THC and CBD, and several mechanisms of action have been proposed for this phytocannabinoid, including inhibition of anandamide and GABA uptake, partial agonism at $CB_{1/2}$receptors, TRPA1 and TRPV1/2 channels, and α2-adrenoceptors, antagonism at 5-HT_{1A} receptors and TRPM8 channels, modulation of phospholipase A_2,COX-1/−2 inhibition, and blockaded of voltage-gated sodium channels (Izzo et al. 2009; dos Santos et al. 2015; Ranieri et al. 2016; Morales et al. 2017). Preclinical studies suggest that at least some of these actions are involved in the analgesic, anti-inflammatory, antibacterial, and anticancer properties of CBG (Izzo et al. 2009; Ranieri et al. 2016; Morales et al. 2017).

3.2.9 CBC

Together with THC and CBD, CBC is one of the most abundant phytocannabinoids, and although it shares a similar pharmacology with THC (inducing hypothermia, sedation, and hypoactivity in animals), it is not euphoriant, and it is 2.5 times more toxic than THC. CBC acts as a TRPA1 agonist and as an inhibitor of anandamide reuptake and MAGL, and there is preclinical evidence that it has anti-inflammatory, analgesic, antidepressant, antibacterial, and anticancer effects (Izzo et al. 2009; Ranieri et al. 2016).

3.2.10 CBDA

CBDA is the acidic form of CBD, which is 95% of the CBD form present in cannabis. CBDA acts as a selective COX-2 inhibitor, a TRPA1 and TRPV1 agonist, a TRPM8 antagonist, and a modulator of the GPR55 receptor, and has showed anti-inflammatory, anticancer, and analgesic actions in preclinical studies (Izzo et al. 2009; Morales et al. 2017).

3.3 Neurochemical and behavioral effects of THC and CBD: Human studies

3.3.1 THC

The action of THC as a partial agonist at $CB_{1/2}$ receptors, but especially at the CB_1 receptor, is its main mechanism of action, being responsible for the characteristic effects of cannabis: euphoria/dysphoria, relaxation/anxiety, and changes in perceptions and thought content/psychotic symptoms. As mentioned above, the CB_1 receptor is high in brain areas related to emotion and cognition, including the amygdala, hippocampus, striatum, and prefrontal cortex. These areas are the neural subtracts of the subjective, emotional, and cognitive effects of THC, including anxiety/relaxation (amygdala), learning/memory (hippocampus), reward processing/motivation (striatum), and euphoria/ "high" (frontal cortex) (Weinstein et al. 2016; Colizzi and Bhattacharyya 2017; Sagar and Gruber 2018).

Studies of acute administration of cannabis or THC to healthy volunteers often report increase in scales measuring "linking", "intoxicated", and "high", but also show impaired cognition and increase in scales measuring anxiety and psychotic symptoms (Bhattacharyya et al. 2009; Fusar-Poli et al. 2009; Morrison et al. 2009; Bhattacharyya et al. 2010; Bhattacharyya et al. 2012; Martin-Santos et al. 2012; Niesink and van Laar 2013; Weinstein et al. 2016; Colizzi and Bhattacharyya 2017; Grimm et al. 2018; Sagar and Gruber 2018), and functional magnetic resonance imaging (fMRI) studies assessing the neural basis of the effects of THC in healthy volunteers suggest that the effects of this compound on fronto-striatal and limbic/paralimbic function are involved in its effects on verbal learning, psychotic symptoms, and emotion processing (Bhattacharyya et al. 2009; Fusar-Poli et al. 2009; Bhattacharyya et al. 2010; Bhattacharyya et al. 2012; Weinstein et al. 2016; Colizzi and Bhattacharyya 2017; Sagar and Gruber 2018). Neuroimaging studies have also shown that acute THC administration stimulates striatal dopamine neurotransmission in healthy human volunteers (Weinstein et al. 2016). However, previous genetic and brain structural and functional characteristics of cannabis users participating in these studies often influence the subjective and cognitive differences among these individuals and controls, suggesting that the observed deficits (when these are observed) are influenced by other factors and not necessarily by cannabis use. Moreover, age and history of cannabis use also influence these results (Weinstein et al. 2016; Sagar and Gruber 2018).

Regarding more prolonged or chronic use, a recent meta-analysis of neuroimaging studies of recreational cannabis users showed that the most consistent functional alterations were decreased activation in the anterior cingulate cortex (ACC) and dorsolateral prefrontal cortex (DL-PFC) and increased activation in the striatum (Weinstein et al. 2016; Sagar and Gruber 2018; Yanes et al. 2018). The ACC and DL-PFC are associated with behavioral control, pain processing, learning and memory, and the striatum with reward and pain processing, social judgments, and attention and inhibition control. Regarding dopaminergic neurotransmission, although acute THC administration stimulates striatal dopamine release in humans, several studies failed to find differences in striatal $D_{2/3}$ dopamine receptor occupancy between regular cannabis users and controls, but regular cannabis use was associated with reduced dopamine transporter (DAT) availability and dopamine synthesis capacity in the striatum (Weinstein et al. 2016).

These functional alterations could be related to the negative effects of cannabis use on reward processing, memory, and executive function, although a recent meta-analysis of cannabis use and cognitive function in adolescents and young adults concluded that previous studies overestimated the magnitude and persistence of the cognitive deficits associated with cannabis use, since these effects are small and of questionable clinical relevance for most individuals (Scott et al. 2018). Moreover, the observed effects are probably reflecting residual effects from acute use or withdrawal symptoms, since they are reduced in studies reporting abstinence periods longer than72 h (Weinstein et al. 2016; Sagar and Gruber 2018; Scott et al. 2018). Indeed, in several studies cannabis users perform similar to nonusing controls in cognitive tests, even when neurofunctional differences are found (and they are not always found) (Weinstein et al. 2016; Sagar and Gruber 2018). Furthermore, previous genetic and brain structural/functional characteristics of cannabis users influence the results of these studies, suggesting that the functional alterations observed are influenced by other factors and are not necessarily caused by cannabis use (Weinstein et al. 2016; Sagar and Gruber 2018).

Structural studies share these same limitations and also report conflicting results, with some studies failing to find differences and others reporting alterations in brain areas rich in CB_1receptors and involved in executive function and memory, including larger cerebellar and striatal volumes, reduced gray matter volume in the hippocampus, and lower white matter integrity (Weinstein et al. 2016; Sagar and Gruber 2018). However, as with functional findings, results from studies assessing brain structure in cannabis users are contradictory and not always correlated to cognitive or psychiatric deficits and are modulated by previous genetic and structural characteristics (Weinstein et al. 2016; Sagar and Gruber 2018; Scott et al. 2018).

In the case of patients using medicinal cannabinoids, it is important to acknowledge that most studies on the effects of cannabis use on cognitive function and on brain structure and function have examined the impact of heavy, chronic, recreational cannabis use (Sagar and Gruber 2018; Scott et al. 2018). Therefore, conclusions from these studies may neither be generalizable to light/moderate use nor to medicinal use. Indeed, recent observational studies suggest that patients using medicinal cannabis to improve anxiety, depression, chronicpain, and sleep, show improvements not only in their mood, quality-of-life, and sleep, but also on executive function and brain activation after starting cannabis treatment (Gruber et al. 2016; Gruber et al. 2018). Specifically, after 3 months of medical cannabis use, patients showed increased activation on the cingulate and frontal cortices during a cognitive task, effects that were not observed while doing the task at baseline (Gruber et al. 2018). These cognitive improvements could be related to the fact that these patients were typically older than recreational users which reduced the use of conventional medication during the study period. The observed improvements in their mood, quality-of-life, and sleep could also have improved their cognitive performance. Furthermore, in contrast to recreational user, patients usually use products with low THC levels and rich in other therapeutic cannabinoids which can counteract some of the undesired effects of THC, such as CBD (Gruber et al. 2016; Gruber et al. 2018).

However, a neuroimaging study with multiple sclerosis patients using smoked cannabis to reduce spasticity and pain observed reduced brain volume in subcortical, medial temporal, and prefrontal regions, which was associated with cognitive impairments in memory and processing speed (Romero et al. 2015). Therefore, further studies are needed to better understand the possible beneficial or deleterious effects of medicinal cannabis and cannabis-based products in different clinical populations. Moreover, as most of these studies only report short treatment periods/follow-ups, the possible long-term effects of these products are largely unknown and should be further investigated.

3.3.2 CBD

As described above (Table 3.2), the mechanisms of action of CBD are not fully understood, and CBD is known as a "promiscuous" compound, since it interacts with several neural systems. For instance, the actions of CBD include not only modulation of the ECS which is independent of $CB_{1/2}$ receptors, but this phytocannabinoid also activates $5\text{-}HT_{1A}$ serotonergic receptors and inhibits the uptake of serotonin, inhibits the uptake of adenosine, noradrenaline, dopamine and GABA, activates TRPV1/2 and TRPA1channels, antagonizes α1-adrenergic and μ-opioid receptors, and stimulates the activity of the inhibitory glycine-receptor, just to mention some of its possible mechanisms (Izzo et al. 2009; dos Santos et al. 2015; Gobira et al. 2015; Campos et al. 2017; Perucca 2017; Crippa et al. 2018; Rodríguez-Muñoz et al. 2018; Schonhofen et al. 2018). It seems that the pharmacological promiscuity of CBD is the reason for the several therapeutic potentials of this compound (Crippa et al. 2018).

These mechanisms of action of CBD make the pharmacology and toxicology of this compound differ from that of THC. Indeed, human studies show that CBD has a good safety and tolerability profile from a physiological and subjective perspective, both after acute and chronic administration, in a wide range of doses (from a single 6 g dose up to 3.5 g/day for 3 months) (Bergamaschi et al. 2011, 2011; Kerstin and Grotenhermen 2017; Crippa et al. 2018; Schoedel et al. 2018; Schonhofen et al. 2018; Taylor et al. 2018). Moreover, CBD does not produce the prototypical euphoriant and cognitive effects of THC and is devoid of abuse liability (Schoedel et al. 2018). Indeed, since the late 1970s, different research groups have shown that CBD counteracts/reduces some of the negative effects of THC, such as increases in anxiety and psychotic symptoms and cognitive deficits (Karniol et al. 1974; Zuardi et al. 1982; Bhattacharyya et al. 2009; Fusar-Poli et al. 2009; Morrison et al. 2009; Bhattacharyya et al. 2010, 2012; Niesink and van Laar 2013; Weinstein et al. 2016; Colizzi and Bhattacharyya 2017; Crippa et al. 2018). Furthermore, naturalistic studies of cannabis users comparing those who use cannabis varieties with low-CBD/high-THC content versus high-CBD/low-THC content showed that users of varieties with high-CBD/low-THC content had attenuated memory impairment and psychotic symptoms compared with users of low-CBD/high-THC content (Morgan et al. 2010, 2011; Colizzi and Bhattacharyya 2017).

Results from neuroimaging studies in humans comparing the subjective, cognitive, and neural effects of CBD with THC show that these phytocannabinoids have opposite effects on the brain (Weinstein et al. 2016; Colizzi and Bhattacharyya 2017; Crippa et al. 2018). For instance, while acute THC administration increases anxiety and psychotic symptoms, intoxication, and sedation, CBD does not induce any of these psychological effects and is indeed associated with reduced subjective anxiety (Martin-Santos et al. 2012; Colizzi and Bhattacharyya 2017; Crippa et al. 2018). Moreover, while the effects of THC on anxiety seem to be regulated by modulation of frontal and parietal brain structures, the anxiolytic effects of CBD were associated with reduced activation and functional connectivity of limbic and paralimbic regions (such as the amygdala and the ACC) during processing of intensely fearful faces (Fusar-Poli et al. 2009). Further, CBD also showed an opposite pattern of subjective effects (psychotic symptoms) and brain activity compared to THC in prefrontal, striatal, and hippocampal function during auditory, visual, and attentional salience processing (Bhattacharyya et al. 2009, 2010, 2012). In a recent study in healthy volunteers, CBD administration significantly increased fronto-striatal connectivity, while no significant difference was observed with THC (Grimm et al. 2018).

More recently, an open-label study of prolonged administration of CBD to regular cannabis users showed that CBD was well tolerated (no impairments on cognition or psychological function) and reduced the euphoria of participants while they smoked cannabis. Moreover, compared to baseline, cannabis users reported less depressive and psychotic symptoms and improved attention and memory, and an apparent recovery of hippocampal volume (Beale et al. 2018; Solowij et al. 2018).

Considering the good safety and tolerability profile of CBD in both healthy volunteers and clinical populations and the already recognized therapeutic indications for this compound (Crippa et al. 2018), the potential neuroprotective effects of CBD should be further assessed in randomized trials with clinical populations with marked cognitive impairments, such as patients with psychosis and Parkinson's Disease. In fact, these studies are already being performed (see below). However, it must be acknowledged that most experimental and clinical studies of CBD administration conducted so far only report short treatment periods and follow-ups. Therefore, long-term effects should be further investigated.

3.4 Approved indications of cannabis-based products, THC and CBD

The information gathered in the next sections was extracted and adapted from following citations: a systematic review from the American Academy of Neurology on the efficacy and safety of cannabis and cannabinoids in the treatment of neurologic disorders, published in 2014 (Koppel et al. 2014), a systematic review and meta-analysis of the efficacy and safety of cannabis and cannabinoids for the treatment of several diseases, published in 2015 (Whiting et al. 2015), epidemiological studies on the characteristics, safety, and efficacy of cannabis-based products (Yassin and Robinson 2017; Abuhasira et al. 2018; Bellnier et al. 2018; Hausman-Kedem et al. 2018; McCoy et al. 2018; Sarid et al. 2018), an open-label study involving the administration of synthetic CBD to cancer patients (Kenyon et al. 2018), articles with regulatory information on cannabinoid medications and products (Abuhasira et al. 2018; Bramness et al. 2018), and a recent narrative/expert review on the same topic (MacCallum and Russo 2018). The main therapeutic indications of cannabis-based products, THC and CBD are summarized in Table 3.3.

3.4.1 Cannabis-based products

In the case of herbal (raw) cannabis and cannabis extracts/oils, there is a great variety of products, indications, and legislations. Medicinal cannabis is permitted in 30 US States and in Canada, the Netherlands, Italy, Germany, Israel, and Brazil (Abuhasira et al. 2018; Bramness et al. 2018). Products include herbal cannabis to be smoked, vaporized, or ingested (as sold in several dispensaries across 30 US States), homemade extracts and oils (as sold in the US and Brazil), and standardized medications (Sativex®, Epidiolex®, TIL-TC150; discussed below). The main indications for medicinal cannabis include chronic pain, sleep disorders, anxiety and mood disorders, posttraumatic stress disorder, Parkinson disease, and epilepsy, with some of these indications lacking assessment in RCTs. Thus, the level of evidence for recommending medicinal use in some indications varies from moderate (e.g., epilepsy, Parkinson's disease) to inconclusive (e.g., anxiety and mood disorders). Moreover, available cannabis-based products are often not standardized and show wide variation in cannabinoid content, which could induce intoxications (in the case of a high THC content) or lack of therapeutic efficacy (in the absence of CBD or THC) (Vandrey et al. 2015; Crippa et al. 2016). Other risk of nonstandardized products is intoxication with more toxic products, such as potent synthetic cannabinoids (Horth et al. 2018).

3.4.2 THC

3.4.2.1 Nausea and vomiting due to chemotherapy

There is conclusive/substantial evidence that THC (Dronabinol®, Cesamet®, Marinol®, Syndros®) and Nabiximols (Sativex®) are an effective treatment for chemotherapy-induced nausea and vomiting. The antiemetic effects of

Table 3.3 Summary of approved indications of cannabis-based products, THC and CBD[a]

Product	Indication	Where it is approved[b]
Herbal cannabis (*Cannabis sp.*, Bedrocan®, Bedrobinol®, Bediol®, Bedica®, Bedrolite®) and nonstandardized cannabis extracts/ oils	Anxiety disorders, AD, ADHD, ALS, appetite and decreasing weight loss associated with HIV/AIDS, cancer (glioma), cancer-associated pain, CD, chemotherapy-associated nausea, chronic pain, clusterheadache, compassion treatment, CUD, dementia, epilepsy, ET, fibromyalgia, glaucoma, IBS, mood disorders, MS, MSA, nonspecific pain, PD, PTSD, PVD, rheumatoid arthritis, sleep disorders, tension headache, tic disorder, TS, ulcerative colitis, etc.[c]	Australia, Brazil, Canada, Germany, Israel, Italy, Lithuania, Netherlands, New Zealand, Portugal, Spain, UK, Uruguay, 30 US States
Dronabinol (Marinol®, Syndros®)/ THC	Appetite and decreasing weight loss associated with HIV/AIDS, nausea and vomiting due to chemotherapy, neuropathicpain, TS	Austria, Belgium, Brazil, Canada, Croatia, Denmark, France, Netherlands, Norway, Romania, Spain, Switzerland, UK, US
Nabilone (Cesamet®, Canemes®)/ THC analog	Nausea and vomiting due to chemotherapy	Austria, Croatia, Denmark, France, Germany, Mexico, UK
Nabiximols (Sativex®)/ THC:CBD (1:1)	MS-associated spasticity and chronic pain	Brazil, Israel, 22 European countries
TIL-TC150	Treatment of intractable seizures in epileptic syndromes (Dravet syndrome)	Canada
CBD (Epidiolex®, Purodiol®)	Treatment of intractable seizures in epileptic syndromes (Dravet and Lennox-Gastaut syndromes), cancer	Brazil, UK, US

[a]Adapted from the following references: Koppel et al. 2014; Whiting et al. 2015; Yassin and Robinson 2017; Abuhasira et al. 2018; Abuhasira et al. 2018; Bellnier et al. 2018; Bramness et al. 2018; Hausman-Kedem et al. 2018; Kenyon et al. 2018; MacCallum and Russo 2018; McCoy et al. 2018; Sarid et al. 2018
[b]Includes licensed medicinal products, nonapproved products prescribed under specific conditions, off-label use, and compassionate prescribing. Several examples of countries are reported, but this list is not exhaustive
[c]Nonexaustive list of indications
AD Alzheimer' disease, *ADHD* attention deficit and hyperactivity disorder, *ALS* amyotrophic lateral sclerosis, *CBD* cannabidiol, *CD* Crohn's disease, *CUD* cannabis use disorder, *ET* essential tremor, *IBS* irritable bowel syndrome, *MS* multiple sclerosis, *MSA* multiple system atrophy, *PD* Parkinson's disease, *PTSD* post-traumatic stress disorder, *PVD* peripheral vascular disease, *THC* tetrahydrocannabinol, *TS* Tourette's syndrome, *UK* United Kingdom, *US* United States

THC are produced by its agonistic action on CB_1 receptors.

3.4.2.2 Appetite and decreasing weight loss associated with HIV/AIDS

There is conclusive/substantial evidence that THC (Dronabinol®, Cesamet®, Marinol®, Syndros®) is an effective treatment for increasing appetite and improves decreasing weight loss associated with HIV/AIDS. The effects of THC on appetite and weight gain are produced by its agonistic action on CB_1 receptors.

3.4.2.3 Multiple sclerosis symptoms (spasticity and chronic pain)

There is conclusive/substantial evidence that Nabiximols (THC:CBD in a 1:1 ratio, Sativex®) is an effective treatment for multiple sclerosis spasticity symptoms and chronic pain. The therapeutic effects of Nabiximols include the analgesic, anti-inflammatory, and sleep-promoting effects of THC and CBD. In the case of THC, these effects are produced by its agonistic action of THC on $CB_{1/2}$ receptors. The mechanisms of action of CBD are not fully understood but seem to be independent of cannabinoid receptors.

3.4.2.4 Chronic pain (neuropathic and cancer pain)

There is moderate evidence that THC (Dronabinol®, Cesamet®, Marinol®, Syndros®) is an effective treatment for chronic neuropathic and cancer pain. The analgesic and anti-inflammatory effects of THC are mediated by its agonistic action on $CB_{1/2}$ receptors.

3.4.3 CBD

3.4.3.1 Antiepileptic

There is conclusive/substantial evidence that purified CBD (Epidiolex®, Purodiol®) is an effective treatment for intractable seizures in epileptic syndromes such as Dravet and Lennox-Gastaut. There is preliminary evidence from an open-label study that a cannabis-extract with purified CBD and THC in a 50:1 CBD:THC ratio (TIL-TC150) is an effective treatment for intractable seizures in children with Dravet syndrome. The antiepileptic mechanisms of action of CBD are not fully understood but seem to be independent of cannabinoid receptors and involve ion channels and G-protein-coupled receptors (see Table 3.2 above).

3.4.3.2 Therapeutic potentials of CBD with moderate/modest evidence from RCTs

In the last decade, accumulating evidence from clinical studies shows that CBD has anxiolytic effects in social anxiety (Bergamaschi et al. 2011, 2011), antipsychotic effects in schizophrenia (Leweke et al. 2012; McGuire et al. 2018) and Parkinson's disease (Zuardi et al. 2009), improvements on well-being and quality of life in Parkinson's disease (Chagas et al. 2014), antiaddictive effects for tobacco and opioid dependence (Morgan et al. 2013; Hurd et al. 2015), and antitumor effects (Kenyon et al. 2018). It is possible that the therapeutic uses of CBD for some of these indications could be regulated in the near future, especially as an antipsychotic drug and for the treatment of some symptoms of Parkinson's disease (Crippa et al. 2018). However, further controlled trials with bigger samples and longer treatment periods are needed to replicate (or refute) most of these results.

3.5 Adverse effects of THC and CBD

The information gathered in the next sections was extracted and adapted from the following citations: Chagas et al. 2014; Koppel et al. 2014; Whiting et al. 2015; Gaston et al. 2017; Perucca 2017; Yassin and Robinson 2017; Abuhasira et al. 2018; Bellnier et al. 2018; Crippa et al. 2018; Hausman-Kedem et al. 2018; Kaufman 2018; Kenyon et al. 2018; Lattanzi et al. 2018; MacCallum and Russo 2018; McCoy et al. 2018; McGuire et al. 2018; Sarid et al. 2018; Schoedel et al. 2018; Schonhofen et al. 2018; Taylor et al. 2018. The main adverse effects of cannabis-based products, THC and CBD are summarized in Table 3.4.

3.5.1 Cannabis-based products and THC

In the last decades, several observational (Gruber et al. 2016; Yassin and Robinson 2017; Abuhasira et al. 2018; Bellnier et al. 2018; de Hoop et al. 2018; Gruber et al. 2018; Hausman-Kedem et al. 2018; McCoy et al. 2018; Mondello et al. 2018; Sarid et al. 2018; Schonhofen et al. 2018) and clinical (open-label and RCTs) studies (Koppel et al. 2014; Whiting et al. 2015) investigated the effects of medicinal cannabis, cannabis-based products, and THC in a diverse group of clinical populations. These products are generally considered safe and well tolerated, at least when they are administered in short treatment periods (weeks, months). Common adverse effects include dizziness, dry mouth, euphoria, nausea, somnolence, drowsiness/fatigue, confusion and disorientation, cough (smoking only), and headache. Less common and rare adverse effects include orthostatic hypotension, ataxia/dyscoordination, anxiety, depression, diarrhea, tachycardia, psychosis/paranoia, hallucinations,

Table 3.4 Summary of the main adverse effects of cannabis-based products, THC and CBD[a]

Product	Adverse effect
Products in which the main adverse reactions are associated with THC	
Herbal cannabis (*Cannabis sp.*, Bedrocan®, Bedrobinol®, Bediol®, Bedica®, Bedrolite®) and nonstandardized cannabis extracts/oils Dronabinol (Marinol®, Syndros®)/THC Nabilone (Cesamet®, Canemes®)/THC analog Nabiximols (Sativex®)/1:1 THC:CBD	*Most common/Common*: Dizziness, euphoria, nausea, somnolence, drowsiness/fatigue, confusion and disorientation, headache, dry mouth, cough/sore throat (smoking/vaporization only) *Less common/Rare*: Orthostatic hypotension, ataxia/dyscoordination, increased appetite, anxiety, depression, diarrhea, tachycardia, psychosis/paranoia, hallucinations, cannabinoid hyperemesis syndrome, seizures
Products in which the main adverse reactions are associated with CBD	
CBD (Epidiolex®, Purodiol®) TIL-TC150/50:1 CBD:THC	*Most common/Common*: Diarrhea, somnolence, nausea, insomnia, fatigue, sedation, decreased appetite, headache, transaminase elevations (with concomitant use of antiepileptics) *Less common/Rare*: Vomiting, fever, lethargy, sleep disorder, seizures, infections, ataxia, rash

[a]This list is not exhaustive. Adapted from the following references: Chagas et al. 2014; Koppel et al. 2014; Whiting et al. 2015; Gaston et al. 2017; Perucca 2017; Yassin and Robinson 2017; Abuhasira et al. 2018; Bellnier et al. 2018; Crippa et al. 2018; Hausman-Kedem et al. 2018; Kaufman 2018; Kenyon et al. 2018; Lattanzi et al. 2018; MacCallum and Russo 2018; McCoy et al. 2018; McGuire et al. 2018; Sarid et al. 2018; Schoedel et al. 2018; Schonhofen et al. 2018; Taylor et al. 2018

CBD cannabidiol, *THC* tetrahydrocannabinol

cannabinoid hyperemesis syndrome, and seizures. Most adverse effects are temporary and are less common with continuous use and titration of these products, since tolerance seems to occur to adverse effects but not necessarily to therapeutic effects (Whiting et al. 2015; Abuhasira et al. 2018; de Hoop et al. 2018; MacCallum and Russo 2018). However, as most of these studies only report short follow-up periods, long-term effects are unknown. Future studies in this area should include longer treatment periods and follow-ups.

3.5.2 CBD

CBD has a good safety and tolerability profile from a physiological and subjective perspective, both after acute and chronic administration in humans, in a wide range of doses (from a single 6 g dose up to 3.5 g/day for 3 months). The most common adverse effects include somnolence, sedation, nausea, diarrhea, headache, changes on appetite, and transaminase elevations. Moreover, CBD does not induce significant effects on cognition, and does not induce tolerance (Bergamaschi et al. 2011, 2011; Colizzi and Bhattacharyya 2017; Gaston et al. 2017; Kerstin and Grotenhermen 2017; Schoedel et al. 2018; Taylor et al. 2018). RCTs of CBD and patients with schizophrenia (McGuire et al. 2018) and Parkinson's disease (Chagas et al. 2014) did not observed differences between placebo and CBD regarding adverse reactions. Indeed, compared with the antipsychotic amisulpride, CBD administration was associated with less extrapyramidal symptoms, weight gain, and prolactin increase (Leweke et al. 2012). A meta-analysis with data from four RCTs of CBD (Epidiolex®) in 550 patients with Lennox-Gastaut or Dravet syndrome showed that CBD was safely administered and produced significant reductions in seizure frequency compared to placebo. CBD administration was associated with a higher rate of adverse effects compared to placebo, but the most common of these effects had a modest clinical relevance and included somnolence, decreased appetite, diarrhea, fatigue, and increased serum aminotransferases (Lattanzi et al. 2018). Less common reactions include vomiting, fever, lethargy, sleep disorder, seizures, infections, and rash (Perucca 2017; Kaufman 2018; Schonhofen et al. 2018).

Thus, CBD seems to show a different profile of adverse reactions depending on the sample, with most studies in healthy volunteers and clinical samples showing no or few adverse effects, except for epileptic syndromes (Kerstin and Grotenhermen 2017; Crippa et al. 2018; Schonhofen et al. 2018). These differences could be related to CBD dose, duration of treatment, differences among samples regarding components of the ECS (e.g., $CB_{1/2}$ receptor expression in different brain areas), and interactions with medications being used concomitantly with CBD. Future clinical studies with bigger samples and in different clinical populations will contribute to a better understanding of the complex pharmacology of CBD.

3.6 Conclusions

Since the early 1980s, THC-based products have been recognized and regulated as a medicines. In the same decade, researchers in different laboratories around the world, including in Brazil, began to show that CBD could antagonize some of the negative effects of THC, such as anxiety, psychotic symptoms, and cognitive deficits. These studies formed the basis for the regulation of Nabiximols as a medicine around the world in the following years. In the mid 1990s and early 2000s, medicinal cannabis programs became active in several countries, and neuroimaging studies started to elucidate the neural basis for the therapeutic and deleterious effects of cannabis and THC and shed light on the difference between these substances and CBD. In the last decade, several legislations included cannabis-based products as regulated medicines, and the research on the possible therapeutic uses of CBD increased significantly.

However, many areas of cannabinoids research still need to be better explored. For instance, the increasing use of nonstandardized herbal cannabis and cannabis oils in the US and other countries for the treatment of several diseases without the appropriate RCTs should be carefully evaluated. Although observational studies with both recreational users and patients suggest that cannabis is not a highly toxic drug when compared with alcohol, heroin, or cocaine, it can have significant psychiatric adverse reactions in a minority of users (e.g., psychosis and cognitive deficits) that should be considered. Observational studies are very important but need to be complemented with RCTs so that the possible therapeutic uses of these products can be done with more information on dosage and adverse effects. The placebo effect can be very powerful in such observational studies, especially in people with difficult-to-treat conditions (e.g., epilepsy, chronic pain) and in a context in which the discussion of cannabis legalization for recreational and medical uses can be very polemic, enforced by commercial interests and the media. This generalization of untested medicinal properties and commercialization of untested and nonstandardized products could have a negative impact in public health, such as poisonings, intoxications, and lack of appropriate treatment.

Thus, more RCTs are needed to explore the effectiveness and safety of herbal cannabis and cannabis oils on specific disorders, and these products need to be standardized for cannabinoid content. It is especially important that these studies include long-term follow-ups.

Moreover, more RCTs should be performed with pure CBD to investigate the possible therapeutic use of this compound in anxiety and mood disorders, substance use disorders, psychotic disorders, Parkinson's disease, epilepsy, and autism. These studies need to be performed not only to establish safety (especially to the developing brain) and dosage, but also to answer the still-to-be-answered question of which products (pure compounds or whole-plant products) are more effective and safer, and for which indications.

Acknowledgements RGS is Fellow of the Programa Nacional de Pós-Doutorado, Brazil (PNPD/CAPES). JECH and JAC are recipients of fellowship awards from Conselho Nacional de Desenvolvimento Científico e Tecnológico (CNPq, Brazil). This review was supported by the Brazilian fund INCT-TM/CNPq/FAPESP (465458/2014-9).

Funding
Research was supported in part by grants from (i) *Fundação de Amparo à Pesquisa do Estado de São Paulo* (FAPESP, Brazil); (ii) *Conselho Nacional de Desenvolvimento Científico e Tecnológico* (CNPq, Brazil); (iii) *Coordenação de Aperfeiçoamento de Pessoal de Nível Superior* (CAPES, Brazil); and (iv) National Institute for Translational Medicine (INCTTM; CNPq, Brazil; INCT-TM/CNPq/FAPESP: 465458/2014–9). JECH and JAC are recipients of fellowship awards from CNPq. JC has a grant from University Global Partnership Network (UGPN)—Global priorities in cannabinoid research excellence.

Conflict of Interest JECH and JAC are coinventors of the patent "Fluorinated CBD compounds, compositions, and uses thereof. Pub. No.: WO/2014/108899. International Application No.: PCT/IL2014/050023" Def. US no. Reg. 62,193,296; 29/07/2015; INPI on 19/08/2015 (BR1120150164927). The University of São Paulo has licensed the patent to Phytecs Pharm (USP Resolution No. 15.1.130002.1.1). The University of São Paulo has an agreement with Prati-Donaduzzi (Toledo, Brazil) to "develop a pharmaceutical product containing synthetic cannabidiol and prove its safety and therapeutic efficacy in the treatment of epilepsy, schizophrenia, Parkinson's disease, and anxiety disorders." JECH and JAC have received travel support from and are medical advisors of BSPG-Pharm. RGS declares no conflicts of interest.

References

Abuhasira R, Schleider LB, Mechoulam R et al (2018) Epidemiological characteristics, safety and efficacy of medical cannabis in the elderly. Eur J Intern Med 49:44–50

Abuhasira R, Shbiro L, Landschaft Y (2018) Medical use of cannabis and cannabinoids containing products - Regulations in Europe and North America. Eur J Intern Med 49:2–6

Andre CM, Hausman JF, Guerriero G (2016) *Cannabis sativa*: the plant of the thousand and one molecules. Front Plant Sci 7:19

Beale C, Broyd SJ, Chye Y et al (2018) Prolonged Cannabidiol Treatment Effects on Hippocampal Subfield Volumes in Current Cannabis Users. Cannabis Cannabinoid Res 3(1):94–107

Bellnier T, Brown GW, Ortega TR (2018) Preliminary evaluation of the efficacy, safety, and costs associated with the treatment of chronic pain with medical cannabis. Ment Health Clin 8(3):110–115

Bergamaschi MM, Queiroz RH, Chagas MH et al (2011) Cannabidiol reduces the anxiety induced by simulated public speaking in treatment-naive social phobia patients. Neuropsychopharmacology 36:1219–1226

Bergamaschi MM, Queiroz RH, Zuardi AW et al (2011) Safety and side effects of cannabidiol, a *Cannabis sativa* constituent. Curr Drug Saf 6(4):237–249

Bhattacharyya S, Crippa JA, Allen P et al (2012) Induction of psychosis by Δ9-tetrahydrocannabinol reflects modulation of prefrontal and striatal function during attentional salience processing. Arch Gen Psychiatry 69 (1):27–36

Bhattacharyya S, Fusar-Poli P, Borgwardt S et al (2009) Modulation of mediotemporal and ventrostriatal function in humans by Delta9-tetrahydrocannabinol: a neural basis for the effects of Cannabis sativa on learning and psychosis. Arch Gen Psychiatry 66(4):442–451

Bhattacharyya S, Morrison PD, Fusar-Poli P et al (2010) Opposite effects of delta-9-tetrahydrocannabinol and cannabidiol on human brain function and psychopathology. Neuropsychopharmacology 5(3):764–774

Bialer M, Johannessen SI, Koepp MJ et al (2018) Progress report on new antiepileptic drugs: a summary of the Fourteenth Eilat Conference on New Antiepileptic Drugs and Devices (EILAT XIV). II. Drugs in more advanced clinical development. Epilepsia 59 (10):1842–1866

Bramness JG, Dom G, Gual A et al (2018) A Survey on the Medical Use of Cannabis in Europe: a Position Paper. Eur Addict Res 24(4):201–205

Campos AC, Fogaça MV, Scarante FF et al (2017) Plastic and Neuroprotective Mechanisms Involved in the Therapeutic Effects of Cannabidiol in Psychiatric Disorders. Front Pharmacol 8:269

Chagas MH, Zuardi AW, Tumas V et al (2014) Effects of cannabidiol in the treatment of patients with Parkinson's disease: an exploratory double-blind trial. J Psychopharmacol 28:1088–1098

Colasanti BK, Lindamood C 3rd, Craig CR (1982) Effects of marijuana cannabinoids on seizure activity in cobalt-epileptic rats. Pharmacol Biochem Behav 16:573–578

Colizzi M, Bhattacharyya S (2017) Does Cannabis Composition Matter? Differential Effects of Delta-9-tetrahydrocannabinol and Cannabidiol on Human Cognition. Curr Addict Rep 4(2):62–74

Consroe P, Wolkin A (1977) Cannabidiol – antiepileptic drug comparisons and interactions in experimentally induced seizures in rats. J Pharmacol Exp Ther 201:26–32

Crippa JA, Crippa AC, Hallak JE et al (2016) Δ9-THC Intoxication by cannabidiol-enriched cannabis extract in two children with refractory epilepsy: full remission after switching to purified cannabidiol. Front Pharmacol 7:359

Crippa JA, Guimarães FS, Campos AC et al (2018) Translational Investigation of the Therapeutic Potential of Cannabidiol (CBD): toward a new age. Front Immunol 9:2009

de Hoop B, Heerdink ER, Hazekamp A (2018) Medicinal cannabis on prescription in The Netherlands: statistics for 2003-2016. Cannabis Cannabinoid Res 3(1):54–55

Di Forti M, Marconi A, Carra E et al (2015) Proportion of patients in south London with first-episode psychosis attributable to use of high potency cannabis: a case-control study. Lancet Psychiatry 2(3):233–238

dos Santos RG, Hallak JE, Leite JP et al (2015) Phytocannabinoids and epilepsy. J Clin Pharm Ther 40 (2):135–143

Englund A, Atakan Z, Kralj A et al (2016) The effect of five day dosing with THCV on THC-induced cognitive, psychological and physiological effects in healthy male human volunteers: a placebo-controlled, double-blind, crossover pilot trial. J Psychopharmacol 30 (2):140–151

Fusar-Poli P, Crippa JA, Bhattacharyya S et al (2009) Distinct effects of {delta}9-tetrahydrocannabinol and cannabidiol on neural activation during emotional processing. Arch Gen Psychiatry 66(1):95–105

Gaston TE, Bebin EM, Cutter GR et al (2017) Interactions between cannabidiol and commonly used antiepileptic drugs. Epilepsia 58(9):1586–1592

Gobira PH, Vilela LR, Gonçalves BD et al (2015) Cannabidiol, a Cannabis sativa constituent, inhibits cocaine-induced seizures in mice: possible role of the mTOR pathway and reduction in glutamate release. Neurotoxicology 50:116–121

Grimm O, Löffler M, Kamping S et al (2018) Probing the endocannabinoid system in healthy volunteers: cannabidiol alters fronto-striatal resting-state connectivity. Eur Neuropsychopharmacol 28(7):841–849

Gruber SA, Sagar KA, Dahlgren MK et al (2016) Splendor in the Grass? A Pilot Study Assessing the Impact of Medical Marijuana on Executive Function. Front Pharmacol 7:355

Gruber SA, Sagar KA, Dahlgren MK et al (2018) The Grass Might Be Greener: medical marijuana patients exhibit altered brain activity and improved executive function after 3 months of treatment. Front Pharmacol 8:983

Hausman-Kedem M, Menascu S, Kramer U (2018) Efficacy of CBD-enriched medical cannabis for treatment of refractory epilepsy in children and adolescents - An observational, longitudinal study. Brain and Development 40(7):544–551

Hill TD, Cascio MG, Romano B et al (2013) Cannabidivarin-rich cannabis extracts are anticonvulsant in mouse and rat via a CB1 receptor-independent mechanism. Br J Pharmacol 170:679–692

Hill AJ, Mercier MS, Hill TD et al (2012) Cannabidivarin is anticonvulsant in mouse and rat. Br J Pharmacol 167:1629–1642

Hill AJ, Weston SE, Jones NA et al (2010) Delta9-Tetrahydrocannabivarin suppresses in vitro epileptiform and in vivo seizure activity in adult rats. Epilepsia 51:1522–1532

Hillig KW (2005) Genetic evidence for speciation in *Cannabis* (Cannabaceae). Genet Resour Crop Evol 52:161–180

Horth RZ, Crouch B, Horowitz BZ et al (2018) Notes from the field: acute poisonings from a synthetic cannabinoid sold as cannabidiol - Utah, 2017-2018. MMWR Morb Mortal Wkly Rep 67(20):587–588

Hurd YL, Yoon M, Manini AF et al (2015) Early phase in the development of cannabidiol as a treatment for addiction: opioid relapse takes initial center stage. Neurotherapeutics 12(4):807–815

Iannotti FA, Hill CL, Leo A et al (2014) Nonpsychotropic plant cannabinoids, cannabidivarin (CBDV) and cannabidiol (CBD), activate and desensitize transient receptor potential vanilloid 1 (TRPV1) channels in vitro: potential for the treatment of neuronal hyperexcitability. ACS Chem Neurosci 5:1131–1141

Izzo AA, Borrelli F, Capasso R et al (2009) Nonpsychotropic plant cannabinoids: new therapeutic opportunities from an ancient herb. Trends Pharmacol Sci 30(10):515–527

Karniol IG, Shirakawa I, Kasinski N et al (1974) Cannabidiol interferes with the effects of delta 9-tetrahydrocannabinol in man. Eur J Pharmacol 28:172–177

Kaufman MB (2018) Pharmaceutical Approval Update. P T 43(9):528–530

Kenyon J, Liu W, Dalgleish A (2018) Report of Objective Clinical Responses of Cancer Patients to Pharmaceutical-grade Synthetic Cannabidiol. Anticancer Res 38(10):5831–5835

Kerstin I, Grotenhermen F (2017) An update on safety and side effects of cannabidiol: a review of clinical data and relevant animal studies. Cannabis Cannabinoid Res 2:139–154

Koppel BS, Brust JC, Fife T et al (2014) Systematic review: efficacy and safety of medical marijuana in selected neurologic disorders: report of the Guideline Development Subcommittee of the American Academy of Neurology. Neurology 82(17):1556–1563

Lattanzi S, Brigo F, Trinka E et al (2018) Efficacy and safety of Cannabidiol in epilepsy: a systematic review and meta-analysis. Drugs 78(17):1791–1804. https://doi.org/10.1007/s40265-018-0992-5

Leweke FM, Piomelli D, Pahlisch F et al (2012) Cannabidiol enhances anandamide signaling and alleviates psychotic symptoms of schizophrenia. Transl Psychiatry 2:e94

MacCallum CA, Russo EB (2018) Practical considerations in medical cannabis administration and dosing. Eur J Intern Med 49:12–19

Martin-Santos R, Crippa JA, Batalla A et al (2012) Acute effects of a single, oral dose of delta9-tetrahydrocannabinol (THC) and cannabidiol (CBD) administration in healthy volunteers. Curr Pharm Des 18(32):4966–4979

McCoy B, Wang L, Zak M et al (2018) A prospective open-label trial of a CBD/THC cannabis oil in dravet syndrome. Ann Clin Transl Neurol 5(9):1077–1088

McGuire P, Robson P, Cubala WJ et al (2018) Cannabidiol (CBD) as an adjunctive therapy in schizophrenia: a multicenter randomized controlled trial. Am J Psychiatry 175:225–231

Mondello E, Quattrone D, Cardia L et al (2018) Cannabinoids and spinal cord stimulation for the treatment of failed back surgery syndrome refractory pain. J Pain Res 11:1761–1767

Morales P, Reggio PH, Jagerovic N (2017) An Overview on Medicinal Chemistry of Synthetic and Natural Derivatives of Cannabidiol. Front Pharmacol 8:422

Morgan CJ, Das RK, Joye A et al (2013) Cannabidiol reduces cigarette consumption in tobacco smokers: preliminary findings. Addict Behav 38(9):2433–2436

Morgan CJ, Gardener C, Schafer G et al (2011) Sub-chronic impact of cannabinoids in street cannabis on cognition, psychotic-like symptoms and psychological well-being. Psychol Med 29:1–10

Morgan CJ, Schafer G, Freeman TP (2010) Impact of cannabidiol on the acute memory and psychotomimetic effects of smoked cannabis: naturalistic study. Br J Psychiatry 197(4):285–290

Morrison PD, Zois V, McKeown DA et al (2009) The acute effects of synthetic intravenous Delta9-tetrahydrocannabinol on psychosis, mood and cognitive functioning. Psychol Med 39(10):1607–1616

Niesink RJ, van Laar MW (2013) Does Cannabidiol Protect Against Adverse Psychological Effects of THC? Front Psych 4:130

Perucca E (2017) Cannabinoids in the treatment of epilepsy: hard evidence at last? J Epilepsy Res 7(2):61–76

Ranieri R, Marasco D, Bifulco M et al (2016) Phytocannabinoids and Cannabimimetic Drugs: recent patents in central nervous system disorders. Recent Pat CNS Drug Discov 10(2):157–177

Rodríguez-Muñoz M, Onetti Y, Cortés-Montero E et al (2018) Cannabidiol enhances morphine antinociception, diminishes NMDA-mediated seizures and reduces stroke damage via the sigma 1 receptor. Mol Brain 11(1):51

Romero K, Pavisian B, Staines WR, Anthony FA (2015) Multiple sclerosis, cannabis, and cognition: A structural MRI study. NeuroImage: Clinical 8:140–147

Rubin R (2018) The Path to the First FDA-Approved Cannabis-Derived Treatment and What Comes Next. JAMA 320(12):1227–1229

Rzepa E, Tudge L, McCabe C (2016) The CB1 Neutral Antagonist Tetrahydrocannabivarin Reduces Default Mode Network and Increases Executive Control Network Resting State Functional Connectivity in Healthy Volunteers. Int J Neuropsychopharmacol 19(2): pyv092

Sagar KA, Gruber SA (2018) Marijuana matters: reviewing the impact of marijuana on cognition, brain structure and function, & exploring policy implications and barriers to research. Int Rev Psychiatry 30 (3):251–267

Sarid N, Zada M, Lev-Ran S et al (2018) Medical cannabis use by hodgkin lymphoma patients: experience of a single center. Acta Haematol 140(4):194–202

Schoedel KA, Szeto I, Setnik B et al (2018) Abuse potential assessment of cannabidiol (CBD) in recreational polydrug users: a randomized, double-blind, controlled trial. Epilepsy Behav 88:162–171

Schonhofen P, Bristot IJ, Crippa JA et al (2018) Cannabinoid-based therapies and brain development: potential harmful effect of early modulation of the endocannabinoid system. CNS Drugs 32(8):697–712. https://doi.org/10.1007/s40263-018-0550-4

Scott JC, Slomiak ST, Jones JD et al (2018) Association of cannabis with cognitive functioning in adolescents and young adults: a systematic review and meta-analysis.

Seeman P (2016) Cannabidiol is a partial agonist at dopamine D2High receptors, predicting its antipsychotic clinical dose. JAMA Psychiatry 75(6):585–595

Solowij N, Broyd SJ, Beale C et al (2018) Therapeutic effects of prolonged cannabidiol treatment on psychological symptoms and cognitive function in regular cannabis users: a pragmatic open-label clinical trial. Cannabis Cannabinoid Res 3:21–34

Taylor L, Gidal B, Blakey G et al (2018) A Phase I, randomized, double-blind, placebo-controlled, single ascending dose, multiple dose, and food effect trial of the safety, tolerability and pharmacokinetics of highly purified cannabidiol in healthy subjects. CNS Drugs 32 (11):1053–1067

Vandrey R, Raber JC, Raber ME et al (2015) Cannabinoid Dose and Label Accuracy in Edible Medical Cannabis Products. JAMA 313(24):2491–2493

Volkow ND, Swanson JM, Evins AE et al (2016) Effects of cannabis use on human behavior, including cognition, motivation, and psychosis: a review. JAMA Psychiat 73(3):292–297

Weinstein A, Livny A, Weizman A (2016) Brain imaging studies on the cognitive, pharmacological and neurobiological effects of cannabis in humans: evidence from studies of adult users. Curr Pharm Des 22 (42):6366–6379

Whiting PF, Wolff RF, Deshpande S (2015) Cannabinoids for medical use: a systematic review and meta-analysis. JAMA 313(24):2456–2473

Yanes JA, Riedel MC, Ray KL et al (2018) Neuroimaging meta-analysis of cannabis use studies reveals convergent functional alterations in brain regions supporting cognitive control and reward processing. J Psychopharmacol 32(3):283–295

Yassin M, Robinson D (2017) Effect of Adding Medical Cannabis Treatment (MCT) to analgesic treatment in patients with low back pain related to fibromyalgia: an observational cross-over single center study. Int J Anesth Pain Med 3:2

Zuardi AW, Crippa JA, Hallak JE et al (2009) Cannabidiol for the treatment of psychosis in Parkinson's disease. J Psychopharmacol 23:979–983

Zuardi AW, Shirakawa I, Finkelfarb E et al (1982) Action of cannabidiol on the anxiety and other effects produced by delta9-THC in normal subjects. Psychopharmacology 76(3):245–250

Emerging Roles of Cannabinoids and Synthetic Cannabinoids in Clinical Experimental Models

4

Paula Morales and Patricia H. Reggio

Abstract

In recent years, an increasing number of investigations has demonstrated the therapeutic potential of molecules targeting the endocannabinoid system. Cannabinoids of endogenous, phytogenic, and synthetic nature have been assessed in a wide variety of disease models ranging from neurological to metabolic disorders. Even though very few compounds of this type have already reached the market, numerous preclinical and clinical studies suggest that cannabinoids are suitable drugs for the clinical management of diverse pathologies.

In this chapter, we will provide an overview of the endocannabinoid system under certain physiopathological conditions, with a focus on neurological, oncologic, and metabolic disorders. Cannabinoids evaluated as potential therapeutic agents in experimental models with an emphasis in the most successful chemical entities and their perspectives towards the clinic will be discussed.

Keywords

Cannabinoids · Clinical studies · Endocannabinoid system · Experimental models · Synthetic cannabinoids

P. Morales (✉) · P. H. Reggio
University of North Carolina, Greensboro, NC, USA
e-mail: phreggio@uncg.edu

4.1 Introduction

Components from the plant *Cannabis Sativa* as well as synthetic derivatives developed by academic and industry researchers have been extensively studied as therapeutics in the past few decades. However, very few have successfully entered the clinical scenario, thus far. Numerous ongoing investigations are trying to decipher the potential of these chemical entities in the treatment of a wide variety of diseases.

A growing number of preclinical studies published in the last years highlight the therapeutic actions of these compounds in different experimental models. Therefore, medical efforts and patient hopes are quite high for the development of cannabinoids as pharmacological agents for metabolic, neurological, or oncologic diseases among others. Presumably, in the near future, this field will greatly benefit patients with otherwise difficult to treat disorders. It is noteworthy that in June 2018, the U.S. Food and Drug Administration *approved* the non-psychoactive phytocannabinoid cannabidiol (CBD, commercialized as Epidiolex®) for the treatment of seizures in children with Lennox–Gastaut and Dravet syndromes (Devinsky et al. 2018, 2019).

Cannabinoids are molecules that target the endocannabinoid system (ECS), which are involved in the regulation of numerous physiological and pathological processes. These compounds may bind or modulate one or various receptors that are part of ECS. Thus far, two

E. Murillo-Rodriguez et al. (eds.), *Cannabinoids and Neuropsychiatric Disorders*, Advances in Experimental Medicine and Biology 1264, https://doi.org/10.1007/978-3-030-57369-0_4

G-protein-coupled receptors (GPCRs) have been identified as the two major cannabinoid receptors CB_1 and CB_2. CB_1 is mostly found in the central nervous system, while CB_2 is predominantly in the immune system among other organs and tissues (Matsuda et al. 1990; Herkenham et al. 1991; Demuth and Molleman 2006). Their endogenous ligands (endocannabinoids) and the enzymes implicated in their biosynthesis and degradation [(fatty acid amide hydrolase (FAAH) and monoacylglycerol lipase (MAGL)] are also part of this intricate system (Mechoulam et al. 1995, 1996; Beltramo et al. 1997; Fu et al. 2011; Marsicano and Chaouloff 2011). Whether additional cannabinoid receptors are part of the ECS still instigates a strong debate (Morales and Reggio 2017). Recent studies have shown that several cannabinoid ligands bind to the receptor GPR55 (Morales and Jagerovic 2016) and GPR18 (McHugh et al. 2010), supporting the idea that they may play an important role in ECS. Moreover, there is extensive evidence indicating that ECS also interacts with a number of established non-CB_1, non-CB_2 GPCRs, ion channels, and nuclear receptors (Pertwee et al. 2010; Morales et al. 2017; Morales and Reggio 2017).

4.1.1 Cannabinoid Classifications

Cannabinoid classifications have been established according to their pharmacology, their molecular structure, or their origin. Attending to the last criterion, cannabinergic compounds can be classified as endogenous (endocannabinoids), phytogenic (phytocannabinoids), and synthetic compounds.

4.2 Endocannabinoids

Endocannabinoids are endogenous lipidic molecules that bind to the cannabinoid receptors mediating retrograde neurotransmission (Wilson and Nicoll 2001). This family of compounds is formed by eicosanoids derived from arachidonic acid and other polyunsaturated fatty acids. Anandamide (AEA) and 2-arachidonoylglycerol (2-AG, Fig. 4.1) are the first endocannabinoids discovered and are most abundant in the human brain (Basavarajappa 2007). AEA partially activates both cannabinoid receptors CB_1 and CB_2, whereas 2-AG fully activates both of them. (Di Marzo et al. 1994; Stella et al. 1997). Other endocannabinoids identified include 2-arachidonoylglyceryl ether (noladin ether, 2-AGE), O-arachidonoyl ethanolamine (virodhamine), and N-arachidonoyl-dopamine (NADA) (Fig. 4.1).

The endocannabinoid tone is sustained by enzymes that synthesize and degrade these eicosanoids. Due to the physiopathological implication of this machinery, diverse drug discovery approaches have explored the modulation of the endocannabinoid tone. Strategies such as inhibition of degrading enzymes, positive allosteric modulation of CB_1 and/or CB_2, and development of endocannabinoid mimetics with a lower affinity towards metabolic enzymes have shown promising results in preclinical models (Pertwee 2005; Di Marzo 2018). Medicinal chemistry programs have developed synthetic analogs of endocannabinoids with structural modifications at key positions following the aforementioned strategies. Instances of this approach are ACEA (arachidonyl-2′-chloroethylamide) or ACPA (arachidonylcyclopropylamide, Fig. 4.1), analogs of AEA with improved CB_1 affinity (Hillard et al. 1999). (*R*)-(+)-Methanandamide (Met-AEA, Fig. 4.1), a methylated AEA derivative, displays the same functional profile at the cannabinoid receptors while being longer-lived because it is more difficult for FAAH to metabolize.

4.3 Phytocannabinoids

To date, over 120 cannabinoids, termed "phytocannabinoids", have been isolated from the Cannabis plant. These compounds bear a benzone-1,3-diol or a benzopyran ring and a hydrophobic alkyl chain. Δ^9-tetrahydrocannabinol (Δ^9-THC) and cannabidiol (CBD, Fig. 4.1) are the most abundant cannabinoids in the plant and the most widely studied. Other

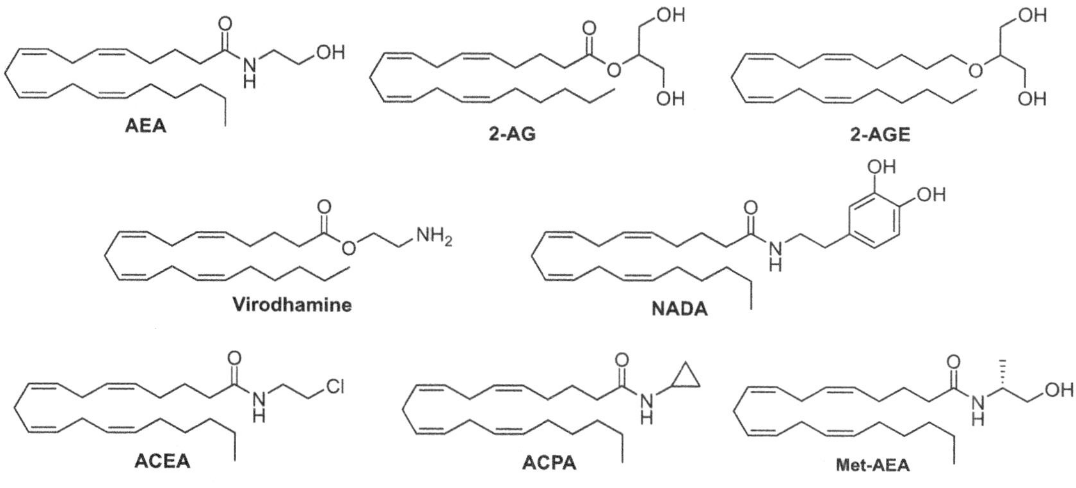

Fig. 4.1 Structures of endogenous cannabinoids and synthetic endocannabinoid derivatives

phytocannabinoids include cannabinol (CBN), cannabigerol (CBG), and cannabichromene (CBC) (Fig. 4.1).

Phytocannabinoids exhibit different activities at the cannabinoid receptors CB_1 and CB_2 (Morales and Reggio 2017). Δ^9-THC has been consistently shown to activate CB_1 and CB_2 with similar potency. Many of the therapeutic effects as well as the psychotropic outcomes of *Cannabis Sativa* are due to this phytocannabinoid. The non-psychoactive plant derivative CBD has also shown pharmacological potential in a wide range of pathologies (Mechoulam et al. 2007). Its functional profile at ECS is quite complex and is currently being investigated by diverse research groups (Morales and Reggio 2019) (Fig. 4.2).

Synthetic cannabinoid derivatives have been developed in the search for improved therapeutics and often trying to dissociate CB1 and CB2 activity. Structure-activity relationship studies of phytocannabinoid analogs have helped to understand the molecular requirements for cannabinoid activity. Derivatization at pharmacophoric positions including the alkyl lipophilic chain, the phenolic, and the pyran ring has resulted in compounds with a cannabinoid selective profile. Widely studied synthetic phytocannabinoid derivatives include CP55,940, HU210, JWH133, and HU308 (Fig. 4.3). CP55,940 and HU210 are very potent CB_1/CB_2 agonists, whereas the deoxy and the methoxy-Δ^9-THC derivatives JWH133 and HU308 are CB_2 agonists with significant selectivity over CB_1 (Huffman 2000). The only structural modification of Δ^9-THC that has led to an approved drug, thus far, is nabilone (Fig. 4.3).

4.4 Synthetic Cannabinoids

The therapeutic relevance of ECS has prompted the identification of numerous synthetic cannabinoid scaffolds. Strategies for the development of cannabimimetic compounds include the design of drugs that selectively activate or block CB_1 or CB_2, molecules that can act as allosteric modulators or biased agonists of these receptors, inhibitors of the metabolic enzymes FAAH or MAGL, as well as the development of compounds acting at peripheral cannabinoid receptors (Morales and Jagerovic 2020). These synthetic cannabinoids aim to provide optimized therapeutic effects and pharmacokinetical profile, while reducing undesirable side actions.

As we will describe in the following sections, numerous synthetic compounds have been used as pharmacological tools or therapeutic agents in different disease models.

The best-known compounds of this synthetic family involve aminoalkyindoles, such as *R*-(+)-WIN55,212–2 (D'Ambra et al. 1992) and

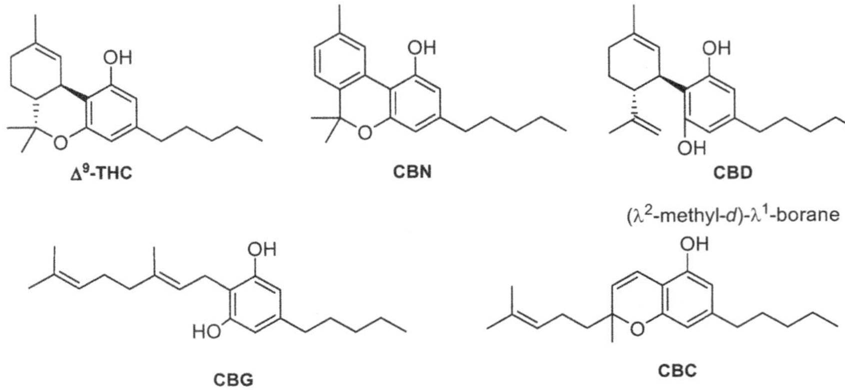

Fig. 4.2 Structures of the most abundant phytocannabinoids

JWH-015 (Fig. 4.4), CB_1/CB_2 and CB_2 agonists, respectively; arylpyrazoles, such as SR141716A (rimonabant) (Rinaldi-Carmona et al. 1994) or AM251 (Fig. 4.4), CB_1 antagonist/inverse agonists; or indole-2-carboxamides such as ORG27569 (Fig. 4.4), identified as the first CB_1 allosteric modulator (Price et al. 2005).

In the following sections, we will describe the ECS upregulation in diverse pathologies to provide an overview of the chemical entities evaluated in experimental disease models. Their potential for further drug development or their progress towards the clinic will be also discussed.

4.5 Cannabinoids in Neuromodulation

ECS has a crucial role in mediating and modulating physiological responses in the central nervous system (CNS). ECS has been shown to be involved in synaptic plasticity and homeostatic processes in the brain. Therefore, it is not surprising that numerous reports have proved the dysregulation of cannabinoid receptor expression under specific neurological disorders providing a therapeutic scenario for the use of cannabinoids.

CB_1 is one of the most abundant GPCRs in CNS, its expression is found particularly high in

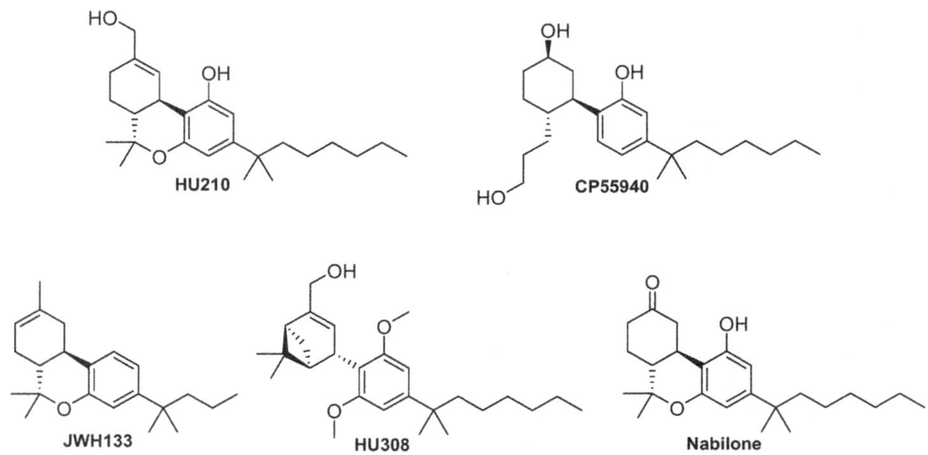

Fig. 4.3 Structures of representative phytocannabinoid synthetic derivatives

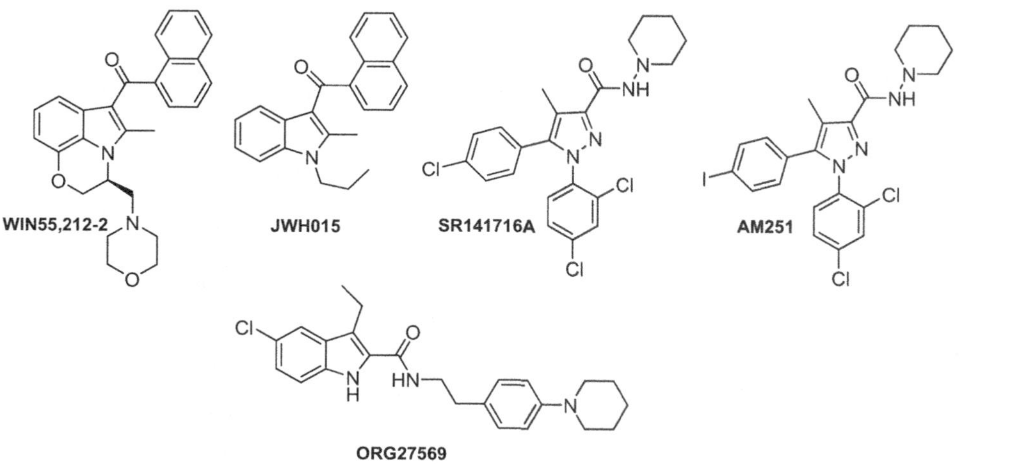

Fig. 4.4 Structure of representative synthetic cannabinoids

the basal ganglia, neocortex, hippocampus, and cerebellum CNS (Herkenham et al. 1991; Marsicano and Kuner 2008). The CB_1 receptors are highly present at the presynaptic and axonal compartments, and thus their function is tightly associated with synaptic activity (Straiker and Mackie 2005). The activation of these receptors has been found to positively affect inwardly rectifying potassium channel conductance, while triggering a decrease in the N-type and P/Q-type voltage-operated calcium channel conductance and to reduce endocannabinoid production. This cascade of events leads to a decrease of neurotransmitter release at excitatory and inhibitory synapses conferring to CB_1 the ability to modulate neurotransmission (Katona et al. 1999; Blázquez et al. 2011). Numerous investigations have demonstrated that the CB_1 receptors exhibit neuroprotective effects against excitotoxicity induced by diverse stimuli (Marsicano et al. 2003). Therefore, multiple pathophysiological events, ranging from neurodegenerative disorders to memory deficits, have been associated with their actions (Kano et al. 2009; Di Marzo et al. 2015).

Moreover, the CB_2 receptors, although initially thought to be peripherally restricted, have been found in particular brain regions offering a very promising therapeutic approach in certain neurological diseases. At a central level, the expression of these receptors is enhanced upon inflammation being mainly localized in the microglia (Fernández-Ruiz et al. 2015). Since neuroinflammatory alterations are associated with several neurological pathologies, the CB_2 receptor agonists offer a promising therapeutic approach for the treatment of these disorders (Roche and Finn 2010; Navarro et al. 2016).

4.5.1 Cannabinoids in Epilepsy

Epilepsy is characterized by an imbalance between excitatory and inhibitory neurotransmitter release and abnormal neuronal electrical activity. Even though, antiepileptic drugs have been shown to limit seizures, over 30% of patients remain pharmacoresistant (Kwan et al. 2011). In this scenario, increasing research demonstrates that the exogenous modulation of ECS offers a promising and effective option for the treatment of refractory epilepsy (Rosenberg et al. 2015; Billakota et al. 2019). Although, the exact molecular mechanisms are still under investigation, the anticonvulsant potential of cannabinoids is supported by their neuromodulatory effects and their ability to inhibit hyperexcitability (Rosenberg et al. 2015).

Diverse phytocannabinoids, including Δ^9-THC, Δ^9-THCA (Δ^9-tetrahydrocannabinolic

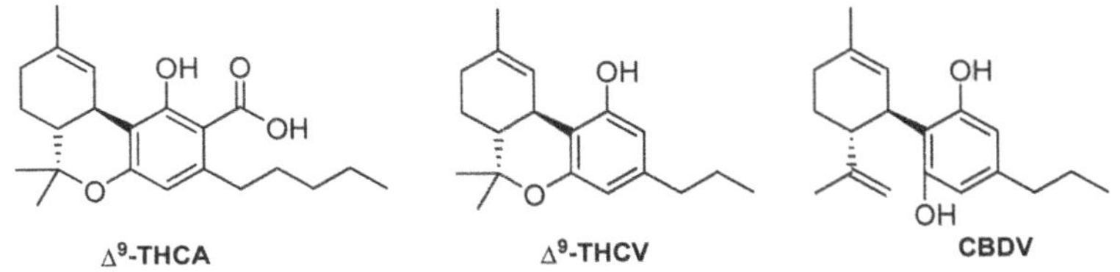

Fig. 4.5 Structures of phytocannabinoids Δ^9-THCA, Δ^9-THCV, and CBDV

acid, Fig. 4.5), Δ^9-THCV (Δ^9-tetrahydrocannabivarin, Fig. 4.5), CBD, and CBDV (cannabidivarin, Fig. 4.5), have shown anticonvulsant effects in different experimental models of seizures. Whereas, very few studies have been reported for the use of Δ^9-THCA, Δ^9-THCV, and CBDV, abundant data support the potential use of Δ^9-THC and CBD for the treatment of epilepsy (Gaston and Friedman 2017).

Most studies have supported the anticonvulsant potential of Δ^9-THC, however, some experiments have revealed mixed or no effects (Rosenberg et al. 2015). Among cannabinoids, the non-psychoactive phytocannabinoid, CBD is currently the best hope for the treatment of refractory epileptic seizures. Its potent anticonvulsant actions have been widely demonstrated in in vitro and in vivo human studies leading to CBD's approval for the management of seizures in children with Lennox–Gastaut and Dravet syndromes (Devinsky et al. 2018, 2019). Placebo-controlled clinical trials revealed that CBD is well-tolerated and does not present side effects on CNS or vital signs (Bergamaschi et al. 2011; Friedman et al. 2019).

The proposed mechanisms of CBD anti-epileptogenic actions include the activation of TRPV1 channels (Bisogno et al. 2001), blockage of T-type voltage-gated calcium channels (VGCC) (Ibeas Bih et al. 2015), and modulation of GPCRs including the cannabinoid receptors CB_1 and CB_2 (Wallace et al. 2001, 2002), GPR55, the adenosine receptors A1 and A2 (Gaston and Friedman 2017), and the serotonin receptors 5-HT1A and 5-HT2A (Sourbron et al. 2016).

Synthetic cannabinoids have also been tested in preclinical seizures models (Rosenberg et al. 2015). FAAH inhibitors such as URB597 and AM404 (Fig. 4.6) did not exert significant anticonvulsant actions in animal models. Likewise, the CB_1 antagonists, including SR141716A and AM251 (Fig. 4.4), were not successful in the assessed models. CB1 agonists, such as WIN55,212–2 (Fig. 4.4) and ACEA (Fig. 4.1), showed anti-seizure effects, although proconvulsive effects were reported in a low percentage of cases (Rosenberg et al. 2015). In fact, one study suggested that the CB_1 agonists may exhibit proconvulsant effects at high doses via TRPV1 activation (Manna and Umathe 2012).

In summary, the activation of ECS exerts antiepileptic effects whereas inhibition of the endogenous cannabinoid machinery does not prevent seizures in reported epilepsy models.

4.5.2 Cannabinoids in Alzheimer's Disease

Alzheimer's disease (AD) is a neurodegenerative disorder that is defined by the progressive deterioration of cognition and memory caused by the formation of β-amyloid plaques and neurofibrillary tangles. Alteration of ECS has been identified in animal models and human *postmortem* samples in the AD brain, especially in the hippocampus and cerebral cortex brain regions severely affected by this disease. AD patients experience a loss of the neuronal CB_1 receptors (Ramírez et al. 2005), while significant upregulation of the CB_2 receptors in microglial cells has been extensively reported (Benito et al. 2003; Aso and Ferrer 2016; López et al. 2018). Additionally, increased 2-AG and elevation of FAAH enzymes have also been associated with the progression of AD pathogenesis (Benito et al. 2003).

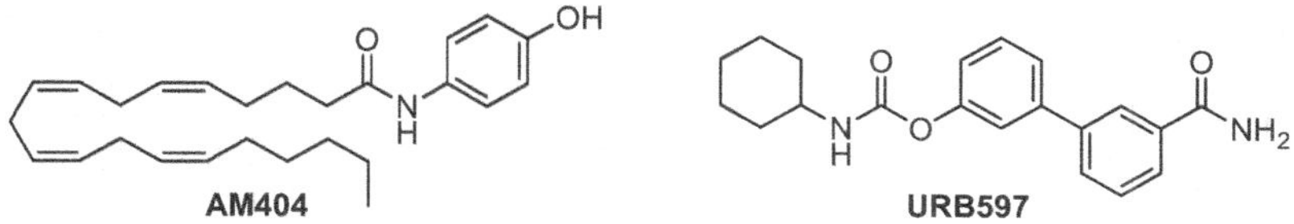

Fig. 4.6 Structures of FAAH inhibitors tested in epilepsy experimental models

The enhanced 2-AG levels along with the increased CB_2 receptors expression in microglial cells have been proposed to exert protective effects against β-amyloid-induced neuroinflammation and neuronal injury (Benito et al. 2003; López et al. 2018). However, the CB_1 receptor downregulation in the hippocampus and basal ganglia may contribute to the destructive inflammatory process experienced by the AD patients (Ramírez et al. 2005). Increased FAAH activity in astrocytes has been associated with the formation of more arachidonic acid, which eventually leads to pro-inflammatory effects.

The exogenous modulation of ECS has shown promising results in preclinical AD models. On the one hand, CB_1 activation has been reported to prevent amyloid β-induced neurotoxicity in vitro (Milton 2002; Benito et al. 2003; Ramírez et al. 2005) and to improve memory deficits and cognitive impairment in diverse animal models (Van Der Stelt et al. 2006; Haghani et al. 2012; Aso et al. 2012). Moreover, the activation of the CB_2 receptors has been reported to attenuate the inflammation associated with AD modulating Aβ aberrant processing (Aso and Ferrer 2016). On the other hand, the inhibition of the endocannabinoid enzymes, FAAH and MAGL, has also been proposed as a potential therapeutic strategy for AD (Benito et al. 2012).

Among the cannabinoids tested in AD experimental models, the most promising results come from the phytocannabinoids Δ^9-THC, CBD, or combinations of both (commercialized as Sativex®) (Fernández-Ruiz et al. 2015). These molecules, and the Δ^9-THC synthetic derivative nabilone (Fig. 4.3), have been shown to counteract specific pathological hallmarks of AD, such as tau and β-amyloid aggregation, leading to cognitive and behavioral improvements. The few clinical trials performed so far confirmed the results observed in the animal models of the disease.[1] However, more controlled trials are needed to evaluate the efficacy of cannabinoids in the management of the different stages of this neurodegenerative disease.

Synthetic cannabinoids with diverse pharmacological profiles have also been tested in AD preclinical models. For instance, CB_2 agonists, such as the naphthoylindole, JWH-015 (Fig. 4.4), or the phytocannabinoid derivatives, JWH-133 (Fig. 4.3), and HU-308 (Fig. 4.3), have been shown to reduce plaque aggregation, thereby exerting anti-inflammatory effects (Aso and Ferrer 2016). Likewise, CB_1/CB_2 mixed agonists including WIN55,212–2 (Fig. 4.4) and HU-210 (Fig. 4.3) have been demonstrated to have the ability to reduce pro-inflammatory markers and improve cognitive performance in the AD models (Ramírez et al. 2005; Martín-Moreno et al. 2011). Although, more studies need to confirm these effects, endocannabinoid reuptake inhibitors, such as VDM11 (Fig. 4.7) or MAGL inhibitors such as JLZ184 (Fig. 4.7), can decrease amyloid neurotoxicity (Van Der Stelt et al. 2006; Chen et al. 2012).

It has been extensively demonstrated that the pleiotropic activity of cannabinoids can target several crucial processes associated with AD. This includes neuroinflammation, β-amyloid and tau aberrant processing, excitotoxicity, or oxidative stress. In a multifactorial disease, such as AD, this offers a promising strategy. Hopefully, results from more clinical trials will shed additional light into this research such that AD patients worldwide can soon benefit from cannabinoid therapy.

[1] Clinical trials: THC in Alzheimer Disease - ClinicalTrials.gov. https://clinicaltrials.gov/ct2/results?cond=Alzheimer+Disease&term=THC&cntry=&state=&city=&dist=. Accessed 7 Oct 2019.

Fig. 4.7 Structure of endocannabinoid reuptake inhibitor VDM11 and MAGL inhibitor JLZ184

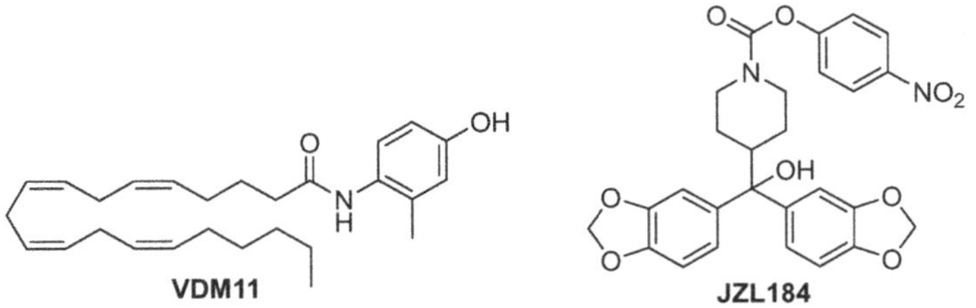

4.5.3 Cannabinoids in Parkinson's Disease

Parkinson's disease (PD) is a long-term degenerative *disorder that mainly affects motor coordination, although non-motor symptoms also appear with the progression of the disease. One of the main pathological hallmarks of PD is cell death in the basal ganglia, especially of dopaminergic neurons.*

As in the previously mentioned neurological disorders, ECS has been shown to be abnormally regulated in this pathology. For instance, the upregulation of the CB_1 receptors has been shown in the basal ganglia of experimental models of PD (Stampanoni Bassi et al. 2017). Moreover, a loss of the neuronal CB_2 receptors was detected in the postmortem tissues of PD patients due to the degeneration of nigrostriatal dopaminergic neurons (García et al. 2015).

Pharmacological cannabinoid strategies to manage PD include activation of CB_2, to control inflammatory events, and blockage of CB_1 receptors, to improve akinesia and reduce motor inhibition. Since one of the main characteristics of PD is high oxidative stress, the experiments reported so far in the PD models have been focused on the use of antioxidant phytocannabinoids. The evaluation of Δ^9-THC (Lastres-Becker et al. 2005), CBD (Lastres-Becker et al. 2005; García-Arencibia et al. 2007; García et al. 2011), and Δ^9-THCV (García et al. 2011) in animal models revealed their ability to reduce parkinsonian motor symptoms. In fact, clinical trials to assess the potential of CBD, nabilone, or Cannabis oils in the PD motor and non-motor symptoms are currently ongoing.[2]

Synthetic cannabinoids such as the potent CB_1/CB_2 receptor agonists WIN55,212–2 (Price et al. 2009; More and Choi 2015) and CP55,940 (Jimenez-Del-Rio et al. 2008) or the AEA synthetic derivative AM404 (García-Arencibia et al. 2007) have been shown to provide neuroprotection in the PD models.

Even though further clinical research is required, the knowledge gained in this field and ongoing clinical efforts point towards a cannabinoid-based neuroprotection for the treatment of PD.

As thoroughly reviewed by others, cannabinoids have been shown to impact many other neurological disease models, such as multiple sclerosis (MS), traumatic brain injury (TBI) or amyotrophic lateral sclerosis (ALS), as well as mental disorders including schizophrenia, anxiety, or depression (Kendall and Yudowski 2017; Aymerich et al. 2018; Ibarra-Lecue et al. 2018; Friedman et al. 2019). Moreover, symptoms associated with these diseases can also be treated with cannabinoid-based medicines, for instance, Sativex® is used for the symptomatic relief of pain and spasticity in adults suffering from MS (Giacoppo et al. 2017).

Even though much more research needs to be conducted, the modulation of ECS is a great therapeutic opportunity for the treatment of several neuropsychiatric and neurodegenerative disorders.

---

[2] Clinical trials: cannabinoids in Parkinson Disease-ClinicalTrials.gov. https://clinicaltrials.gov/ct2/results?cond=Parkinson+Disease&term=cannabis&cntry=&state=&city=&dist=. Accessed 3 Oct 2019.

4.6 Cannabinoids in Cancer

The ability of Cannabis to prevent nausea and vomiting, stimulate appetite, and reduce pain has been widely demonstrated. Therefore, cannabinoids have been successfully used in the treatment of specific cancer chemotherapy side effects (Abrams and Guzman 2015).

A few decades ago, dronabinol (Marinol®) and nabilone (Cesamet®) were approved to treat emesis and nausea induced by antitumor agents (Tramèr et al. 2001). However, they are only prescribed in certain countries upon failure of conventional anti-emetics (Sharkey et al. 2014).

Extensive research has demonstrated the palliative potential of cannabinoids for cancer patients. For instance, Δ^9-THC acts as an appetite stimulant increasing food intake in rodents. Clinical trials confirmed this orexigenic effect in the management of cancer anorexia (Jatoi et al. 2002; Berry and Mechoulam 2002; Walsh et al. 2003). Moreover, the ability of cannabinoids in reducing chemotherapy-induced pain has also been reported. Δ^9-THC and synthetic analogs have shown to act as potent analgesic drugs in diverse clinical trials highlighting their beneficial role in the treatment of cancer pain (Campbell et al. 2001; Iversen and Chapman 2002; Mantyh et al. 2002). Actually, Sativex® can be currently prescribed in certain countries to reduce pain in adults with advanced tumors (Pertwee 2009; Fallon et al. 2017).

Preclinical data indicate that peripheral neuropathies associated with cancer treatment can also be ameliorated upon cannabinoid administration (Guindon et al. 2014). Synthetic agonists such as the aminoalkylindole WIN55,212–2, diminishes mechanical and cold allodynia in rodent models of paclitaxel (Pascual et al. 2005), vincristine (Rahn et al. 2007), and cisplatin-evoked neuropathy (Vera et al. 2007). Moreover, CBD is able to reduce doxorubicin-induced cardiomyopathies (Hao et al. 2015) and cisplatin-induced nephrotoxicity (Pan et al. 2009).

Besides their palliative potential, cannabinoids have exhibited antitumor effects in numerous in vitro and in vivo experimental models of cancer (Guzmán 2003; Chakravarti et al. 2014; Velasco et al. 2016). Since the early 2000s, a growing body of research has evidenced the potential of cannabinoids in the reduction of tumor growth and progression in diverse cancer models (Galve-Roperh et al. 2000; Guzmán et al. 2002; Guzmán 2003; Carracedo et al. 2006; Sarfaraz et al. 2008; Velasco et al. 2012).

ECS alterations have also been detected in cancer physiopathology. Abnormal expression of the ECS components in neoplasms compared with healthy tissues has been detected (Guzmán 2003; Caffarel et al. 2006; Malfitano et al. 2011; Velasco et al. 2012). These data can be tumor type-specific and therefore, studies need to determine how ECS is regulated in each cancer type (Malfitano et al. 2011; Velasco et al. 2016). In specific cancer types, such as glioblastoma (Schley et al. 2009) or specific breast tumors (Qamri et al. 2009; Caffarel et al. 2010), increased CB_2 receptor levels have been shown. Other tumors, including gastric carcinoma (Miyato et al. 2009) or rhabdomyosarcoma (Oesch et al. 2009) are characterized by the overexpression of the CB_1 receptor. Upregulated expression of both CB_1 and CB_2 has also been detected in acute myeloid leukemia (Joseph et al. 2004) malignant astrocytomas (Stella 2010), pancreatic cancer (Carracedo et al. 2006), and hepatocellular carcinoma (Giuliano et al. 2009) among others. Levels of endocannabinoids, AEA and 2-AG, have also been shown to differ between cancer cells and their normal counterparts in specific tumors (Bifulco et al. 2006). Upregulation of the putative cannabinoid receptor, GPR55, has also been observed in cells of diverse cancer types including breast adenocarcinoma, squamous skin cell carcinoma, or gliomas (Oka et al. 2010; Andradas et al. 2011; Leyva-Illades and Demorrow 2013; Pérez-Gómez et al. 2013). GPR55 expression has been shown to correlate with proliferation and thus, it has been proposed as a novel oncology biomarker with a potential prognostic value (Henstridge et al. 2011). Expression of GPR55-CB_2 heterodimers has also been reported in human tumors (Moreno et al. 2014; Balenga et al. 2014).

Even if further research is required to clarify the intricate role of this complex system in the course of oncological processes, there is no doubt that cannabinoids are useful drugs for the management of cancer and related symptoms.

As in previously described diseases, thus far, preclinical and clinical studies on cannabinoids as antitumor agents have been mainly focused on understanding the mechanism of action of Δ^9-THC and CBD (Pellati et al. 2018; Hinz and Ramer 2019). Δ^9-THC has shown antiproliferative effects in diverse cancer types including glioblastoma, prostate, breast, colon, pancreatic, lymphoma, or lung among others (Fowler 2015; Fraguas-Sánchez et al. 2016). Mechanisms of this antitumor action include the CB receptor-dependent and independent pathways (Powles et al. 2005). Moreover, CBD has been widely proved to reduce tumor growth via proapoptotic actions in numerous cancer cell lines (Hinz and Ramer 2019). The anticancer effects of CBD have been suggested to be mediated by several targets, including COX-2, 5-LOX, PPARγ, TRPV2, mTOR, and the p38 MAPK pathway (Ligresti 2006; Hinz and Ramer 2019). Clinical trials are trying to unravel the antitumor potential of phytocannabinoids (such as Δ^9-THC) alone or in combination with benchmark chemotherapeutic agents in different types of cancer. Guzmán et al. developed the first clinical trial to further explore the antitumor actions of cannabinoids in cancer patients. This pilot trial investigated the effects of Δ^9-THC on nine patients with recurrent glioblastoma multiforme. The preliminary results attained from this study suggest a reduction in tumor growth upon Δ^9-THC administration (Guzmán et al. 2006). Ongoing clinical trials are trying to decipher the potential antitumor role of cannabinoids.[3]

Even if phytocannabinoids are in the forefront towards the clinic, many other cannabinoids with antitumor properties have been reported in the literature (Morales and Jagerovic 2019). For instance, the well-known aminoalkylindole WIN55,212–2 is able to decrease cell proliferation and migration in models of different cancer types, hepatocellular carcinoma (Xu et al. 2015), neuroblastoma (Müller et al. 2017), myeloma (Barbado et al. 2017), renal carcinoma (Khan et al. 2018), prostate (Morell et al. 2016), or gastric cancer (Xian et al. 2016) among them.

Moreover, it is worth highlighting the anticancer potential of cannabinoid quinones. Oxidized derivatives of phytocannabinoids cannabidiol (HU-331, Fig. 4.8), Δ^8-THC (HU-336, Fig. 4.8) and cannabinol (HU-345, Fig. 4.8) were effective in reducing tumor growth in mice cancer models (Kogan et al. 2004). However, their biological activity was attributed to their quinone structure independently of their cannabinoid character, since they do not modulate the cannabinoid receptors (Kogan et al. 2006, 2007). *Para-* and *ortho-* quinones of chromenopyrazoles were also reported as antitumor agents (Morales et al. 2013, 2015). These compounds were able to reduce cancer proliferation through mechanisms that involve the cannabinoid receptors. For instance, *para*-quinones PM49 (Fig. 4.8) was able to reduce prostate cancer in vitro and in vivo (Morales et al. 2013). 1,4-naphthoquinone derivatives, such as 3a (Fig. 4.8), have also been reported to inhibit tumor proliferation. GPR55 has been proposed as the target through which they exhibit their antitumor effects (Badolato et al. 2019).

Currently, the use of cannabinoids is limited to the management of chemotherapy-induced side effects. Nevertheless, the aforementioned preclinical data clearly evidence the antitumor potential of cannabinoids. Hopefully, further clinical data can soon confirm the therapeutic potential of cannabinoids in the treatment of cancer.

4.7 Cannabinoids in Metabolic Disorders

ECS has been recognized to play a crucial role in the regulation of metabolic events, particularly in energy balance, food intake, and lipid metabolism (Scherma et al. 2014; Williams et al. 2015). This

[3]Clinical trials: cannabinoids in Cancer-ClinicalTrials.gov. https://clinicaltrials.gov/ct2/results?cond=Cancer&term=cannabinoid&cntry=&state=&city=&dist= Accessed 29 June 2020.

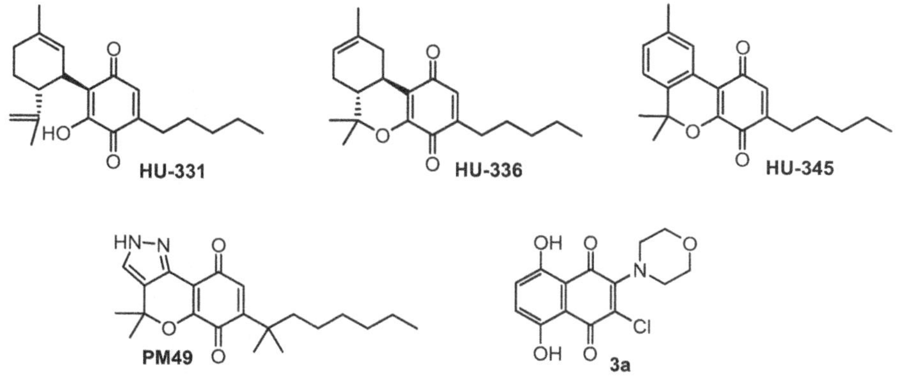

Fig. 4.8 Structures of quinones related to cannabinoids with reported antitumor potential

system has shown dysregulation in metabolic pathologies including obesity. For instance, the increased levels of circulating endocannabinoids (Blüher et al. 2006; Matias et al. 2006) and upregulation of the CB_1 receptors have been observed in obese rodents and human obesity (Murdolo et al. 2007; Pagano et al. 2007). In this disorder, ECS dysregulation has been reported, not only in CNS but also at the peripheral level, in diverse organs including the pancreas, liver, and adipose tissues.

It is well-known that ECS activation induces orexigenic effects (Rossi et al. 2018), therefore, the inhibition of the CB_1 receptors has been considered as a potential strategy for the management of obesity and metabolic syndrome. In fact, the CB_1 antagonist/inverse agonist rimonabant (SR141716A, Fig. 4.4, commercialized as Acomplia®), was approved in certain European countries in 2006 for the management of obesity (Després et al. 2006). The anti-obesity effects of this drug were accompanied by the undesired effects such as depression, anxiety, headache, and suicidal thoughts forcing its withdrawal from the clinic, a couple of years later. Numerous research projects from academia and the pharmaceutical industry were centered on the development of CB_1 receptor antagonists, however, the psychiatric side effects of rimonabant led to a significant decrease in the continuation of this approach (Serrano et al. 2012; Silvestri and Di Marzo 2012; Sharma et al. 2015; Yadav and Murumkar 2018; Amato et al. 2019).

Other pharmacological strategies targeting ECS, but without severe psychiatric side effects, have been attempted. Peripherally restricted CB_1 antagonists, such as URB447 and AM6545 (Fig. 4.9), have shown promising results in the control of fat intake and obesity (DiPatrizio et al. 2011; Argueta and DiPatrizio 2017).

Moreover, molecules acting preferentially via the CB_2 receptors have shown efficacy in a rat model of alcoholic hepatic steatosis by decreasing the liver/body weight ratio and hepatic triglyceride content (Lotersztajn et al. 2008, 2011). The inhibitors of the enzymes involved in the degradation of endocannabinoids, such as FAAH inhibitors, has also shown potential for the regulation of energy balance (Balsevich et al. 2018). However, this approach should be taken with caution, since the FAAH inhibitor BIA 10–2474 (Fig. 4.9) caused severe neurotoxicity in a phase I clinical trial probably due to off-target effects (Van Esbroeck et al. 2017).

Despite the clinical failures obtained so far, ECS still represents a very promising pharmacological target to treat metabolic disorders.

4.8 Conclusions

It has been widely demonstrated that compounds targeting ECS, particularly CB_1 and/or CB_2, have therapeutic potential for the clinical management of numerous diseases. These include neurological disorders, metabolic pathologies, cancer, or

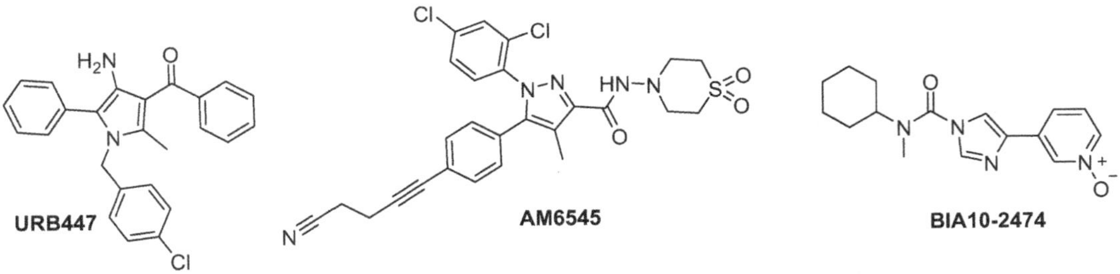

Fig. 4.9 Structures of CB_1 antagonists URB447 and AM6545 and FAAH inhibitor BIA 10–2474

symptoms such as inflammatory and neuropathic pain. However, just a few of these diseases can be treated with cannabinoid-based medicines currently (Table 4.1).

Even though CB_1/CB_2 agonists are in the forefront of clinical research for neuroprotection or cancer treatment, there is an increasing interest in exploiting novel pharmacological approaches (Picone and Kendall 2015). CB_2 selective agonists or peripherally restricted CB_1/CB_2 agonists exhibit a promising therapeutic potential for treating various pathologies, while avoiding the adverse psychotropic effects related to the modulation of CB_1 in the brain (Dhopeshwarkar and Mackie 2014). CB_1 and/or CB_2 antagonists or inverse agonists, as well as, allosteric cannabinoid ligands are also emerging and may prove useful in the treatment of certain diseases (Picone and Kendall 2015; Vemuri and Makriyannis 2015). Biased cannabinoid agonists can also fine-tune the therapeutic effects, while minimizing side effects associated with other receptor pathways (Morales et al. 2018; Al-zoubi et al. 2019). Even though

Table 4.1 Representative cannabinoids that have been reported to exhibit therapeutic potential in specific diseases

Molecule	Disease	Development stage	References
Δ^9-THC	Epilepsy	Preclinical	(Rosenberg et al. 2015)
	AD	Clinical	(Fernández-Ruiz et al. 2015) See footnote 1
	PD	Clinical	See footnote 2
	Cancer	Clinical	(Guzmán et al. 2006)
Δ^9-THCA	Epilepsy	Preclinical	(Gaston and Friedman 2017)
Δ^9-THCV	Epilepsy	Preclinical	(Hill et al. 2010)
	PD	Preclinical	(García et al. 2011)
CBD	Epilepsy	In the market[a]	(Devinsky et al. 2018, 2019)
	AD	Preclinical[b]	(Martín-Moreno et al. 2011)
	PD	Clinical	See footnote 2
	Cancer	Clinical	(Ligresti 2006; Hinz and Ramer 2019)
CBDV	Epilepsy	Preclinical	(Hill et al. 2012)
Nabilone	PD	Clinical	See footnote 2
	Cancer	In the market[c]	(Sharkey et al. 2014)
WIN55,212–2	AD	Preclinical	(Martín-Moreno et al. 2011)
	PD	Preclinical	(Price et al. 2009; More and Choi 2015)
SR141716A	Obesity	Withdrawn from the market[d]	(Després et al. 2006)
AM404	PD	Preclinical	(García-Arencibia et al. 2007)
BIA 10–2474	Obesity	Failed in clinical trials	(Van Esbroeck et al. 2017)

[a]Approved as Epidiolex®
[b]Clinical trials currently recruiting
[c]Approved as Cesamet®
[d]Commercialized as Acomplia®

phytocannabinoids are way closer to the bedside, some of the aforementioned synthetic cannabinoids may provide advantages in the treatment of specific pathologies. Nonetheless, more preclinical and especially clinical research needs to be done in this field.

References

Abrams DI, Guzman M (2015) Cannabis in cancer care. Clin Pharmacol Ther 97:575–586. https://doi.org/10.1002/cpt.108

Al-zoubi R, Morales P, Reggio PH (2019) Structural insights into CB1 receptor biased signaling. Int J Mol Sci 20:1837. https://doi.org/10.3390/ijms20081837

Amato G, Khan NS, Maitra R (2019) A patent update on cannabinoid receptor 1 antagonists (2015-2018). Expert Opin Ther Pat 29:261–269. https://doi.org/10.1080/13543776.2019.1597851

Andradas C, Caffarel MM, Pérez-Gómez E et al (2011) The orphan G protein-coupled receptor GPR55 promotes cancer cell proliferation via ERK. Oncogene 30:245–252. https://doi.org/10.1038/onc.2010.402

Argueta DA, DiPatrizio NV (2017) Peripheral endocannabinoid signaling controls hyperphagia in western diet-induced obesity. Physiol Behav 171:32–39. https://doi.org/10.1016/j.physbeh.2016.12.044

Aso E, Ferrer I (2016) CB2 cannabinoid receptor as potential target against Alzheimer's disease. Front Neurosci 10:1–10. https://doi.org/10.3389/fnins.2016.00243

Aso E, Palomer E, Juvés S et al (2012) CB1 agonist ACEA protects neurons and reduces the cognitive impairment of AβPP/PS1 mice. J Alzheimers Dis 30:439–459. https://doi.org/10.3233/JAD-2012-111862

Aymerich MS, Aso E, Abellanas MA et al (2018) Cannabinoid pharmacology/therapeutics in chronic degenerative disorders affecting the central nervous system. Biochem Pharmacol 157:67–84. https://doi.org/10.1016/j.bcp.2018.08.016

Badolato M, Carullo G, Caroleo MC et al (2019) Discovery of 1,4-Naphthoquinones as a new class of Antiproliferative agents targeting GPR55. ACS Med Chem Lett acsmedchemlett 10(4):402–406. https://doi.org/10.1021/acsmedchemlett.8b00333

Balenga NA, Martínez-Pinilla E, Kargl J et al (2014) Heteromerization of GPR55 and cannabinoid CB2 receptors modulates signaling. Br J Pharmacol 171:5387–5406. https://doi.org/10.1111/bph.12850

Balsevich G, Sticht M, Bowles NP et al (2018) Role for fatty acid amide hydrolase (faah) in the leptin-mediated effects on feeding and energy balance. Proc Natl Acad Sci U S A 115:7605–7610. https://doi.org/10.1073/pnas.1802251115

Barbado MV, Medrano M, Caballero-Velázquez T et al (2017) Cannabinoid derivatives exert a potent anti-myeloma activity both in vitro and in vivo. Int J Cancer 140:674–685. https://doi.org/10.1002/ijc.30483

Basavarajappa BS (2007) Neuropharmacology of the endocannabinoid signaling system-molecular mechanisms, biological actions and synaptic plasticity. Curr Neuropharmacol 5:81–97. https://doi.org/10.2174/157015907780866910

Beltramo M, Stella N, Calignano A et al (1997) Functional role of high-affinity anandamide transport, as revealed by selective inhibition. Science 277:1094–1097. https://doi.org/10.1126/science.277.5329.1094

Benito C, Núñez E, Tolón RM et al (2003) Cannabinoid CB2 receptors and fatty acid amide hydrolase are selectively overexpressed in Neuritic plaque-associated glia in Alzheimer's disease brains. J Neurosci 23:11136–11141. https://doi.org/10.1523/JNEUROSCI.23-35-11136.2003

Benito C, Tolõn RM, Castillo AI et al (2012) β-Amyloid exacerbates inflammation in astrocytes lacking fatty acid amide hydrolase through a mechanism involving PPAR-α, PPAR-γ and TRPV1, but not CB1 or CB2 receptors. Br J Pharmacol 166:1474–1489. https://doi.org/10.1111/j.1476-5381.2012.01889.x

Bergamaschi MM, Queiroz RHC, Zuardi AW, Crippa JAS (2011) Safety and side effects of cannabidiol, a Cannabis sativa constituent. Curr Drug Saf 6:237–249

Berry EM, Mechoulam R (2002) Tetrahydrocannabinol and endocannabinoids in feeding and appetite. Pharmacol Ther 95:185–190. https://doi.org/10.1016/S0163-7258(02)00257-7

Bifulco M, Laezza C, Pisanti S, Gazzerro P (2006) Cannabinoids and cancer: pros and cons of an antitumour strategy. Br J Pharmacol 148:123–135. https://doi.org/10.1038/sj.bjp.0706632

Billakota S, Devinsky O, Marsh E (2019) Cannabinoid therapy in epilepsy. Curr Opin Neurol 32:220–226. https://doi.org/10.1097/WCO.0000000000000660

Bisogno T, Hanuš L, De Petrocellis L et al (2001) Molecular targets for cannabidiol and its synthetic analogues: effect on vanilloid VR1 receptors and on the cellular uptake and enzymatic hydrolysis of anandamide. Br J Pharmacol 134:845–852. https://doi.org/10.1038/sj.bjp.0704327

Blázquez C, Chiarlone A, Sagredo O et al (2011) Loss of striatal type 1 cannabinoid receptors is a key pathogenic factor in Huntington's disease. Brain 134:119–136. https://doi.org/10.1093/brain/awq278

Blüher M, Engeli S, Klöting N et al (2006) Dysregulation of the peripheral and adipose tissue endocannabinoid system in human abdominal obesity. Diabetes 55:3053–3060. https://doi.org/10.2337/db06-0812

Caffarel MM, Andradas C, Mira E et al (2010) Cannabinoids reduce ErbB2-driven breast cancer progression through Akt inhibition. Mol Cancer 9:196–206. https://doi.org/10.1186/1476-4598-9-196

Caffarel MM, Sarrió D, Palacios J et al (2006) Delta9-tetrahydrocannabinol inhibits cell cycle progression in human breast cancer cells through Cdc2 regulation.

Cancer Res 66:6615–6621. https://doi.org/10.1158/0008-5472.CAN-05-4566
Campbell FA, Tramèr MR, Carroll D et al (2001) Are cannabinoids an effective and safe treatment option in the management of pain? A qualitative systematic review. BMJ 323:13–16
Carracedo A, Gironella M, Lorente M et al (2006) Cannabinoids induce apoptosis of pancreatic tumor cells via endoplasmic reticulum stress-related genes. Cancer Res 66:6748–6755. https://doi.org/10.1158/0008-5472.CAN-06-0169
Chakravarti B, Ravi J, Ganju RK (2014) Cannabinoids as therapeutic agents in cancer : current status and future implications. Oncotarget 5:5852–5872
Chen R, Zhang J, Wu Y et al (2012) Monoacylglycerol lipase is a therapeutic target for Alzheimer's disease. Cell Rep 2:1329–1339. https://doi.org/10.1016/j.celrep.2012.09.030
D'Ambra TE, Estep KG, Bell MR et al (1992) Conformationally restrained analogs of pravadoline: nanomolar potent, enantioselective, (aminoalkyl)indole agonists of the cannabinoid receptor. J Med Chem 35:124–135. https://doi.org/10.1021/jm00079a016
Demuth DG, Molleman A (2006) Cannabinoid signalling. Life Sci 78:549–563. https://doi.org/10.1016/j.lfs.2005.05.055
Després J-P, Lemieux I, Alméras N (2006) Contribution of CB1 blockade to the management of high-risk abdominal obesity. Int J Obes (Lond) 30(Suppl 1):S44–S52. https://doi.org/10.1038/sj.ijo.0803278
Devinsky O, Nabbout R, Miller I et al (2019) Long-term cannabidiol treatment in patients with Dravet syndrome: an open-label extension trial. Epilepsia 60:294–302. https://doi.org/10.1111/epi.14628
Devinsky O, Patel AD, Cross JH et al (2018) Effect of Cannabidiol on drop seizures in the Lennox–Gastaut syndrome. N Engl J Med 378:1888–1897. https://doi.org/10.1056/NEJMoa1714631
Dhopeshwarkar A, Mackie K (2014) CB2 cannabinoid receptors as a therapeutic target - what does the future hold? Mol Pharmacol 86:430–437. https://doi.org/10.1124/mol.114.094649
Di Marzo V (2018) New approaches and challenges to targeting the endocannabinoid system. Nat Rev Drug Discov 17(9):623–639. https://doi.org/10.1038/nrd.2018.115
Di Marzo V, Fontana A, Cadas H et al (1994) Formation and inactivation of endogenous cannabinoid anandamide in central neurons. Nature 372:686–691. https://doi.org/10.1038/372686a0
Di Marzo V, Stella N, Zimmer A (2015) Endocannabinoid signalling and the deteriorating brain. Nat Rev Neurosci 16:30–42. https://doi.org/10.1530/ERC-14-0411.Persistent
DiPatrizio NV, Astarita G, Schwartz G et al (2011) Endocannabinoid signal in the gut controls dietary fat intake. Proc Natl Acad Sci U S A 108:12904–12908. https://doi.org/10.1073/pnas.1104675108
Fallon MT, Lux EA, Mcquade R et al (2017) Sativex oromucosal spray as adjunctive therapy in advanced cancer patients with chronic pain unalleviated by optimized opioid therapy : two double-blind , randomized, placebo-controlled phase 3 studies. Br J Pain 11:119–133. https://doi.org/10.1177/2049463717710042
Fernández-Ruiz J, Romero J, Ramos JA (2015) Endocannabinoids and neurodegenerative disorders: Parkinson's disease, Huntington's chorea, Alzheimer's disease, and others. In: Pertwee RG (ed) Endocannabinoids. Springer International Publishing, Cham, pp 233–259
Fowler CJ (2015) Delta9-tetrahydrocannabinol and cannabidiol as potential curative agents for cancer: a critical examination of the preclinical literature. Clin Pharmacol Ther 97:587–596. https://doi.org/10.1002/cpt.84
Fraguas-Sánchez AI, Fernández-Carballido A, Torres-Suárez AI (2016) Phyto-, endo- and synthetic cannabinoids: promising chemotherapeutic agents in the treatment of breast and prostate carcinomas. Expert Opin Investig Drugs 25:1311–1323. https://doi.org/10.1080/13543784.2016.1236913
Friedman D, French JA, Maccarrone M (2019) Safety, efficacy, and mechanisms of action of cannabinoids in neurological disorders. Lancet Neurol 18:504–512. https://doi.org/10.1016/S1474-4422(19)30032-8
Fu J, Bottegoni G, Sasso O et al (2011) A catalytically silent FAAH-1 variant drives anandamide transport in neurons. Nat Neurosci 15:64–69. https://doi.org/10.1038/nn.2986
Galve-Roperh I, Sánchez C, Cortés ML et al (2000) Antitumoral action of cannabinoids: involvement of sustained ceramide accumulation and extracellular signal-regulated kinase activation. Nat Med 6:313–319. https://doi.org/10.1038/73171
García MC, Cinquina V, Palomo-Garo C et al (2015) Identification of CB2 receptors in human nigral neurons that degenerate in Parkinson's disease. Neurosci Lett 587:1–4. https://doi.org/10.1016/j.neulet.2014.12.003
García C, Palomo-Garo C, García-Arencibia M et al (2011) Symptom-relieving and neuroprotective effects of the phytocannabinoid Δ 9-THCV in animal models of Parkinson's disease. Br J Pharmacol 163:1495–1506. https://doi.org/10.1111/j.1476-5381.2011.01278.x
García-Arencibia M, González S, de Lago E et al (2007) Evaluation of the neuroprotective effect of cannabinoids in a rat model of Parkinson's disease: importance of antioxidant and cannabinoid receptor-independent properties. Brain Res 1134:162–170. https://doi.org/10.1016/j.brainres.2006.11.063
Gaston TE, Friedman D (2017) Pharmacology of cannabinoids in the treatment of epilepsy. Epilepsy Behav 70:313–318. https://doi.org/10.1016/j.yebeh.2016.11.016

Giacoppo S, Bramanti P, Mazzon E (2017) Sativex in the management of multiple sclerosis-related spasticity: an overview of the last decade of clinical evaluation. Mult Scler Relat Disord 17:22–31. https://doi.org/10.1016/j.msard.2017.06.015

Giuliano M, Pellerito O, Portanova P et al (2009) Apoptosis induced in HepG2 cells by the synthetic cannabinoid WIN: involvement of the transcription factor PPARgamma. Biochimie 91:457–465. https://doi.org/10.1016/j.biochi.2008.11.003

Guindon J, Lai Y, Takacs SM et al (2014) Alterations in endocannabinoid tone following chemotherapy-induced peripheral neuropathy: effects of endocannabinoid deactivation inhibitors targeting fatty-acid amide hydrolase and monoacylglycerol lipase in comparison to reference analgesics following. Pharmacol Res 67:94–109. https://doi.org/10.1016/j.phrs.2012.10.013.Alterations

Guzmán M (2003) Cannabinoids: potential anticancer agents. Nat Rev Cancer 3:745–755. https://doi.org/10.1038/nrc1188

Guzmán M, Duarte MJ, Blázquez C et al (2006) A pilot clinical study of Delta9-tetrahydrocannabinol in patients with recurrent glioblastoma multiforme. Br J Cancer 95:197–203. https://doi.org/10.1038/sj.bjc.6603236

Guzmán M, Sánchez C, Galve-Roperh I (2002) Cannabinoids and cell fate. Pharmacol Ther 95:175–184. https://doi.org/10.1016/S0163-7258(02)00256-5

Haghani M, Shabani M, Javan M et al (2012) CB1 cannabinoid receptor activation rescues amyloid ß-induced alterations in behaviour and intrinsic electrophysiological properties of rat hippocampal CA1 pyramidal Neurones. Cell Physiol Biochem 29:391–406. https://doi.org/10.1159/000338494

Hao E, Mukhopadhyay P, Cao Z et al (2015) Cannabidiol protects against doxorubicin-induced cardiomyopathy by modulating mitochondrial function and biogenesis. Mol Med 21:38–45. https://doi.org/10.2119/molmed.2014.00261

Henstridge CM, Balenga NAB, Kargl J et al (2011) Minireview: recent developments in the physiology and pathology of the lysophosphatidylinositol-sensitive receptor GPR55. Mol Endocrinol 25:1835–1848. https://doi.org/10.1210/me.2011-1197

Herkenham M, Lynn AB, Johnson MR et al (1991) Characterization and localization of cannabinoid receptors in rat brain: a quantitative in vitro autoradiographic study. J Neurosci 11:563–583

Hill A, Mercier M, Hill T et al (2012) Cannabidivarin is anticonvulsant in mouse and rat. Br J Pharmacol 167:1629–1642. https://doi.org/10.1111/j.1476-5381.2012.02207.x

Hill AJ, Weston SE, Jones NA et al (2010) Δ9-Tetrahydrocannabivarin suppresses in vitro epileptiform and in vivo seizure activity in adult rats. Epilepsia 51:1522–1532. https://doi.org/10.1111/j.1528-1167.2010.02523.x

Hillard CJ, Manna S, Greenberg MJ et al (1999) Synthesis and characterization of potent and selective agonists of the neuronal cannabinoid receptor (CB1). J Pharmacol Exp Ther 289:1427–1433

Hinz B, Ramer R (2019) Anti-tumour actions of cannabinoids. Br J Pharmacol 176:1384–1394. https://doi.org/10.1111/bph.14426

Huffman JW (2000) The search for selective ligands for the CB2 receptor. Curr Pharm Des 6:1323–1337

Ibarra-Lecue I, Pilar-Cuéllar F, Muguruza C et al (2018) The endocannabinoid system in mental disorders: evidence from human brain studies. Biochem Pharmacol 157:97–107. https://doi.org/10.1016/j.bcp.2018.07.009

Ibeas Bih C, Chen T, Nunn AVW et al (2015) Molecular targets of Cannabidiol in neurological disorders. Neurotherapeutics 12(4):699–730. https://doi.org/10.1007/s13311-015-0377-3

Iversen L, Chapman V (2002) Cannabinoids: a real prospect for pain relief? Curr Opin Pharmacol 2:50–55

Jatoi A, Windschitl HE, Loprinzi CL et al (2002) Dronabinol versus megestrol acetate versus combination therapy for cancer-associated anorexia: a north central cancer treatment group study. J Clin Oncol 20:567–573

Jimenez-Del-Rio M, Daza-Restrepo A, Velez-Pardo C (2008) The cannabinoid CP55,940 prolongs survival and improves locomotor activity in Drosophila melanogaster against paraquat: implications in Parkinson's disease. Neurosci Res 61:404–411. https://doi.org/10.1016/j.neures.2008.04.011

Joseph J, Niggemann B, Zaenker KS, Entschladen F (2004) Anandamide is an endogenous inhibitor for the migration of tumor cells and T lymphocytes. Cancer Immunol Immunother 53:723–728. https://doi.org/10.1007/s00262-004-0509-9

Kano M, Ohno-Shosaku T, Hashimotodani Y et al (2009) Endocannabinoid-mediated control of synaptic transmission. Physiol Rev 89:309–380. https://doi.org/10.1152/physrev.00019.2008

Katona I, Sperlágh B, Sík A et al (1999) Presynaptically located CB1 cannabinoid receptors regulate GABA release from axon terminals of specific hippocampal interneurons. J Neurosci 19:4544–4558

Kendall DA, Yudowski GA (2017) Cannabinoid receptors in the central nervous system: their signaling and roles in disease. Front Cell Neurosci 10:294. https://doi.org/10.3389/fncel.2016.00294

Khan MI, Soboci AA, Brodaczewska KK et al (2018) Involvement of the CB 2 cannabinoid receptor in cell growth inhibition and G0 / G1 cell cycle arrest via the cannabinoid agonist WIN 55, 212 – 2 in renal cell carcinoma. BMC Cancer 18:583

Kogan NM, Bl C, Alvarez L et al (2006) A cannabinoid Quinone inhibits angiogenesis by targeting vascular endothelial cells. Mol Pharmacol 70:51–59. https://doi.org/10.1124/mol.105.021089.cardiotoxic

Kogan NM, Rabinowitz R, Levi P et al (2004) Synthesis and antitumor activity of Quinonoid derivatives of cannabinoids. J Med Chem 47:3800–3806

Kogan NM, Schlesinger M, Peters M et al (2007) A cannabinoid anticancer quinone, HU-331, is more potent and less cardiotoxic than doxorubicin: a comparative in vivo study. J Pharmacol Exp Ther 322:646–653

Kwan P, Schachter SC, Brodie MJ (2011) Drug-Resistant Epilepsy. N Engl J Med 365:919–926. https://doi.org/10.1056/NEJMra1004418

Lastres-Becker I, Molina-Holgado F, Ramos JA et al (2005) Cannabinoids provide neuroprotection against 6-hydroxydopamine toxicity in vivo and in vitro: relevance to Parkinson's disease. Neurobiol Dis 19:96–107. https://doi.org/10.1016/j.nbd.2004.11.009

Leyva-Illades D, Demorrow S (2013) Orphan G protein receptor GPR55 as an emerging target in cancer therapy and management. Cancer Manag Res 5:147–155. https://doi.org/10.2147/CMAR.S35175

Ligresti A (2006) Antitumor activity of plant cannabinoids with emphasis on the effect of Cannabidiol on human breast carcinoma. J Pharmacol Exp Ther 318:1375–1387. https://doi.org/10.1124/jpet.106.105247

López A, Aparicio N, Pazos MR et al (2018) Cannabinoid CB2 receptors in the mouse brain: relevance for Alzheimer's disease. J Neuroinflammation 15:158. https://doi.org/10.1186/s12974-018-1174-9

Lotersztajn S, Teixeira-Clerc F, Julien B et al (2008) CB2 receptors as new therapeutic targets for liver diseases. Br J Pharmacol 153:286–289. https://doi.org/10.1038/sj.bjp.0707511

Lotersztajn S, Teixeira-Clerc F, Mallat A, Louvet A (2011) Selective CB2 receptor agonists for use in the prevention or treatment of alcoholic liver disease

Malfitano AM, Ciaglia E, Gangemi G et al (2011) Update on the endocannabinoid system as an anticancer target. Expert Opin Ther Targets 15:297–308. https://doi.org/10.1517/14728222.2011.553606

Manna SSS, Umathe SN (2012) Involvement of transient receptor potential vanilloid type 1 channels in the pro-convulsant effect of anandamide in pentylenetetrazole-induced seizures. Epilepsy Res 100:113–124. https://doi.org/10.1016/j.eplepsyres.2012.02.003

Mantyh PW, Clohisy DR, Koltzenburg M, Hunt SP (2002) Molecular mechanisms of cancer pain. Nat Rev Cancer 2:201–209. https://doi.org/10.1038/nrc747

Marsicano G, Chaouloff F (2011) Moving bliss: a new anandamide transporter. Nat Neurosci 15:5–6. https://doi.org/10.1038/nn.3011

Marsicano G, Goodenough S, Monory K et al (2003) CB1 cannabinoid receptors and on-demand defense against Excitotoxicity. Science (80-) 302:84–88. https://doi.org/10.1126/science.1088208

Marsicano G, Kuner R (2008) Anatomical distribution of receptors, ligands and enzymes in the brain and in the spinal cord: circuitries and neurochemistry. In: Cannabinoids and the brain. Springer US, Boston, MA, pp 161–201

Martín-Moreno AM, Reigada D, Ramírez BG et al (2011) Cannabidiol and other cannabinoids reduce microglial activation in vitro and in vivo : relevance to Alzheimer ' s disease. Mol Pharmacol 79:964–973. https://doi.org/10.1124/mol.111.071290.Alzheimer

Matias I, Gonthier MP, Orlando P et al (2006) Regulation, function, and dysregulation of endocannabinoids in models of adipose and β-pancreatic cells and in obesity and hyperglycemia. J Clin Endocrinol Metab 91:3171–3180. https://doi.org/10.1210/jc.2005-2679

Matsuda LA, Lolait SJ, Brownstein MJ et al (1990) Structure of a cannabinoid receptor and functional expression of the cloned cDNA. Nature 346:561–564. https://doi.org/10.1038/346561a0

McHugh D, Hu SSJ, Rimmerman N et al (2010) N-arachidonoyl glycine, an abundant endogenous lipid, potently drives directed cellular migration through GPR18, the putative abnormal cannabidiol receptor. BMC Neurosci 11:44–56. https://doi.org/10.1186/1471-2202-11-44

Mechoulam R, Ben Shabat S, Hanus L et al (1996) Endogenous cannabinoid ligands--chemical and biological studies. J Lipid Mediat Cell Signal 14:45–49

Mechoulam R, Ben-Shabat S, Hanus L et al (1995) Identification of an endogenous 2-monoglyceride, present in canine gut, that binds to cannabinoid receptors. Biochem Pharmacol 50:83–90. https://doi.org/10.1016/0006-2952(95)00109-D

Mechoulam R, Peters M, Murillo-Rodriguez E, Hanus LO (2007) Cannabidiol--recent advances. Chem Biodivers 4:1678–1692. https://doi.org/10.1002/cbdv.200790147

Milton NGN (2002) Anandamide and noladin ether prevent neurotoxicity of the human amyloid-β peptide. Neurosci Lett 332:127–130. https://doi.org/10.1016/S0304-3940(02)00936-9

Miyato H, Kitayama J, Yamashita H et al (2009) Pharmacological synergism between cannabinoids and paclitaxel in gastric cancer cell lines. J Surg Res 155:40–47. https://doi.org/10.1016/j.jss.2008.06.045

Morales P, Blasco-Benito S, Andradas C et al (2015) Selective, nontoxic CB2 cannabinoid o-quinone with in vivo activity against triple-negative breast cancer. J Med Chem 58:2256–2264. https://doi.org/10.1021/acs.jmedchem.5b00078

Morales P, Goya P, Jagerovic N (2018) Emerging strategies targeting CB2 cannabinoid receptor: biased agonism and allosterism. Biochem Pharmacol 157:8–17. https://doi.org/10.1016/j.bcp.2018.07.031

Morales P, Hurst DP, Reggio PH (2017) Molecular targets of the Phytocannabinoids: a complex picture. Prog Chem Org Nat Prod 103:103–131. https://doi.org/10.1007/978-3-319-45541-9_4

Morales P, Jagerovic N (2016) Advances towards the discovery of GPR55 ligands. Curr Med Chem 23:2087–2100

Morales P, Jagerovic N (2019) Antitumor cannabinoid Chemotypes: structural insights. Front Pharmacol 10:621. https://doi.org/10.3389/fphar.2019.00621

Morales P, Jagerovic N (2020) Novel approaches and current challenges with targeting the endocannabinoid system. Expert Opin Drug Discov 00:1–14. https://doi.org/10.1080/17460441.2020.1752178

Morales P, Reggio PH (2017) An update on non-CB1, non-CB2 cannabinoid related G-protein-coupled receptors. Cannabis Cannabinoid Res 2:265–273. https://doi.org/10.1089/can.2017.0036

Morales P, Reggio PH (2019) CBD: a new Hope? ACS Med Chem Lett 10:694–695. https://doi.org/10.1021/acsmedchemlett.9b00127

Morales P, Vara D, Goméz-Cañas M et al (2013) Synthetic cannabinoid quinones: preparation, in vitro antiproliferative effects and in vivo prostate antitumor activity. Eur J Med Chem 70:111–119

More SV, Choi D-K (2015) Promising cannabinoid-based therapies for Parkinson's disease: motor symptoms to neuroprotection. Mol Neurodegener 10:17. https://doi.org/10.1186/s13024-015-0012-0

Morell C, Bort A, Vara D et al (2016) The cannabinoid WIN 55,212-2 prevents neuroendocrine differentiation of LNCaP prostate cancer cells. Prostate Cancer Prostatic Dis 19:248–257. https://doi.org/10.1038/pcan.2016.19

Moreno E, Andradas C, Medrano M et al (2014) Targeting CB2-GPR55 receptor Heteromers modulates cancer cell signaling. J Biol Chem 289:21960–21972. https://doi.org/10.1074/jbc.M114.561761

Müller L, Radtke A, Decker J et al (2017) The synthetic cannabinoid WIN 55,212-2 elicits death in human cancer cell lines. Anticancer Res 37:6341–6345

Murdolo G, Kempf K, Hammarstedt A et al (2007) Insulin differentially modulates the peripheral endocannabinoid system in human subcutaneous abdominal adipose tissue from lean and obese individuals. J Endocrinol Investig 30:RC17–RC21. https://doi.org/10.1007/BF03347440

Navarro G, Morales P, Rodríguez-Cueto C et al (2016) Targeting cannabinoid CB2 receptors in the central nervous system. Medicinal chemistry approaches with focus on neurodegenerative disorders. Front Neurosci 10:1–11. https://doi.org/10.3389/fnins.2016.00406

Oesch S, Walter D, Wachtel M et al (2009) Cannabinoid receptor 1 is a potential drug target for treatment of translocation-positive rhabdomyosarcoma. Mol Cancer Ther 8:1838–1845. https://doi.org/10.1158/1535-7163.MCT-08-1147

Oka S, Kimura S, Toshida T et al (2010) Lysophosphatidylinositol induces rapid phosphorylation of p38 mitogen-activated protein kinase and activating transcription factor 2 in HEK293 cells expressing GPR55 and IM-9 lymphoblastoid cells. J Biochem 147:671–678. https://doi.org/10.1093/jb/mvp208

Pagano C, Pilon C, Calcagno A et al (2007) The endogenous cannabinoid system stimulates glucose uptake in human fat cells via phosphatidylinositol 3-kinase and calcium-dependent mechanisms. J Clin Endocrinol Metab 92:4810–4819. https://doi.org/10.1210/jc.2007-0768

Pan H, Mukhopadhyay P, Rajesh M et al (2009) Cannabidiol attenuates cisplatin-induced nephrotoxicity by decreasing oxidative/nitrosative stress, inflammation, and cell death. J Pharmacol Exp Ther 328:708–714. https://doi.org/10.1124/jpet.108.147181

Pascual D, Goicoechea C, Suardíaz M, Martín MI (2005) A cannabinoid agonist, WIN 55,212-2, reduces neuropathic nociception induced by paclitaxel in rats. Pain 118:23–34. https://doi.org/10.1016/j.pain.2005.07.008

Pellati F, Borgonetti V, Brighenti V et al (2018) Cannabis sativa L. and nonpsychoactive cannabinoids: their chemistry and role against oxidative stress, inflammation, and cancer. Biomed Res Int 2018:1–15. https://doi.org/10.1155/2018/1691428

Pérez-Gómez E, Andradas C, Flores JM et al (2013) The orphan receptor GPR55 drives skin carcinogenesis and is upregulated in human squamous cell carcinomas. Oncogene 32:2534–2542. https://doi.org/10.1038/onc.2012.278

Pertwee RG (2005) The therapeutic potential of drugs that target cannabinoid receptors or modulate the tissue levels or actions of endocannabinoids. AAPS J 7: E625–E654. https://doi.org/10.1208/aapsj070364

Pertwee RG (2009) Emerging strategies for exploiting cannabinoid receptor agonists as medicines. Br J Pharmacol 156:397–411. https://doi.org/10.1111/j.1476-5381.2008.00048.x

Pertwee RG, Howlett AC, Abood ME et al (2010) International Union of Basic and Clinical Pharmacology. LXXIX. Cannabinoid receptors and their ligands: beyond CB_1 and CB_2. Pharmacol Rev 62:588–631. https://doi.org/10.1124/pr.110.003004

Picone RP, Kendall D (2015) Minireview: from the bench, toward the clinic: therapeutic opportunities for cannabinoid receptor modulation. Mol Endocrinol 29:801–813. https://doi.org/10.1210/me.2015-1062

Powles T, Te Poele R, Shamash J et al (2005) Cannabis-induced cytotoxicity in leukemic cell lines: the role of the cannabinoid receptors and the MAPK pathway. Blood 105:1214–1221. https://doi.org/10.1182/blood-2004-03-1182

Price MR, Baillie GL, Thomas A et al (2005) Allosteric modulation of the cannabinoid CB1 receptor. Mol Pharmacol 68:1484–1495. https://doi.org/10.1124/mol.105.016162.view

Price DA, Martinez AA, Seillier A et al (2009) WIN55, 212-2, a cannabinoid receptor agonist, protects against nigrostriatal cell loss in the 1-methyl-4-phenyl-1,2,3,6-tetrahydropyridine mouse model of Parkinson's disease. Eur J Neurosci 29:2177–2186. https://doi.org/10.1111/j.1460-9568.2009.06764.x

Qamri Z, Preet A, Nasser MW et al (2009) Synthetic cannabinoid receptor agonists inhibit tumor growth and metastasis of breast cancer. Mol Cancer Ther 8:3117–3129. https://doi.org/10.1158/1535-7163.MCT-09-0448

Rahn EJ, Makriyannis A, Hohmann AG (2007) Activation of cannabinoid CB1 and CB2 receptors suppresses neuropathic nociception evoked by the chemotherapeutic agent vincristine in rats. Br J Pharmacol 152:765–777. https://doi.org/10.1038/sj.bjp.0707333

Ramírez BG, Blázquez C, Gómez Del Pulgar T et al (2005) Prevention of Alzheimer's disease pathology by cannabinoids: Neuroprotection mediated by blockade of microglial activation. J Neurosci 25:1904–1913. https://doi.org/10.1523/JNEUROSCI.4540-04.2005

Rinaldi-Carmona M, Barth F, Héaulme M et al (1994) SR141716A, a potent and selective antagonist of the brain cannabinoid receptor. FEBS Lett 350:240–244

Roche M, Finn DP (2010) Brain CB2 receptors: implications for neuropsychiatric disorders. Pharmaceuticals 3:2517–2533. https://doi.org/10.3390/ph3082517

Rosenberg EC, Tsien RW, Whalley BJ, Devinsky O (2015) Cannabinoids and epilepsy. Neurotherapeutics 12(4):747–768. https://doi.org/10.1007/s13311-015-0375-5

Rossi F, Punzo F, Umano GR et al (2018) Role of cannabinoids in obesity. Int J Mol Sci 19:2690. https://doi.org/10.3390/ijms19092690

Sarfaraz S, Adhami VM, Syed DN et al (2008) Cannabinoids for cancer treatment: progress and promise. Cancer Res 68:339–342. https://doi.org/10.1158/0008-5472.CAN-07-2785

Scherma M, Fattore L, Paola Castelli M et al (2014) The role of the Endocannabinoid system in eating disorders: neurochemical and Behavioural preclinical evidence. Curr Pharm Des 20:2089–2099

Schley M, Ständer S, Kerner J et al (2009) Predominant CB2 receptor expression in endothelial cells of glioblastoma in humans. Brain Res Bull 79:333–337. https://doi.org/10.1016/j.brainresbull.2009.01.011

Serrano A, Pavon FJ, Suarez J et al (2012) Obesity and the Endocannabinoid system: is there still a future for CB1 antagonists in obesity? Curr Obes Rep 1:216–228. https://doi.org/10.1007/s13679-012-0031-x

Sharkey KA, Darmani NA, Parker LA (2014) Regulation of nausea and vomiting by cannabinoids and the endocannabinoid system. Eur J Pharmacol 722:134–146. https://doi.org/10.1016/j.ejphar.2013.09.068

Sharma MK, Murumkar PR, Barmade MA et al (2015) A comprehensive patents review on cannabinoid 1 receptor antagonists as antiobesity agents. Expert Opin Ther Pat 25:1–24. https://doi.org/10.1517/13543776.2015.1064898

Silvestri C, Di Marzo V (2012) Second generation CB1 receptor blockers and other inhibitors of peripheral endocannabinoid overactivity and the rationale of their use against metabolic disorders. Expert Opin Investig Drugs 21:1309–1322. https://doi.org/10.1517/13543784.2012.704019

Sourbron J, Schneider H, Kecskés A et al (2016) Serotonergic modulation as effective treatment for Dravet syndrome in a Zebrafish mutant model. ACS Chem Neurosci 7:588–598. https://doi.org/10.1021/acschemneuro.5b00342

Stampanoni Bassi M, Sancesario A, Morace R et al (2017) Cannabinoids in Parkinson's disease. Cannabis Cannabinoid Res 2:21–29. https://doi.org/10.1089/can.2017.0002

Stella N (2010) Cannabinoid and cannabinoid-like receptors in microglia, astrocytes, and astrocytomas. Glia 58:1017–1030. https://doi.org/10.1002/glia.20983

Stella N, Schweitzer P, Piomelli D (1997) A second endogenous cannabinoid that modulates long-term potentiation. Nature 388:773–778. https://doi.org/10.1038/42015

Straiker A, Mackie K (2005) Depolarization-induced suppression of excitation in murine autaptic hippocampal neurones. J Physiol 569:501–517. https://doi.org/10.1113/jphysiol.2005.091918

Tramèr MR, Carroll D, Campbell FA et al (2001) Cannabinoids for control of chemotherapy induced nausea and vomiting: quantitative systematic review. Br Med J 323:16–21

Van Der Stelt M, Mazzola C, Esposito G et al (2006) Endocannabinoids and β-amyloid-induced neurotoxicity in vivo: effect of pharmacological elevation of endocannabinoid levels. Cell Mol Life Sci 63:1410–1424. https://doi.org/10.1007/s00018-006-6037-3

Van Esbroeck ACM, Janssen APA, Cognetta AB et al (2017) Activity-based protein profiling reveals off-target proteins of the FAAH inhibitor BIA 10-2474. Science (80-) 356:1084–1087. https://doi.org/10.1126/science.aaf7497

Velasco G, Hernández-Tiedra S, Dávila D, Lorente M (2016) The use of cannabinoids as anticancer agents. Prog Neuro-Psychopharmacol Biol Psychiatry 64:259–266. https://doi.org/10.1016/j.pnpbp.2015.05.010

Velasco G, Sánchez C, Guzmán M (2012) Towards the use of cannabinoids as antitumour agents. Nat Rev Cancer 12:436–444. https://doi.org/10.1038/nrc3247

Vemuri VK, Makriyannis A (2015) Medicinal chemistry of cannabinoids. Clin Pharmacol Ther 97:553–558. https://doi.org/10.1111/cpt.115

Vera G, Chiarlone A, Cabezos PA et al (2007) WIN 55,212-2 prevents mechanical allodynia but not alterations in feeding behaviour induced by chronic cisplatin in the rat. Life Sci 81:468–479. https://doi.org/10.1016/j.lfs.2007.06.012

Wallace MJ, Martin BR, DeLorenzo RJ (2002) Evidence for a physiological role of endocannabinoids in the modulation of seizure threshold and severity. Eur J Pharmacol 452:295–301. https://doi.org/10.1016/S0014-2999(02)02331-2

Wallace MJ, Wiley JL, Martin BR, DeLorenzo RJ (2001) Assessment of the role of CB1 receptors in cannabinoid anticonvulsant effects. Eur J Pharmacol 428:51–57. https://doi.org/10.1016/S0014-2999(01)01243-2

Walsh D, Nelson KA, Mahmoud FA (2003) Established and potential therapeutic applications of cannabinoids in oncology. Support Care Cancer 11:137–143. https://doi.org/10.1007/s00520-002-0387-7

Williams CM, Whalley BJ, McCabe C (2015) Cannabinoids and appetite (dys)regulation. In: Fattore L (ed) Cannabinoids in neurologic and mental disease. Elsevier, pp 315–339

Wilson RI, Nicoll RA (2001) Endogenous cannabinoids mediate retrograde signalling at hippocampal synapses. Nature 410:588–592. https://doi.org/10.1038/35069076

Xian X, Huang L, Zhang B et al (2016) WIN 55,212-2 inhibits the epithelial Mesenchymal transition of gastric cancer cells via COX-2 signals. Cell Physiol Biochem 39:2149–2157. https://doi.org/10.1159/000447910

Xu D, Wang J, Zhou Z et al (2015) Cannabinoid WIN55, 212-2 induces cell cycle arrest and inhibits the proliferation and migration of human BEL7402 hepatocellular carcinoma cells. Mol Med Rep 12:7963–7970. https://doi.org/10.3892/mmr.2015.4477

Yadav MR, Murumkar PR (2018) Advances in patented CB1 receptor antagonists for obesity. Pharm Pat Anal 7:169–173. https://doi.org/10.4155/ppa-2018-0020

Cannabis and Depression

5

Daniel Feingold and Aviv Weinstein

Abstract

There is a growing body of evidence pointing to the co-occurrence of cannabis use and depression. There is also some evidence that the use of cannabis may lead to the onset of depression; however, strong evidence points to the inverse association; i.e. that depression may lead to the onset or increase in cannabis use frequency. Observational and epidemiological studies have not indicated a positive long-term effect of cannabis use on the course and outcome of depression. The association between cannabis use and depression may be stronger among men during adolescence and emerging adulthood and stronger in women during midlife. There is an indication for potential genetic correlation contributing to the comorbidity of cannabis dependence and major depression, namely that serotonin (5-HT) may mediate such association and there is also evidence for specific risk alleles for cannabis addiction. There is preclinical evidence that alteration in the endocannabinoid system could potentially benefit patients suffering from depression. However, the issue of using cannabis as an anti-depressant is at an early stage of examination and there is little evidence to support it. Finally, there has been little support to the notion that selective serotonin reuptake inhibitors (SSRIs) may be effective in decreasing depressive symptoms or rates of substance use in adolescents treated for depression and a co-occurring substance use disorder. In conclusion, despite methodological limitations, research in the past decades has broadened our knowledge on the association between cannabis use and depression from epidemiological, neurological, genetic, and pharmacological perspectives.

Keywords

Cannabis · Mental illness · Mood disorders · Major Depressive Disorder · Bipolar Disorder · Anxiety Disorders

D. Feingold (✉) · A. Weinstein
Department of Behavioral Sciences, Ariel University, Ariel, Israel

5.1 Introduction

Major depressive disorder (MDD) is a psychiatric condition characterized by continuous low mood and loss of interest or pleasure in enjoyable activities. The past-year prevalence of MDD has been estimated to be 4.7% globally and it is considered as one of the most severe mood disorders (Ferrari et al. 2013), associated with high mortality due to suicide and a major functional impairment caused directly and indirectly. In 2010, MDD was the most disabling mental disorder worldwide, accounting for more than 40% of the disability-adjusted life years

E. Murillo-Rodriguez et al. (eds.), *Cannabinoids and Neuropsychiatric Disorders*, Advances in Experimental Medicine and Biology 1264, https://doi.org/10.1007/978-3-030-57369-0_5

(a combination of premature mortality and disability) caused by mental illness (Whiteford et al. 2013). Therefore, extensive effort has been made throughout the years in order to identify risk factors associated with the onset of MDD and its clinical course, including the literature published concerning the effect of substance use and substance use disorders (Moore et al. 2007; Whiteford et al. 2013).

Following tobacco and alcohol, cannabis is the most commonly used addictive substance, with the estimated worldwide past-year prevalence rate of 3–4.5% (Degenhardt et al. 2011; United Nations Office on Drugs and Crime 2012). The number of cannabis users continues to increase globally by roughly 16% between 2006–2016, currently estimated at around 190 million worldwide (WHO 2016). In accordance with an incline in the number of cannabis users, the prevalence of the past-year diagnostic and statistical manual of mental disorders (DSM)-IV cannabis use disorder (i.e. cannabis abuse or cannabis dependence) was 1.4% and 8.5% in 2005 (Stinson et al. 2006), while in 2013, the prevalence of DSM-5 cannabis use disorder (CUD) was 2.54% (Hasin et al. 2016).

In this chapter, we will first review evidence on the co-occurrence of depression and cannabis use reported in cross-sectional studies. We will then present data on the longitudinal association between cannabis use and depression, exploring two inverse causal pathways. We will then review the possible contribution of age and gender to the association between cannabis use and depression, as well as newly emerging evidence on possible genetic and neurological factors that may underline this association. Finally, we will review the preclinical and clinical evidence for the use of cannabis as an antidepressant, and pharmacological treatment for comorbid cannabis use disorder and depression.

5.2 Co-Occurrence of Cannabis Use and Depression

5.2.1 Cannabis Use among Individuals with Depression

With a growing number of studies reporting on the co-occurrence of cannabis use and depression, Degenhardt et al. (2003) concluded in an early review that "there is increasing evidence that regular cannabis use and depression occur together more often than we might expect by chance" (p. 1497). Results from the United States National Comorbidity Survey have indicated that more than half of the individuals with MDD reported lifetime use of cannabis (Chen et al. 2002). Data from the national epidemiological survey on alcohol and related conditions (NESARC) focusing on adults with past-year MDD or dysthymia ($N = 6534$) have indicated that the past-year prevalence of cannabis use among these individuals was 10%, with nearly equally distributed between regular users (those using cannabis at least once a week; 4.5%) and occasional users (using less than weekly; 5.4%) (Aspis et al. 2015). According to data from the National Survey on Drug Use and Health, the past-year prevalence of cannabis use among adolescents with depression was substantially higher compared to adult population, estimated at 25%, compared to only 12% among those without depression (SAMHSA 2007).

Concerning the co-occurrence of MDD and CUDs, a recent study encompassing more than 28,000 cannabis users indicated that past year major depressive episode was associated with the increased number of DSM-IV cannabis dependence criteria, regardless of cannabis frequency of use (Dierker et al. 2018). In this study, participants with depression were significantly more likely to use cannabis than they intended to and spent much time on acquiring cannabis, using it or recovering from the effect of cannabis use compared to those without depression. They were also more likely to have continued to use cannabis despite negative consequences, repeatedly failed in efforts to stop or reduce cannabis use, have important activities in life superseded by cannabis use and needed an increasing amount of cannabis in order to obtain its effect.

5.2.2 Depression among Cannabis Users

According to data from the Epidemiologic Catchment Area study, more than half of the individuals who qualify for a lifetime diagnosis of DSM-IV CUD had a concurrent diagnosis of mental illness (Regier et al. 1990). According to a Dutch survey, the three-year incidence of MDD among cannabis users was 11.7% compared to 5% among nonusers (Van Laar et al. 2007). Data from NEASRC indicated that lifetime and past-year CUD were associated with a nearly three-fold increase in the risk for the past-year diagnosis of MDD (Odds Ratio (OR) = 2.8, 95% Confidence Interval (CI) = 2.33–3.41 for past-year use, and OR = 2.6, 95% CI = 2.26–2.95 for lifetime use) (Hasin et al. 2016). Odds for concurrent MDD were even higher among adolescents, with a study among 14–17 years old Europeans indicating that the lifetime prevalence of MDD was higher among individuals with lifetime cannabis use (OR = 2.7, 95% Confidence Interval (CI) = 1.6–4.4) and those with lifetime CUD (OR = 4.7, 95% CI = 2.3–9.4) compared to those who did not use cannabis (Wittchen et al. 2007).

In conclusion: Cross-sectional studies have indicated that depression and cannabis use tend to co-occur.

5.3 Cannabis Use and Depression: Longitudinal Evidence

Longitudinal studies allow for further interpreting the cross-sectional association between cannabis use and depression by addressing two inverse temporal hypotheses. The first addresses the extent to which cannabis use is associated with a future onset of MDD or an incline in depressive symptoms. While earlier studies suggested that baseline cannabis use was associated with a higher risk for future MDD (Bovasso 2001; Fergusson and Horwood 1997), accumulating evidence from latter studies has indicated that cannabis users, including heavy cannabis users, were, in fact, not more likely to be diagnosed with MDD over the course of time compared to nonusers (Degenhardt et al. 2013; Georgiades and Boyle 2007). A meta-analysis focusing on longitudinal evidence in the effect of cannabis use on depression has suggested that cannabis use, and particularly heavy cannabis use, may be associated with a mild significant increased risk for developing depression (AOR = 1.17, 95% CI = 1.05–1.30 for any cannabis use and AOR = 1.62, 95% CI = 1.21–2.16 for heavy cannabis use) (Lev-Ran et al. 2013). However, the authors concluded that findings should be regarded with caution, given lack of consistency in terms of statistical adjustment for possible confounding variables implemented in these studies.

Two population-based longitudinal studies have supported the notion that when taking into account of such confounders, cannabis use may not elevate the risk for depression. In a Swedish population-based study, crude analyzes have indicated that cannabis use at the baseline was associated with greater odds for consequent depression (Rate Ratio (RR) = 1.22, 05% CI = 1.06–1.42), yet after controlling for baseline confounders significance was not maintained (RR = 0.99, 95% CI = 0.82–1.17) (Danielsson et al. 2015). In another study that implied a similar methodology, based on Waves 1 and 2 of NESARC, following the control for baseline confounders cannabis use even daily, was not associated with increased incidence MDD at a follow-up (Feingold et al. 2015). Two separate reports, published in Europe by the World Health Organization (WHO, 2016) and in the U.S. by the National Academies of Sciences, Engineering, and Medicine (2017), have concluded that evidence concerning the effect of cannabis use and CUD on depression incidence is limited and warrant further attention, with the latter stating that "cannabis use does not appear to increase the likelihood of developing depression" (p. 12–1).

An inverse notion has focused on the possible contribution of depression to the initiation or increase in cannabis use in the future. The notion that substance use may be triggered by low mood has been reported in several clinical observations (Khantzian and Albanese 2008), retrospective studies (Gruber et al. 1997), and exploratory studies (Ogborne et al. 2000). It has long been suggested that individuals suffering from psychological distress may 'self-medicate' their negative effect by using substances (Khantzian 1985), yet this notion has received only little support in longitudinal research (Degenhardt et al. 2003). In a population-based survey, approximately 9% of the individuals, who qualified for a diagnosis of MDD have reported using drugs or misusing prescription medication for the purpose of relieving depressive symptoms (Bolton et al. 2009); however, cannabis use was not specifically addressed.

A longitudinal study by Feingold et al. (2015) has indicated that among lifetime cannabis abstainers, baseline MDD predicted the initiation of cannabis use throughout a three-year period (Adjusted Odds Ratio (AOR) = 1.72, 95% CI = 1.10 2.69). However, this was not replicated in the study by Danielsson et al. (2015), in which following control for additional illicit drug use, baseline depression was not associated with higher odds for cannabis use initiation at the follow-up (RR = 1.24, 95% CI 0.99–1.54).

In conclusion: There is a weak evidence pointing that cannabis use may lead to the onset of depression; however, there is strong evidence pointing to the inverse association; i.e. that depression may lead to the initiation or increase in frequency of cannabis use.

5.4 Cannabis Use and Depression: Contributing Factors

5.4.1 Age and Gender as Possible Moderators of the Association between Cannabis Use and Depression

Cannabis use is highly prevalent among adolescents and may be prevalent as early as by age 13 (EMMDDA 2017). Initial estimations suggest that nearly 14 million adolescents aged 15–16 use cannabis each year, equivalent to the rate of nearly 6% (WHO 2016). The relatively high availability of cannabis, along with low-harm perception associated with cannabis, makes it one of the most common substances used during adolescence. Based on national monitoring data collected in 2012, the prevalence of cannabis peaks at about 20 years of age, with a general decrease in through age 25 and above (SAMHSA 2014), suggesting that cannabis use may decrease with age and its increasing responsibilities (work, family, and so on). In recent years, it has been suggested that the association between cannabis use and depression may peak at younger age. For example, cannabis use at adolescence was associated with more depressive symptoms compared to nonusers at ages 13–18 (Kaasbøll et al. 2018) and frequency of past-year cannabis use was associated with more current depressive symptoms at ages 16–19 (Leadbeater et al. 2018). However, the association between cannabis use and depression may decrease in its magnitude with time. For example, at age 18 and above, no differences across age were found between the frequency of cannabis use and number of depressive symptoms (Leadbeater et al. 2018). In addition, regular cannabis use at ages 14, 16, and 21 was not associated with increased risk for developing MDD by age 33 (Guttmannova et al. 2017). An integration of four Australian longitudinal cohorts has indicated that cannabis use prior to age 17, even at a daily level, was not associated with increased odds for depression by age 30 (AOR = 1.02, 95% CI = 0.52–2.01) (Silins et al. 2014). These findings suggest a gradually decreasing time-varying effect of cannabis use on depression.

A recently published 40-years follow-up study of adolescents has evaluated the effects of cannabis use on the odds for developing MDD, taking into account the age of cannabis use initiation and frequency of cannabis use. Adjusted analyzes suggested that early cannabis use initiation (age < 18) was significantly associated with increased odds for consequent MDD compared to nonusers, both among frequent (AOR = 8.83, 95% CI = 1.29–70.79) and infrequent users (AOR = 2.41, 95% CI = 1.22–4.76). However, late cannabis use initiation (age > 27) was not significantly associated with higher odds for the onset of MDD at a follow-up, for both frequent (AOR = 0.68, 95% CI = 0.10–2.65) and infrequent users (AOR = 2.23, 95% CI = 0.26–14.94) compared to nonusers (Schoeler et al. 2018). Additional analyzes have indicated that the early initiation of cannabis use predicted a more rapid onset of MDD, regardless of cannabis use frequency, implying that age at the first cannabis use may play a significant role in the duration to onset of MDD (Fig. 5.1).

Exploring the inverse notion that MDD may lead to the onset of cannabis use, particularly at younger ages, yield conflicting results. On one hand, integrated finding from four Australian cohorts has indicated a moderate positive association between baseline frequent cannabis use and depression scores at a follow-up (Horwood et al. 2012). In this study, the association between cannabis use frequency and future depression has been found to decrease with age, peaking at age 15 and declining at age 30. In another study, a one standard deviation increase in cumulative depression symptoms counts between ages 12 and 15 (defined as the sum of depression symptoms during this time) was associated with a 150% risk for qualifying for a DSM-IV CUD (abuse or dependence) at age 18 (Rhew et al. 2017). On the other hand, in an 18-month follow-up of Chilean ninth-graders, baseline depression was associated with an increased use of cannabis at the follow-up (AOR = 1.21, 95% CI = 1.09–1.34), yet significance was not obtained after controlling for additional sociodemographic and clinical variables (AOR = 1.08, 95% CI = 0.94–1.23) (Stapinski et al. 2016).

There is some indication that the association between cannabis use and depression varies across gender. For example, between ages 18 and 25, the association between cannabis use and depressive symptoms was stronger among men compared to women (Leadbeater et al. 2018). This finding may be attributed to heavier use of cannabis, yet it also may be attributed to higher impulsivity, more sensation seeking, and tendency for avoidant coping strategies presented by men at this age. Notably, in the same study at age 25 and above, the association between cannabis use and depressive symptoms was stronger among women compared to men, suggesting that at these ages, women present a "telescoped" trajectory from cannabis use to pathological use.

In conclusion: Cannabis use is more likely to increase the risk for depression at younger age. The association between cannabis use and depression may be stronger among men at adolescence and emerging adulthood, yet stronger among women during midlife.

5.5 The Effect of Cannabis Use on the Course of Depression

5.5.1 The Effect on Natural Outcome of Depression: Population-Based Studies

Unfortunately, there is little data concerning the effect of cannabis use on the severity and course of depression is scarce. An early study indicated that cannabis use may be associated with an elevated feeling of dysphoria among individuals with depression (Ablon and Goodwin 1974). Participants with lifetime depression and current CUD were likely to experience depression and sadness while intoxicated and were reluctant to report experience of happiness or euphoria (Arendt et al. 2007). In a study focusing on individuals with MDD or dysthymia, women using cannabis regularly (at least weekly) reported of lower mental quality of life compared

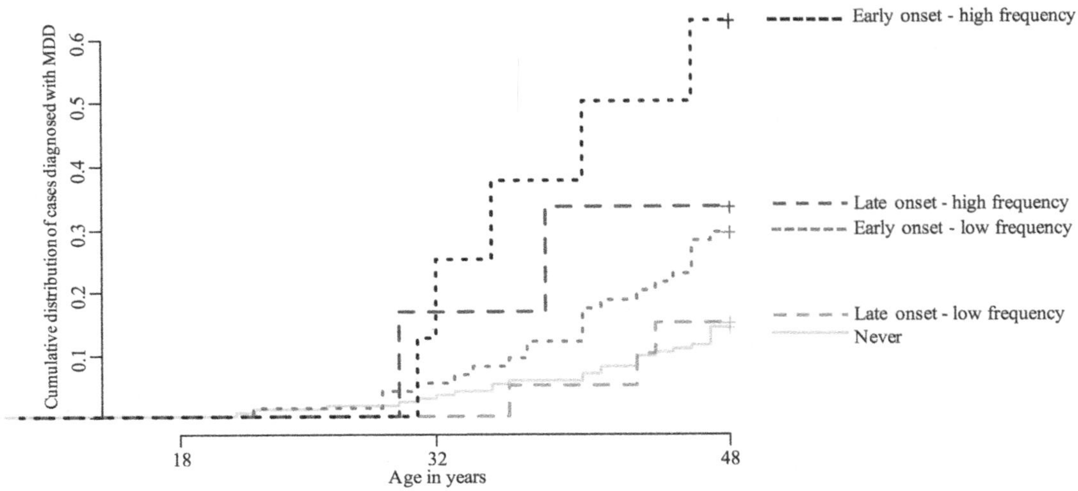

Fig. 5.1 Survival curves of cannabis use pattern and time until diagnosis of MDD. This plot illustrates the cumulative odds of developing the major depressive disorder (MDD) from age 18 to 48. Individuals with early onset of cannabis use exhibited a more rapid onset of MDD, regardless of cannabis use frequency (Schoeler et al. 2018)

to women who did not use cannabis (Aspis et al. 2015).

Several longitudinal studies have suggested that cannabis use may alter the course of illness among individuals with depression. Otten and Engels (2013) have reported that individuals who used cannabis presented more depressive symptoms through adolescence. In another study based on the NESARC sample, individuals who qualified for a baseline diagnosis of MDD ($N = 2300$) were followed throughout a three-year period. Results have indicated that cannabis use and CUD throughout the course of the study was associated with more symptoms of depression at the follow-up, specifically anhedonia, insomnia or hypersomnia, changes in body weight, and psychomotor problems (Feingold et al. 2017). However, the authors have concluded that for the most part, preliminary differences in sociodemographic and clinical factors rather than cannabis use per se were responsible for poorer clinical and functional outcomes of depression. Notably, the results from this study have not indicated a positive effect of cannabis use on the course and outcome of depression, suggesting that "self-medication" may not be effective.

There is evidence suggesting that cannabis use, and frequent cannabis use, in particular, may result in a reduced efficiency of pharmacological treatment for depression in clinical populations (Bricker et al. 2007). In another study, among 300 psychiatric outpatients receiving treatment for depression, baseline cannabis use predicted more suicide ideation, less treatment utilization, less improvement in depressive symptoms, and poorer quality of life compared to nonusers at a 12-month follow-up (Bahorik et al. 2018).

In recent years, technological and methodological advances allow for a more sensitive investigation of the effect of cannabis use on the depressed mood. For example, Cuttler et al. (2018) have analyzed data from the Strainprint™ app, designed to allow users of medical cannabis to track changes in their affective symptoms in relation to different cannabis dosing and chemotypes. Exploring more than 3000 contacts made by app users, participants reported a reduction in depressive symptoms from before to after using cannabis in 89.3% of tracked sessions. No gender differences were found in the magnitude of this change, yet a greater reduction in symptom rating was observed among individuals using low-THC/high-CBD strains of cannabis compared to those who used high-THC/low-CBD strains.

Notably, analyzing sessions made by users over time in this study, authors have reported that participants' rates of baseline depression (right before using cannabis) significantly increased with time/sessions (Cuttler et al. 2018). This may indicate that the effects of cannabis use on depression may act in two inverse paths, decreasing depressive symptoms on the short run, but increasing baseline depression in the long run. This notion is supported by findings from a study in which during 28-days of monitored period, abstinent adolescent cannabis users reported a significant decline in the level of depression compared to nonusers (Jacobus et al. 2017).

In conclusion: Observational and epidemiological studies did not indicate a positive long-term effect of cannabis use on the outcome of depression.

5.5.2 Genetics and the Neurological Basis of the Association between Cannabis Use and Depression

5.5.2.1 Genetic Studies on CUD

There is evidence for genetic associations in CUD shown in epidemiological and in clinical studies (see Benyamina et al. 2016 for review). Epidemiological studies have investigated the roles of environmental and genetic factors in cannabis use disorders and in the progression from experimentation to CUD. Studies on CUD have shown correlations between parents and children that range from 0.3 to 0.47; among siblings, these figures are between 0.39 and 0.47 (Agrawal and Lynskey 2006; Meller et al. 1988; Verweij et al. 2010). Large-scale twin studies have also estimated the role of genetic, environmental, and shared environmental factors Verweij et al. (2010). Thus, genetic factors would seem to contribute significantly to progression to CUD.

Genetic studies have examined many genes that could be associated with CUD. One gene that was investigated was catechol-O-methyltransferase (COMT), which regulates dopamine (Batalla et al. (2014), AKT_1 polymorphism (rs 1,130,253) (Bhattacharyya et al. 2014), DRD_2, (Colizzi et al. 2015; Taurisano et al. 2014), and the cannabinoid receptor 1 gene (CNR1) (Agrawal et al. 2009).

Genome-wide association studies (GWAS) have identified single nucleotide polymorphisms (SNPs) that show an association between schizophrenia, depression, and CUD. Sherva et al. (2016) have studied 14,754 participants and they have found three SNPs rs143244591, rs146091982 (SLC35G1), and rs77378271 in the CUB and Sushi multiple domain 1 gene (CSMD1) that were associated with CUD. They also found genes that were affecting both MDD and CUD and risk for schizophrenia. This is the first study that has identified specific cannabis addiction risk alleles and potential genetic factors contributing to the comorbidity of cannabis dependence with major depression and schizophrenia.

5.5.2.2 Specific Genetic Studies on the Association between Cannabis Use and Depression

A major study inking cannabis use and depression by Lynskey et al. (2004) has measured correlations between early cannabis use and lifetime cannabis dependence and MDD, and suicidal ideation and attempts for suicide. They have investigated 311 same-sex twins who differ in their early start of cannabis use (before 17) and 277 same-sex twins discordant for cannabis dependence. They have found that cannabis-dependent individuals compared with their twins who were not cannabis dependent had higher odds ratio of suicidal ideation and suicide attempts (2.5 to 2.9 times, respectively). Cannabis was also associated with higher risks for MDD in nonidentical twins. Those who initiated cannabis use, prior to age 17 years, had elevated rates of subsequent suicide attempt (OR = 3.5, 95% CI = 1.4–8.6]) but not of MDD or suicidal ideation. The risk of cannabis dependence was associated with early MDD and suicidal ideation in nonidentical twins who differed in cannabis use (discordant) but not in identical twins who differed in cannabis use discordant dizygotic twins. This evidence supports the suggestion that the comorbidity of cannabis dependence and MDD (but not suicidal behavior) has both genetic and environmental vulnerability factors. Early-onset cannabis use may be a predisposition factor for MDD or it may share genetic and environmental predisposition,

The relationship between cannabis use and depression and the short allele of the 5-hydroxytryptamine (serotonin) transporter gene-linked polymorphic region (5-HTTLPR) genotype in 310 adolescents over 4 years has been studied by Otten and Engels (2013). Cannabis use was associated with an increase in depressive symptoms over time but only in those who had the short allele of the 5-HTTLPR genotype. This is further evidence for the genetic contribution to the co-occurrence of cannabis use and depression. Finally, Hodgson et al. (2017) have studied 1284 Mexican-Americans from 75 large multi-generation families and an additional 57 genetically unrelated spouses. A linkage peak for major depression on Chromosome 22 and a peak for cannabis use on Chromosome 21 was found. Chromosome 11 had a linkage peak that affected both cannabis use and MDD as well as an SNP 20 kb upstream of NCAM1 (rs7932341) that was associated with both disorders.

In conclusion: there seems to be a common genetic association between cannabis use and MDD, which is found in twins and family studies located on Chromosome 11.

5.5.3 The Use of Medical Cannabis for Treating Depression

Medical conditions such as chronic pain multiple sclerosis, post-traumatic stress disorder, and Parkinson's disease have been recently treated with medical cannabis (Amato et al. 2017; Borgelt et al. 2013; Pertwee 2001). In order to understand the potential therapeutic effects of medical cannabis, it is mandatory to learn about the interactions of cannabinoids and other neurotransmitters (Pertwee 2014; Cohen et al. 2019).

5.5.3.1 Pre-Clinical Studies on Cannabis as an Anti-Depressant

Preclinical studies have shown that cannabis may be therapeutically effective for depression (Scherma et al. 2018). The agonistic effect of cannabinoids to the central CB_1 receptors (CB_1Rs) may mediate this effect. Shearman et al. (2003) have evaluated the CB_1R modulation of antidepressant-like effects. They have administered mice with the CB_1R inverse agonist AM251 and tested on the tail-suspension test (TST) and on the forced swim test (FST). On both tests, AM251 has significantly reduced immobility without an increase in motor activity in the open field indicating and antidepressant effect. Inverse cannabinoid agonism of CB_1R may be therefore useful for the regulation of mood. Furthermore, a low dose of a CB_1R agonist WIN55,212–2 has a potential antidepressant effect in the rat forced swim test (Bambico et al. 2007). This effect was blocked by the CB_1R antagonist rimonabant and also by pretreatment with the 5-HT-depleting agent parachlorophenylalanine. CB_1R agonists may therefore have antidepressant effects and they modulate 5-HT neuronal activity via the medial prefrontal cortex in rats.

CB_1R density and function, as well as CB_1 messenger RNA (mRNA) levels, were high in the dorsolateral prefrontal cortex of depressed humans after suicide found in postmortem studies, (Hungund et al. 2004; Choi et al. 2012). However, patients suffering from major depression have not shown any alterations in CB_1R mRNA and protein levels in the dorsal prefrontal cortex (Eggan et al. 2010). Depressed patients treated with serotonin selective reuptake inhibitors (SSRIs) showed a reduced expression of CB_1R in the anterior cingulate cortex (Koethe et al. 2007), suggesting that an elevated CB_1 has antidepressant properties..

An association between polymorphisms in the CNR_1 gene and increased vulnerability to develop a depressive episode following exposure to life stress was shown by Juhasz et al. (2009), and increased risk of resistance to the effects of antidepressants (Domschke et al. 2008).

Susceptibility to bipolar disorder and MDD may be associated with CNR_1 and FAAH gene polymorphisms (Monteleone et al. 2010). Finally, the CNR_1 gene may be involved in the development of MDD and in the effects of Citalopram, an SSRI (Mitjans et al. 2013). *In conclusion*, deficient endocannabinoid signaling may be associated with depression; and therefore, activating the endocannabinoid system could be an effective treatment for MDD but the mechanism of elevated CB_1 as an anti-depressant needs further examination.

5.5.3.2 Clinical Studies on Cannabis as an Anti-Depressant

Cannabis users often use cannabis as self-medication for depression and manic symptoms (Grinspoon and Bakalar 1998; Ashton et al. 2005). This is supported by five cases described by Gruber et al. (1996). Individuals who have used cannabis occasionally or even daily have lower levels of depressive symptoms than those who have never tried cannabis (Denson and Earleywine 2006). Depressed patients also have used cannabis to improve their sleep they developed (Babson et al. 2013). There are seven cross-sectional studies that showed clear evidence of an improvement in depressed mood by cannabis (Walsh et al. 2017). In conclusion, the evidence so far is anecdotal and it relies heavily on single case studies and cross-sectional studies and there are no placebo-controlled clinical trials that show that cannabis is useful for the treatment of depression.

5.5.4 Discussion

This issue of using cannabis as an antidepressant is at an early stage of examination and there is little evidence to support it. Psychiatric outpatients who used medical cannabis showed worse mental and physical health function compared with nonusers. Nonmedical cannabis correlated with increased suicidal ideation and mental health problems and fewer psychiatry visits (Bahorik et al. 2018). Nonmedical cannabis over time correlated with lowered improvement in depression symptoms and suicidal ideation. Cannabis use in depressed patients can prevent improvement in depressive symptoms and impair patients' treatment.

The evidence in favor of cannabis treatment for anxiety and mood disorders relies on few single-dose studies with a small sample size and flawed design (Turna et al. 2017). Furthermore, there are other factors such as interactions with other medications, frequency of use, dose of cannabis, time of use, way of delivery, and characterization of patients who may have influenced the results (D'souza and Ranganathan 2015). It remains to be investigated whether cannabis should be used alone or together with other medications, what patients should be treated, should it be prescribed only to those who do not respond to other medications (such as in the case of pain for example), and whether cannabis is efficient in the long term in comparison with SSRIs, considering its long-term side effects (D'souza and Ranganathan 2015).

5.6 Pharmacological Treatment for Individuals with Comorbid Depression and CUD

5.6.1 Pharmacological Treatment for CUD

There is an increase in the number of patients who are treated for CUD and most patients find it difficult to achieve and maintain abstinence from cannabis use. Hence, there is an urgent need to find medications for the treatment of CUD (see Weinstein and Gorelick 2011; Gorelick 2016 for review). Currently, no medication has been approved for the treatment of CUD. Medications have been tested for their ability to alleviate withdrawal symptoms, influence endogenous cannabinoids, or those that have been used for drug abuse treatment and other psychiatric conditions (Weinstein and Gorelick 2011). Four trials have been documented for the treatment of specific intoxication symptoms, seven trials for withdrawal, and 12 phase II trials for CUD (Gorelick 2016). The only effective medications were gabapentin and N-acetylcysteine (in adolescents). Three trials of antidepressants for CUD with comorbid depression revealed inconsistent results.

In conclusion: There is a need for double-blind placebo-controlled clinical trials in order to test the clinical efficacy of medications for the treatment for CUD.

5.6.2 Treatment of Patients with Comorbidity of Cannabis Dependence and Depression

There are few studies evaluating the use of antidepressant medication for the treatment of comorbid CUD and depression. Preliminary findings in 13 patients treated with fluoxetine an SSRI antidepressant (20–40 mg daily), showed a reduction in cannabis and alcohol dependence and depressive symptoms (Cornelius et al. (2005). However, after a five-year follow-up of 10 patients, although symptoms of cannabis and alcohol dependence have been reduced and academic ability has improved, clinical depression remained. A double-blind, placebo-controlled study using fluoxetine (20 mg daily) for 12 weeks to treat 70 adolescents and young adults with comorbid MDD and CUD showed no greater efficacy than placebo for treating either the depressive symptoms or the cannabis-related symptoms (Cornelius et al. 2010). The negative findings may be due to a small sample size, limited medication efficacy, or high efficacy of the psychosocial treatment.

Finally, a randomized eight-week double-blind and placebo-controlled study of fluoxetine in 29 male and five female adolescents with depressive illness and a comorbid substance use disorder showed no difference in depression ratings between patients treated with fluoxetine and placebo, nor was there any differences in positive urine samples for cannabis (Findling et al. 2009). This study has also suffered from the limitations of a small sample size and a high placebo response rate, limited dose of fluoxetine, and inclusion of participants with depressive disorders other than the major depressive disorder.

In conclusion: SSRIs, such as fluoxetine, do not show improved efficacy in treating depressive symptoms or treatment of CUD (indicated by clean urine samples) in adolescents with comorbid depression and CUD.

5.7 Summary

When exploring the association between cannabis use and depression, one should take into account several methodological limitations. The study of cannabinoids often does not allow for a precise measure of dose, frequency, or chemical composition of the substance used by participants (Mariani et al. 2011), while epidemiological studies also indiscriminately use different definitions of cannabis use, thus standardization is seldom achieved. Likewise, the definition of depression may vary according to the classification method used, i.e. the diagnostic and statistical manual of mental disorders (DSM) (American Psychiatric Association 2013) or the international classification of diseases (ICD) (World Health Organization 1992), and according to the method of assessment used (i.e. clinical assessment, semi-structured interviews, questionnaires, and so on). Furthermore, depression is often defined as the presence of depressive symptoms rather than a full diagnosis.

Despite these limitations, research in the past decades has shed light on the association between cannabis use and depression from epidemiological, neurological, genetic, and pharmacological perspectives. Given the growing prevalence of both cannabis use and depression globally, integrative research is needed in order to comprehend the possible pathways through which these factors interact. In addition, in light of the debate on the legalization of cannabis, further research should assess the direct and indirect effects of cannabis use on co-occurring physical and mental disorders, including depression.

References

Ablon SL, Goodwin FK (1974) High frequency of dysphoric reactions to tetrahydrocannabinol among depressed patients. Am J Psychiatry 131(4):448–453

Agrawal A, Lynskey MT (2006) The genetic epidemiology of cannabis use, abuse and dependence. [research support, N.I.H., extramural review]. Addiction 101 (6):801–812. https://doi.org/10.1111/j.1360-0443.2006.01399.x

Agrawal A, Wetherill L, Dick DM, Xuei X, Hinrichs A, Hesselbrock V, Kramer J, Nurnberger JI, Schuckit M, Bierut LJ, Edenberg HJ, Foroud T (2009) Evidence for association between polymorphisms in the cannabinoid receptor 1 (CNR1) gene and cannabis dependence. Am J Med Genet B Neuropsychiatr Genet 150B(5):736–740

Amato L, Minozzi S, Mitrova Z, Parmelli E, Saulle R, Cruciani F, Davoli M (2017) Systematic review of safeness and therapeutic efficacy of cannabis in patients with multiple sclerosis, neuropathic pain, and in oncological patients treated with chemotherapy. Epidemiol Prev 41(5–6):279–293

American Psychiatric Association (2013) Diagnostic and statistical manual of mental disorders, 5th edn. American Psychiatric Publishing, Arlington, VA

Arendt M, Rosenberg R, Fjordback L, Brandholdt J, Foldager L, Sher L, Munk-Jorgensen P (2007) Testing the self-medication hypothesis of depression and aggression in cannabis-dependent subjects. [research support, non-U.S. Gov't]. Psychol Med 37 (7):935–945. https://doi.org/10.1017/S0033291706009688

Ashton CH, Moore PB, Gallagher P, Young AH (2005) Cannabinoids in bipolar affective disorder: a review and discussion of their therapeutic potential. J Psychopharmacol 19:293–300

Aspis I, Feingold D, Weiser M, Rehm J, Shoval G, Lev-Ran S (2015) Cannabis use and mental health-related quality of life among individuals with depressive disorders. Psychiatry Res 230(2):341–349. https://doi.org/10.1016/j.psychres.2015.09.014

Babson KA, Boden MT, Bonn-Miller MO (2013) Sleep quality moderates the relation between depression symptoms and problematic cannabis use among medical cannabis users. Am J Drug Alcohol Abuse 39:211–216

Bahorik AL, Sterling SA, Campbell CI, Weisner C, Ramo D, Satre DD (2018) Medical and non-medical marijuana use in depression: longitudinal associations with suicidal ideation, everyday functioning, and psychiatry service utilization. J Affect Disord 241:8–14. https://doi.org/10.1016/j.jad.2018.05.065

Bambico FR, Katz N, Debonnel G, Gobbi G (2007) Cannabinoids elicit antidepressant-like behavior and activate serotonergic neurons through the medial prefrontal cortex. J Neurosci 27(43):11700–11711

Batalla A, Soriano-Mas C, Lopez-Sola M, Torrens M, Crippa JA, Bhattacharyya S, Blanco-Hinojo L, Fagundo AB, Harrison BJ, Nogue S, de la Torre R, Farre M, Pujol J, Martin-Santos R (2014) Modulation of brain structure by catechol-O-methyltransferase Val (158) met polymorphism in chronic cannabis users. Addict Biol 19:722–732

Benyamina A, Karila L, Lafaye G, Blecha L (2016) Genetic influences in cannabis use disorder and psychosis: dopamine and beyond. Curr Pharm Des 22 (42):6392–6396. https://doi.org/10.2174/1381612822666160831095707

Bhattacharyya S, Iyegbe C, Atakan Z, Martin-Santos R, Crippa JA, Xu X, Williams S, Brammer M, Rubia K, Prata D, Collier DA, McGuire PK (2014) Protein kinase B (AKT1) genotype mediates sensitivity to cannabis-induced impairments in psychomotor control. Psychol Med 44:3315–3328

Bolton JM, Robinson J, Sareen J (2009) Self-medication of mood disorders with alcohol and drugs in the National Epidemiologic Survey on alcohol and related conditions. J Affect Disord 115(3):367–375. https://doi.org/10.1016/j.jad.2008.10.003

Borgelt LM, Franson KL, Nussbaum AM, Wang GS (2013) Thepharmacologic and clinical effects of medical cannabis. Pharmacotherapy 33(2):195–209

Bovasso GB (2001) Cannabis abuse as a risk factor for depressive symptoms. [research support, U.S. Gov't, non-P.H.S. research support, U.S. Gov't, P.H.S.]. Am J Psychiatry 158(12):2033–2037

Bricker JB, Russo J, Stein MB, Sherbourne C, Craske M, Schraufnagel TJ, Roy-Byrne P (2007) Does occasional cannabis use impact anxiety and depression treatment outcomes?: results from a randomized effectiveness trial. [comparative study multicenter study randomized controlled trial]. Depress Anxiety 24(6):392–398. https://doi.org/10.1002/da.20248

Chen CY, Wagner FA, Anthony JC (2002) Marijuana use and the risk of major depressive episode. Epidemiological evidence from the United States National Comorbidity Survey. [research support, non-U.S. Gov't research support, U.S. Gov't, P.H.S.]. Soc Psychiatry Psychiatr Epidemiol 37(5):199–206. https://doi.org/10.1007/s00127-002-0541-z

Choi K, Le T, McGuire J, Xing G, Zhang L, Li H, Ursano RJ (2012) Expression pattern of the cannabinoid receptor genes in the frontal cortex of mood disorder patients and mice selectively bred for high and low fear. J Psychiatr Res 46(7):882–889

Cohen K, Weizman A, Weinstein A (2019) Modulatory effects of cannabinoids on brain neurotransmission. Eur J Neurosci 50:2322–2345. https://doi.org/10.1111/ejn.14407

Colizzi M, Iyegbe C, Powell J, Ursini G, Porcelli A, Bonvino A, Taurisano P, Romano R, Masellis R, Blasi G, Morgan C, Aitchison K, Mondelli V, Luzi S, Kolliakou A, David A, Murray RM, Bertolino A, Di Forti M (2015) Interaction between functional genetic variation of DRD2 and cannabis use on risk of psychosis. Schizophr Bull 41(5):1171–1182

Cornelius JR, Buksteinb OG, Douaihy AB, Clark DB, Chung TA, Daley DC, Wood DS, Brown SJ (2010) Double-blind fluoxetine trial in comorbid MDD–CUD youth and young adults. Drug Alcohol Depend 112:39–45

Cornelius JR, Clark DB, Bukstein OG, Bimaher B, Salloum IM, Brown SA (2005) Acute phase and five year follow up atusy of fluoxetine in adolescents with major depression and comorbid substance abuse disorder: a review. Addict Behav 30:1824–1833

Cuttler C, Spradlin A, McLaughlin RJ (2018) A naturalistic examination of the perceived effects of cannabis on negative affect. J Affect Disord 235:198–205. https://doi.org/10.1016/j.jad.2018.04.054. Epub 2018 Apr 6

D'souza DC, Ranganathan M (2015) Medical marijuana: is the cart before the horse? JAMA 313(24):2431–2432

Danielsson AK, Lundin A, Agardh E, Allebeck P, Forsell Y (2015) Cannabis use, depression and anxiety: a 3-year prospective population-based study. J Affect Disord 193:103–108. https://doi.org/10.1016/j.jad.2015.12.045

Degenhardt L, Bucello C, Calabria B, Nelson P, Roberts A, Hall W, McLaren J (2011) What data are available on the extent of illicit drug use and dependence globally? Results of four systematic reviews. [research support, non-U.S. Gov't review]. Drug Alcohol Depend 117(2–3):85–101. https://doi.org/10.1016/j.drugalcdep.2010.11.032

Degenhardt L, Coffey C, Romaniuk H, Swift W, Carlin JB, Hall WD, Patton GC (2013) The persistence of the association between adolescent cannabis use and common mental disorders into young adulthood. [research support, non-U.S. Gov't]. Addiction 108(1):124–133. https://doi.org/10.1111/j.1360-0443.2012.04015.x

Degenhardt L, Hall W, Lynskey M (2003) Exploring the association between cannabis use and depression. [review]. Addiction 98(11):1493–1504

Denson TF, Earleywine M (2006) Decreased depression in marijuana users. Addict Behav 31:738–742

Dierker L, Selya A, Lanza S, Li R, Rose J (2018) Depression and marijuana use disorder symptoms among current marijuana users. Addict Behav 76:161–168. https://doi.org/10.1016/j.addbeh.2017.08.013

Domschke K, Dannlowski U, Ohrmann P, Lawford B, Bauer J, Kugel H, Heindel W, Young R, Morris P, Arolt V et al (2008) Cannabinoid receptor 1 (CNR1) gene: impact on antidepressant treatment response and emotion processing in major depression. Eur Neuropsychopharmacol 18(10):751–759

Eggan SM, Stoyak SR, Verrico CD, Lewis DA (2010) Cannabinoid CB1 receptor immunoreactivity in the prefrontal cortex: comparison of schizophrenia and major depressive disorder. Neuropsychopharmacology 35:2060–2071

EMMDDA (2017) European drug report 2017. European Monitoring Centre for Drugs and Drug Addiction, Luxembourg

Feingold D, Rehm J, Lev-Ran S (2017) Cannabis use and the course and outcome of major depressive disorder: a population based longitudinal study. Psychiatry Res 251:225–234

Feingold D, Weiser M, Rehm J, Lev-Ran S (2015) The association between cannabis use and mood disorders: a longitudinal study. J Affect Disord 172:211–218

Fergusson DM, Horwood LJ (1997) Early onset cannabis use and psychosocial adjustment in young adults. [comparative study research support, non-U.S. Gov't]. Addiction 92(3):279–296

Ferrari AJ, Somerville AJ, Baxter AJ, Norman R, Patten SB, Vos T, Whiteford HA (2013) Global variation in the prevalence and incidence of major depressive disorder: a systematic review of the epidemiological literature. Psychol Med 43(03):471–481. https://doi.org/10.1017/S0033291712001511

Findling RL, Pagano ME, McNamara NK, Stansbrey RJ, Faber JE, Lingler J, Reed MD (2009) The short-term safety and efficacy of fluoxetine in depressed adolescents with alcohol and cannabis use disorders: a pilot randomized placebo-controlled trial. Child Adolesc Psychiatry Ment Health 3(1):11. https://doi.org/10.1186/1753-2000-3-11

Georgiades K, Boyle MH (2007) Adolescent tobacco and cannabis use: young adult outcomes from the Ontario child health study. [research support, non-U.S. Gov't]. J Child Psychol Psychiatry 48(7):724–731. https://doi.org/10.1111/j.1469-7610.2007.01740.x

Gorelick DA (2016) Pharmacological treatment of cannabis-related disorders: a narrative review. Curr Pharm Des 22(42):6409–6419. https://doi.org/10.2174/1381612822666160822150822

Grinspoon L, Bakalar JB (1998) The use of cannabis as a mood stabilizer in bipolar disorder: anecdotal evidence and the need for clinical research. J Psychoactive Drugs 30:171–177

Gruber AJ, Pope HG, Brown ME (1996) Do patients use marijuana as an antidepressant? Depression 4:77–80

Gruber AJ, Pope HG, Oliva P (1997) Very long-term users of marijuana in the United States: a pilot study. Subst Use Misuse 32(3):249–264

Guttmannova K, Kosterman R, White HR, Bailey JA, Lee JO, Epstein M, Hawkins JD (2017) The association between regular marijuana use and adult mental health outcomes. [research support, N.I.H., extramural]. Drug Alcohol Depend 179:109–116. https://doi.org/10.1016/j.drugalcdep.2017.06.016

Hasin DS, Kerridge BT, Saha TD, Huang B, Pickering R, Smith SM, Grant BF (2016) Prevalence and correlates of DSM-5 cannabis use disorder, 2012-2013: findings from the National Epidemiologic Survey on alcohol and related conditions–III. Am J Psychiatr 173 (6):588–599

Hodgson K, Almasy L, Knowles EE, Kent JW, Curran JE, Dyer TD, Göring HH, Olvera RL, Woolsey MD, Duggirala R, Fox PT, Blangero J, Glahn DC (2017) The genetic basis of the comorbidity between cannabis use and major depression. Addiction 112(1):113–123

Horwood LJ, Fergusson DM, Coffey C, Patton GC, Tait R, Smart D, Hutchinson DM (2012) Cannabis and

depression: an integrative data analysis of four Australasian cohorts. Drug Alcohol Depend 126 (3):369–378
Hungund BL, Vinod KY, Kassir SA, Basavarajappa BS, Yalamanchili R, Cooper TB, Mann JJ, Arango V (2004) Upregulation of CB1 receptors and agonist-stimulated [35S]GTPγS binding in the prefrontal cortex of depressed suicide victims. Mol Psychiatry 9:184–190
Jacobus J, Squeglia LM, Escobar S, McKenna BM, Hernandez MM, Bagot KS, Huestis MA (2017) Changes in marijuana use symptoms and emotional functioning over 28-days of monitored abstinence in adolescent marijuana users. Psychopharmacology 234 (23–24):3431–3442
Juhasz G, Chase D, Pegg E, Downey D, Toth ZG, Stones K, Platt H, Mekli K, Payton A, Elliott R et al (2009) CNR1 gene is associated with high neuroticism and low agreeableness and interacts with recent negative life events to predict current depressive symptoms. Neuropsychopharmacology 34:2019–2027
Kaasbøll C, Hagen R, Gråwe RW (2018) Population-based associations among cannabis use, anxiety, and depression in Norwegian adolescents. J Child Adolesc Subst Abuse 27(4):238–243
Khantzian EJ (1985) The self-medication hypothesis of addictive disorders: focus on heroin and cocaine dependence. [case reports research support, U.S. Gov't, P.H.S. review]. Am J Psychiatry 142 (11):1259–1264
Khantzian EJ, Albanese MJ (2008) Understanding addiction as self medication: finding Hope behind the pain. Rowman & Littlefield Publishers, Inc., New York
Koethe D, Llenos IC, Dulay JR, Hoyer C, Torrey EF, Leweke FM, Weis S (2007) Expression of CB1 cannabinoid receptor in the anterior cingulate cortex in schizophrenia, bipolar disorder, and major depression. J Neural Transm 114:1055–1063
Leadbeater BJ, Ames ME, Linden-Carmichael AN (2018) Age-varying effects of cannabis use frequency and disorder on symptoms of psychosis, depression, and anxiety in adolescents and adults. Addiction
Lev-Ran S, Roerecke M, Le Foll B, George TP, McKenzie K, Rehm J (2013) The association between cannabis use and depression: a systematic review and meta analysis of longitudinal studies. Psychol Med:1–14. https://doi.org/10.1017/S0033291713001438
Lynskey MT, Glowinski AL, Todorov AA, Bucholz KK, Madden PA, Nelson EC, Statham DJ, Martin NG, Heath AC (2004) Major depressive disorder, suicidal ideation, and suicide attempt in twins discordant for cannabis dependence and early-onset cannabis use. Arch Gen Psychiatry 61(10):1026–1032
Mariani JJ, Brooks D, Haney M, Levin FR (2011) Quantification and comparison of marijuana smoking practices: blunts, joints, and pipes. Drug Alcohol Depend 113(2–3):249–251
Meller WH, Rinehart R, Cadoret RJ, Troughton E (1988) Specific familial transmission in substance abuse. Int J Addict 23(10):1029–1039
Mitjans M, Serretti A, Fabbri C, Gastó C, Catalán R, Fañanás L, Arias B (2013) Screening genetic variability at the CNR1 gene in both major depression etiology and clinical response to citalopram treatment. Psychopharmacology 227:509–519
Monteleone P, Bifulco M, Maina G, Tortorella A, Gazzerro P, Proto MC, Di Filippo C, Monteleone F, Canestrelli B, Buonerba G et al (2010) Investigation of CNR1 and FAAH endocannabinoid gene polymorphisms in bipolar disorder and major depression. Pharmacol Res 61:400–404
Moore TH, Zammit S, Lingford-Hughes A, Barnes TR, Jones PB, Burke M, Lewis G (2007) Cannabis use and risk of psychotic or affective mental health outcomes: a systematic review. [research support, non-U.S. Gov't review]. Lancet 370(9584):319–328. https://doi.org/10.1016/S01406736(07)61162-3
National Academies of Sciences, E., & Medicine (2017) The health effects of cannabis and cannabinoids: the current state of evidence and recommendations for research. National Academies Press
Ogborne AC, Smart RG, Weber T, Birchmore-Timney C (2000) Who is using cannabis as a medicine and why: an exploratory study. J Psychoactive Drugs 32 (4):435–443
Otten R, Engels RC (2013) Testing bidirectional effects between cannabis use and depressive symptoms: moderation by the serotonin transporter gene. Addict Biol 18(5):826–835. https://doi.org/10.1111/j.1369-1600.2011.00380.x
Pertwee RG (2001) Cannabinoid receptors and pain. Prog Neurobiol 63(5):569–611
Pertwee R (2014) Handbook of cannabis. Oxford University Press, Oxford
Regier DA, Farmer ME, Rae DS, Locke BZ, Keith SJ, Judd LL, Goodwin FK (1990) Comorbidity of mental disorders with alcohol and other drug abuse. Results from the epidemiologic catchment area (ECA) study. [research support, U.S. Gov't, P.H.S.]. JAMA 264 (19):2511–2518
Rhew IC, Fleming CB, Vander Stoep A, Nicodimos S, Zheng C, McCauley E (2017) Examination of cumulative effects of early adolescent depression on cannabis and alcohol use disorder in late adolescence in a community-based cohort. Addiction 112 (11):1952–1960
SAMHSA. (2007) Results from the 2006 National Survey on drug use and health: National Findings. Rockville, MD
SAMHSA. (2014) Results from the 2013 National Survey on drug use and health: summary of National Findings Rockville, MD
Scherma, M, Masia P, Deidda M, Fratta W, Tanda G, Fadda P (2018) New perspectives on the use of cannabis in the treatment of psychiatric disorders. Medicines

(Basel), 5(4). pii: E107. doi: https://doi.org/10.3390/medicines5040107

Schoeler T, Theobald D, Pingault J-B, Farrington DP, Coid JW, Bhattacharyya S (2018) Developmental sensitivity to cannabis use patterns and risk for major depressive disorder in mid-life: findings from 40 years of follow-up. Psychol Med:1–8

Shearman LP, Rosko KM, Fleischer R, Wang J, Xu S, Tong XS, Rocha BA (2003) Antidepressant-like and anorectic effects of the cannabinoid CB1 receptor inverse agonist AM251 in mice. Behav Pharmacol 14 (8):573–582

Sherva R, Wang Q, Kranzler H, Zhao H, Koesterer R, Herman A, Gelernter J (2016) Genome-wide association study of cannabis dependence severity, novel risk variants, and shared genetic risks. JAMA Psychiat 73 (5):472–480. https://doi.org/10.1001/jamapsychiatry.2016.0036

Silins E, Horwood LJ, Patton GC, Fergusson DM, Olsson CA, Hutchinson DM, Mattick RP (2014) Young adult sequelae of adolescent cannabis use: an integrative analysis. Lancet Psychiatry 1(4):286–293. https://doi.org/10.1016/S2215-0366(14)70307-4

Stapinski LA, Montgomery AA, Araya R (2016) Anxiety, depression and risk of cannabis use: examining the internalising pathway to use among Chilean adolescents. Drug Alcohol Depend 166:109–115

Stinson FS, Ruan WJ, Pickering R, Grant BF (2006) Cannabis use disorders in the USA: prevalence, correlates and co-morbidity. [research support, N.I.H., extramural research support, N.I.H., intramural]. Psychol Med 36(10):1447–1460. https://doi.org/10.1017/S0033291706008361

Taurisano P, Romano R, Mancini M, Giorgio AD, Antonucci LA, Fazio L, Rampino A, Quarto T, Gelao B, Porcelli A, Papazacharias A, Ursini G, Caforio G, Masellis R, Niccoli-Asabella A, Todarello O, Popolizio T, Rubini G, Blasi G, Bertolino A (2014) Prefronto-striatal physiology is associated with schizotypy and is modulated by a functional variant of DRD2. Front Behav Neurosci 8:235

Turna J, Patterson B, Van Ameringen M (2017) Is cannabis treatment for anxiety, mood, and related disorders ready for prime time? Depress Anxiety 34 (11):1006–1017

United Nations Office on Drugs and Crime (2012) World drug report 2012. United Nations, New York

Van Laar M, van Dorsselaer S, Monshouwer K, de Graaf R (2007) Does cannabis use predict the first incidence of mood and anxiety disorders in the adult population? [research support, non-U.S. Gov't]. Addiction 102 (8):1251–1260. https://doi.org/10.1111/j.1360-0443.2007.01875.x

Verweij KJ, Zietsch BP, Lynskey MT, Medland SE, Neale MC, Martin NG, Vink JM (2010) Genetic and environmental influences on cannabis use initiation and problematic use: a meta-analysis of twin studies. Addiction 105(3):417–430. https://doi.org/10.1111/j.1360-0443.2009.02831.x

Walsh Z, Gonzalez R, Crosby K, Thiessen MS, Carroll C, Bonn-Miller MO (2017) Medical cannabis and mental health: a guided systematic review. Clin Psychol Rev 51:15–29

Weinstein AM, Gorelick DA (2011) Pharmacological treatment of cannabis dependence. Curr Pharm Des 17(14):1351–1358

Whiteford HA, Degenhardt L, Rehm J, Baxter AJ, Ferrari AJ, Erskine HE, Vos T (2013) Global burden of disease attributable to mental and substance use disorders: findings from the global burden of disease study 2010. [research support, N.I.H., extramural research support, non-U.S. Gov't review]. Lancet 382 (9904):1575–1586. https://doi.org/10.1016/S0140-6736(13)61611-6

Wittchen HU, Frohlich C, Behrendt S, Gunther A, Rehm J, Zimmermann P, Perkonigg A (2007) Cannabis use and cannabis use disorders and their relationship to mental disorders: a 10-year prospective-longitudinal community study in adolescents. [research support, N.I.H., extramural research support, non-U.S. Gov't]. Drug Alcohol Depend 88(Suppl 1):S60–S70. https://doi.org/10.1016/j.drugalcdep.2006.12.013

World Health Organization (1992) ICD 10 classifications of mental and Behavioural disorder: clinical descriptions and diagnostic guidelines. World Health Organization, Geneva

World Health Organization (2016) The health and social effects of nonmedical cannabis use. WHO, Geneva, CH

Recent Advances in the Potential of Cannabinoids for Neuroprotection in Alzheimer's, Parkinson's, and Huntington's Diseases

6

Catalina Pérez-Olives, Rafael Rivas-Santisteban, Jaume Lillo, Gemma Navarro, and Rafael Franco

Abstract

Three prevalent neurodegenerative diseases, Parkinson's, Alzheimer's, and Huntington's are in need of symptomatic relief of slowing disease progression or both. This chapter focuses on the potential of cannabinoids to afford neuroprotection, i.e. avoid or retard neuronal death. The neuroprotective potential of cannabinoids is known from the work in animal models and is mediated by the two cannabinoid receptors (CB_1/CB_2) and eventually, by their heteromers, GPR55, orphan receptors (GPR3/GPR6/GPR12/GPR18), or PPARγ. Now, there is the time to translate the findings into patients. The chapter takes primarily into account advances since 2016 and addresses the issue of proving neuroprotection in humans. One recent discovery is the existence of activated microglia with neuroprotective phenotype; cannabinoids are good candidates to skew phenotype, especially via glial CB_2 receptors (CB_2R), whose targeting has, a priori, less side effects those targeting the CBs_1 receptor (CB_1R), which are expressed in both neurons and glia. The fact that a cannabis extract (SativexTM) is approved for human therapy, such that cannabis use will likely be legalized in many countries and different possibilities that cannabinoid pharmacology suggests a successful route of cannabinoids (natural or synthetic) all the way to be approved and used in the treatment of neurodegeneration.

Keywords

Neurodegenerative diseases · Dementia · Drug discovery · Fatty acid amide hydrolase · Heteromers · Therapy · Microglia · Nootropics

Abbreviations

AD	Alzheimer's disease
CB1R	cannabinoid CB_1 receptor
CB_2R	cannabinoid CB_2 receptor
CBD	cannabidiol

C. Pérez-Olives
Molecular Neurobiology laboratory, Department of Biochemistry and Molecular Biomedicine, Universitat de Barcelona, Barcelona, Spain

R. Rivas-Santisteban · J. Lillo · R. Franco (✉)
Molecular Neurobiology laboratory, Department of Biochemistry and Molecular Biomedicine, Universitat de Barcelona, Barcelona, Spain

Centro de Investigación en Red, Enfermedades Neurodegenerativas (CIBERNED). Instituto de Salud Carlos III, Madrid, Spain
e-mail: rfranco@ub.edu

G. Navarro (✉)
Centro de Investigación en Red, Enfermedades Neurodegenerativas (CIBERNED). Instituto de Salud Carlos III, Madrid, Spain

Department Biochemistry and Physiology. Faculty of Pharmacy and Food Sciences, Universitat de Barcelona, Barcelona, Spain

E. Murillo-Rodriguez et al. (eds.), *Cannabinoids and Neuropsychiatric Disorders*, Advances in Experimental Medicine and Biology 1264, https://doi.org/10.1007/978-3-030-57369-0_6

CNS	central nervous system
FAAH	fatty acid amide hydrolase
GPCR	G-protein-coupled receptor
GPRn	orphan GPCR number "n"
MDS-UPDRS	Scale for non-motor symptoms in parkinsonian patients
PD	Parkinson's disease
PET	positron emission tomography
PPARγ	peroxisome proliferator-activated receptor γ
THC	Δ^9-tetrahydrocannabinol

6.1 Introduction

The history of medicines derived from drugs of abuse is fairly interesting. In the case of natural cannabinoids, i.e. those derived from *Cannabis sativa*, there has been a huge delay in the approval of "medical cannabis" despite controversy on whether or not Cannabis consumption leads serious dependence. The first cannabinoids approved for therapeutic use were synthetic derivatives of natural compounds: nabilone and dronabinol; they are used for a wide-range of illnesses but mainly to stop nausea and vomiting associated, for instance, with chemotherapy; they are also used to combat anorexia (Fraguas-Sánchez and Torres-Suárez 2018). Following the long-standing and well-known relaxed legislation existing in Holland, Cannabis sativa consumption is now approved in Uruguay, Canada, and several states of EEUU. It is likely that more and more countries will approve ad hoc legislation to allow consumption, not only for recreational purposes but also for medicinal uses. In fact, it is well known that patients of a huge variety of diseases have stuck to the intake of natural cannabinoids: some cases are related to the diseases of CNS, for instance, Parkinson's disease (PD), some others are related to sclerosis as patients report symptom improvements. Cancer patients even at advanced stages report improvement in cancer-associated pain. Furthermore, medication based on natural cannabinoids has been approved. To our knowledge, there are two cannabinoid-based medicines: Sativex™/nabiximols and Epidiolex™. Epidiolex consists of cannabidiol (CBD), one of the main components of Cannabis sativa, dissolved in sesame oil. Interestingly, Sativex™ is one of the few plant extracts that have been approved as a medicine. It contains several compounds but with enrichment in cannabidiol (CBD) and Δ^9-tetrahydrocannabinol (THC). Sativex™, whose content in CBD and THC is similar, is prescribed for spasticity associated with multiple sclerosis. Rimonabant, a synthetic cannabinoid receptor antagonist was approved for weight loss but was retired after serious adverse events (Carai et al. 2006; Sam et al. 2011). Enhanced cannabinoid action may be afforded by inhibiting the enzyme that degrades endocannabinoids, fatty acid amide hydrolase (FAAH) (Benito et al. 2003; Goncalves et al. 2008; Celorrio et al. 2016) (see below). Unfortunately, a clinical trial using an inhibitor of FAAH led to the death of healthy volunteers (van Esbroeck et al. 2017; Kaur et al. 2018). Despite the issue was independent of enzyme inhibition, this fact has led to some reluctance to develop therapeutic drugs acting on those enzymes. In summary, it is likely that in the future more cannabinoids, natural or synthetic, may be approved for different diseases. Meanwhile, practitioners face the issue of "prescribing" cannabis for patients with neurodegenerative diseases in these countries or in the US States, where consumption is allowed; a review on sources presenting the pros and cons of "medical cannabis" use in patients (Noel 2017) and an account of "appropriate" dosing based on currently approved medicines (MacCallum and Russo 2018) are available. We here report the potential of cannabinoids (natural or synthetic) in neuroprotection related to the three more prevalent neurodegenerative diseases (Alzheimer's, Parkinson's, and Huntington's). Very solid reports have been provided in the last two decades (see (Fernández-Ruiz et al. 2015; Mannucci et al. 2017; Cilia 2018; Fraguas-Sánchez and Torres-Suárez 2018) for review) and, therefore, this chapter focuses on research performed, since 2016.

6.2 Receptors that Respond to Cannabinoids

As of today, two receptors that mediate the physiological effects of cannabinoids are cannabinoid CB_1 and CB_2 receptors, which belong to the G-protein-coupled receptors (GPCRs) superfamily (https://www.guidetopharmacology.org/). In addition, they interact to each other to form CB_1 and CB_2 heteromers of proved physiological relevance and with therapeutic potential as the CB_1 and CB_2 receptors themselves (Callén et al. 2012; Sierra et al. 2015; Navarro et al. 2018, 2018)

The orphan GPCR, GPR55, was at first considered as a third cannabinoid receptor (Ryberg et al. 2007). Although this possibility has not reached consensus and GPR55 may be the receptor of lysophosphatidylinositol, it is known that cannabinoids regulate GPR55 action. In addition, GPR55 may form heteromers with CB_1 receptors or with CB_2 receptors (Kargl et al. 2012; Balenga et al. 2014; Martínez-Pinilla et al. 2014; García-Gutiérrez et al. 2018). Actually, data indicate that GPR55 may be a target for PD (Celorrio et al. 2017) but, besides complex pharmacology, there are few available tools; therefore, it lacks behind the CB_1 and CB_2 receptors in the race to find anti-neurodegenerative drugs.

CBD, at high concentrations, activate cannabinoid receptors. Recently, CBD has also been reported as an allosteric modulator of these receptors (Laprairie et al. 2015; Martínez-Pinilla et al. 2017). Interestingly, CBD behaves as an inverse agonist of some orphan GPCRs such as GPR3, GPR6, and GPR12 (Laun and Song 2017; Laun et al. 2019), which share a high degree of homology with the cannabinoid receptors (Morales et al. 2018). GPR18, another orphan of GPCR that may be regulated by cannabinoids, may interact with CB_2 (CB_2R) but not with the CB_1 receptor (CB_1R) (Reyes-Resina et al. 2018)

6.3 Addressing Neuroprotection in Humans

Addressing neuroprotection is not easy. Even in the case of laboratory animal models of neurodegenerative diseases, the demonstration that a given drug is neuroprotective poses difficulties. In addition, symptom improvements (in animal models) are quite often considered as neuroprotection and this interpretation is wrong. Yet, preclinical research has led to candidates that seem really neuroprotective, i.e. prevent neuronal death, and cannabinoids are among them.

Demonstrating neuroprotection in humans is a serious concern as there is not any "technique" that can prove it. Food and Drug Administration has no special rules that could serve to address this issue. Furthermore, clinical trials, by concept, and also by the pressure of pharmaceutical companies, are limited in time. Demonstrating neuroprotection requires time and requires safe drugs in chronic administration. In summary, patients are in urgent need of protocols to address neuroprotection. In our understanding, this requires new protocols and the use of drugs that are already considered as safe in chronic usage or of complements that are considered as "generally recognized safe" and are commercially available. This specific issue applied to another promising drug class, antagonists of A_{2A} receptors, has been widely discussed elsewhere (Franco and Navarro 2018). Longitudinal studies are likely required in either i) healthy individuals taking memory enhancers (nootropics) for years and looking for the age of appearance of neurodegenerative signs or ii) patients taking additional medication with a "safe" drug (already approved or provisionally approved on the basis of compassionate drug use) and measuring disease progression using ad hoc scores (Franco et al. 2019). In either case, cannabinoids are candidates that deserve to be tested.

Another advantage of cannabinoids is related to the relatively recent development of PET tracers. Especially, relevant are those that are able to detect CB_2R in the brain of living humans (healthy individuals or patients); the very recent papers on tracer development prove the interest of in vivo picturing this receptor (Attili et al. 2019; Kallinen et al. 2019). On the one hand, it is considered that PET for CB_2R gives relevant hints for the neuroinflammation extent (see (Kho et al. 2017) for background). On the other hand, it is considered that reducing neuroinflammation in patients reflects less neurodegeneration and hence, reduced the progression of the disease (see (Spinelli et al. 2017) for review). Very importantly, and as pointed out by (Janssen et al. 2018), it would be instrumental to develop a PET tracer for the neuroprotective M2 microglial phenotype. Such a tracer could be a biomarker for neuroinflammation and/or for assessing neuroprotection in humans.

Another issue is related to dosage. Cannabinoids may act in a hormetic-like fashion, i.e. qualitatively different depending on the dose (Calabrese and Baldwin 2002; Calabrese and Rubio-Casillas 2018). Taking a simple example, CBD at high concentrations activates cannabinoid receptors, whereas CBD at lower doses behaves as a negative allosteric modulator (Martínez-Pinilla et al. 2017). See below (section on AD) for another example involving THC (Calabrese and Rubio-Casillas 2018).

6.4 Potential of Cannabinoids in Parkinson's Disease

Parkinsonian patients are in need of drugs that delay the progression of the disease, i.e. preventing the death of dopaminergic neurons in the *substantia nigra*. Although there are efficacious interventions to address symptoms, they are not exempt of undesirable effects, mainly dyskinesias, i.e. involuntary movements arising after long periods of chronic pharmacological treatment. There is evidence that cannabinoids may be useful for neuroprotection but also for addressing symptoms and for reducing the chances to suffer from dyskinesias. There are other aspects of the disease, particularly those are known as non-motor symptoms. A recent protocol has been disclosed to address the safety and efficacy of nabilone in a cohort of approximately 38 patients entering into a randomized, placebo-controlled, double-blind clinical study. The primary outcome will be the MDS-UPDRS score and the results are expected by the end of 2019 (Peball et al. 2019).

Confirming data in animal and cell models analysis of post-mortem samples and positron emission tomography (PET) in living patients shows that cannabinoid signaling is altered in Parkinson's disease and that cannabinoid receptors exist in the brain regions susceptible of targeting by therapeutic drugs (Cilia 2018). The expression of CB_1R and endocannabinoid synthesizing/degrading enzymes is also altered in the basal ganglia as a consequence of side effects levodopa treatment, more precisely during the active phase of dyskinesia (Rojo-Bustamante et al. 2018). Accordingly, the CB_1 and CB_2 receptors (individually or forming heteromers with other GPCRs) are potential targets of drugs aimed at affording neuroprotection.

At present, the evidence for efficacy in humans is scarce. The authors for a systematic review on Medical Cannabis and Neurodegenerative and Psychiatric indicated that: "*Evaluation of these low-quality trials, as rated on the Cochrane risk of bias tools, was challenged by methodological issues such as inadequate description of allocation concealment, blinding and underpowered sample size. More adequately powered controlled trials that examine the long and short term efficacy, safety and tolerability of cannabis for medical use, and the mechanisms underpinning the therapeutic potential are warranted*". (Lim et al. 2017). In what concerns PD-related pain, prospects are already good; a meta-analysis considering >25 clinical trials (randomized) with idiopathic parkinsonian patients showed that greater pain reductions were achieved with safinamide but followed by cannabinoids and opioids (Qureshi et al. 2018). Therefore, cannabinoids are equipotent as one of the most potent pain relievers, opioids, but with the advantage that cannabinoids have fewer side effects.

Starting with the seminal discoveries of Rafael Mechoulam (Gaoni and Mechoulam 1964; Mechoulam et al. 1970; Mechoulam and Parker 2013) in the cannabinoid field, Israelian laboratories and hospitals have significantly contributed to find evidence for cannabinoid clinical potential. A human-based report by laboratories in this country indicates that cannabinoids are efficacious in "*reducing tremor, dyskinesia, rigidity and pain, and improving sleep*" (Katz et al. 2017). The authors add that medical cannabis may be useful in dementia " *although clinical data are still inadequate"*.

As they are often altered in neurodegenerative models, research on the mitochondrial metabolism and mitochondrial biogenesis is gaining momentum in the field. For instance, THC upregulates proteins involved in biogenesis to MPP^+ toxicity in a dopamine transporter-positive cell line (SH-SY5Y) (Zeissler et al. 2016). The mechanisms are dependent on upregulating a PPARγ co-activator 1α, PGC-1α, and a mitochondrial transcription factor, TFAM.

The potential of glia and cannabinoid receptors in glia merits special attention in PD but also in AD (see below). In the rotenone model of PD, a phytocannabinoid, β-caryophyllene reduces, among other, glial activation and this leads to the protection of dopaminergic neurons (Ojha et al. 2016). While these results indicate that glial activation may be detrimental, as it was currently thought, they contrast with those reporting that targeting the CB_2R reduces the progression of motor symptoms in the LRRK2-transgenic mice (Palomo-Garo et al. 2016). It is interesting that the authors want to correlate without taking glia into account. Indeed, CB_2R expression in CNS neurons (mainly restricted to the globus pallidus and cerebellum) could not explain the results in the LRRK2 mice; therefore, the effects are likely due by CB_2R expressed in the activated microglia (see the section on AD, below). In any case, it would be good to know the status of the glia in the LRRK2 mice and the expression of activation markers and of cannabinoid receptors. As of today, there are no hits in Pubmed for "LRRK2-transgenic mice" and "microglia". However, CB_2R in the glia has already been suggested as a pharmacological target against PD-related inflammation (Gómez-Gálvez et al. 2016). The authors were even able to find the upregulation of the receptors in the glial cells of patients using post-mortem samples. It should be noted that VCE-003.2, which is a synthetic quinone derivative of cannabigerol, provided (in a PPARγ receptor-dependent way) benefits against inflammation-driven neuronal deterioration in a PD model in (García et al. 2018).

As also commented below, it is needed to establish not only the target but the pharmacological properties of the "therapeutic" cannabinoid, i.e. whether more efficacious one will be an agonist, an inverse agonist, or an allosteric modulator. In this sense, it is informative the case of the patient displaying mild cognitive impairments and living independent but who became seriously affected when nabilone was administered. To know the reasons of such psychosis, exacerbation would help in designing the most appropriate type of cannabinoid to address PD (Udow et al. 2018).

Familial early-onset PD may be caused by mutations in the (PINK1) gene, which codes for PTEN-induced putative kinase 1. (Madeo et al. 2016) reported in PINK1 knockout mice a CB_1 receptor dysfunction that was dependent on dopaminergic transmission. The usefulness of such finding in terms of PD therapy will require further experimental effort.

6.5 Potential of Cannabinoids in Alzheimer's Disease

Cannabinoids are among the myriad of drugs that transgenic models are efficacious reducing the pathological hallmarks of Alzheimer's disease (AD) but that, unfortunately, have failed to reach the patient (Franco and Cedazo-Minguez 2014). The handicap is double, i.e. apart from the difficulty in assessing neuroprotection in humans, it turns out that transgenic Alzheimer's disease models do not display neuronal death. Accordingly, these models serve more for hereditary cases (around 10% of total cases) and less for sporadic non-hereditary cases (around 90% of total cases). Can cannabinoids on delaying disease progression? Part of the answer came from analogies, i.e. if cannabinoids are seemingly neuroprotective in other neurodegenerative diseases they may be also efficacious in Alzheimer's patients. Recently, the efficacy of some cannabinoids (individually administered or co-administered) has been tested in a so-called phenotypic screening platform that "*recapitulate proteotoxicity, loss of trophic support, oxidative stress, energy loss, and inflammation*" (Schubert et al. 2019). Compounds were also assayed for "*their ability to remove intraneuronal amyloid*" (Schubert et al. 2019). The conclusion of synergism upon coadministration is notable (Sativex™/nabiximols is, in fact, a mixture of compounds), but the conclusion that the agonists of the CB_1 receptors are affording neuroprotection (Schubert et al. 2019) must be taken with caution as i) previous data do not support this view and ii) phenotypic platforms may not be suitable to measure neuroprotection in this disease. In fact, in one of the newest transgenic models with quicker cognitive impairment onset, the *triple* 3xTg-AD mice, desensitization of the CB_1 receptor may be a "*plausible strategy to improve behavior alterations associated with genetic risk factors for developing Alzheimer's disease*" (Llorente-Ovejero et al. 2018). Other recent results on cannabinoid action on animal models of Alzheimer's disease are provided below.

In what symptoms are concerned, the use of cannabinoids has been suggested to reduce agitation and/or the aggressive behavior found in some patients (Liu et al. 2015). A clinical trial to know the efficacy of a synthetic cannabinoid, nabilone, on agitation in moderate-to-severe Alzheimer's disease (Ruthirakuhan et al. 2019) has seemingly been completed in March 2019 (https://clinicaltrials.gov/ct2/show/NCT02351882) though no results have been posted. A recent meta-analysis based on double-blind, placebo-controlled trials have retrieved six studies with a total of 251 cases; the conclusion is that the results are inconclusive in what concerns aggression or agitation (Ruthirakuhan et al. 2019).

Cannabidiol, which has recently reported as an allosteric modulator of the CB_1 and CB_2 receptors (Laprairie et al. 2015; Martínez-Pinilla et al. 2017; Navarro et al. 2018), may activate peroxisome proliferator-activated receptor γ (PPARγ) and via the Wnt/β-catenin pathway, may reduce the oxidative stress and neuroinflammation associated with the disease. The authors propose cannabidiol as a drug to combat Alzheimer's disease (Vallée et al. 2017). The modulation of genes in the mesenchymal stem cells suggests that cannabidiol leads to an expression pattern that could be more beneficial with any efficacious anti-Alzheimer's disease therapy (Libro et al. 2016).

Rats with intracerebroventricularly administered okadaic acid appear as a model of Alzheimer's disease as they present, in some brain regions, pathological hallmarks (phosphorylated tau and ß-amyloid) and display cognition deficits. Consistent with the potentially relevant role of activated microglia in what concerns neuroprotection, a selective CB_2 receptor agonist was effective in reducing cognitive impairment, neurodegeneration and neuroinflammation (Çakır et al. 2019). The potential of the receptor as a target for neuroprotection is reinforced by the detection of memory impairment and of Tau pathology in the CB_2 receptor knockout mice. Animals presented Tau hyperphosphorylation, on a decrease of AMPK activity and mitochondrial dysfunction (Wang et al. 2018).

Classical activation of microglia has been considered as detrimental but this view has changed. In fact, two different phenotypes arise from the activation of resting M0 microglia such as M1 of proinflammatory and M2 of neuroprotective (see (Franco and Fernández-Suárez 2015) for review). A recent discovery has linked the activated microglia to neuroprotection in Alzheimer's disease. We found that the primary cultures of microglia from a transgenic AD mouse model present the activated phenotype with an important regulatory role of cannabinoids via cannabinoid receptors and receptor heteromers (Navarro et al. 2018). These results in animals that, unlike human patients, do not present any neuronal death leads to the suggestive hypothesis that microglia may be neuroprotective and prevent neuronal death and consequently, the progression of the disease. Results from the effects of ß-amyloid in a novel immortalized microglial cell line (McCarthy et al. 2016) may help in designing drugs leading to microglial M2 phenotype skewing.

In addition, it should be noted that blood flow is important for any neurodegenerative condition. In this sense, both functional impairments that may be regulated by ligands acting at the cannabinoid receptors have been detected in the vessels from a transgenic AD model (Navarro-Dorado et al. 2016).

The negative regulation of ß-amyloid-activated astroglial hemichannels is seen as a neuroprotective mechanism exerted by cannabinoids (Gajardo-Gómez et al. 2017).

Synthetic cannabinoids constituted by indazolylketones are postulated to be potential to combat Alzheimer's disease as some of the generated compounds are able to target the CB_2 receptor to inhibit ß-secretase 1 (the enzyme that participates in the production of the ß-amyloid toxic peptide) and to inhibit butyryl cholinesterase (one of the enzymes that degrade a neurotransmitter reduced in the disease: acetylcholine) (González-Naranjo et al. 2019).

Finally, an interesting hypothesis has been emitted to explain the biphasic effects of THC that is able to alter short-term memory, while it improves neurological function in old animals and in animal models of Alzheimer's disease in which the compound prevents neurodegenerative processes. This paradox may be reconciled by one of the properties of hormetic mechanisms, namely different effects depending on the dose (Calabrese and Rubio-Casillas 2018). Interestingly, the benefits of THC on cognitive deficits in transgenic Alzheimer's disease mice models do not happen in healthy aging of wild-type animals (Aso et al. 2016). In addition, it should be noted that the metabolism of an endocannabinoid, 2-arachidonoylglycerol, is altered by different aggregates of ß-amyloid (Pascual et al. 2017), thus suggesting that endocannabinoid metabolism is altered in patients.

6.6 Potential of Cannabinoids in Huntington's Disease

Huntington's disease is caused by neuronal death due to an abnormal protein, consisting of huntingtin with long poly-glutamine expansions. Therefore, it is hereditary and the altered gene contains CAG expansions that may be generated during spermatogenesis. Recent results have shown that torsional stress promote the generation of CAG expansions in the gene coding for huntingtin (Simard et al. 2017).

Cannabinoids may be neuroprotective in this disease as it has been demonstrated in the R6/2 rodent model of the disease. Neuroprotection was achieved by a Sativex™-like combination of compounds, although the motor performance was not improved. The neuroprotective effect was deduced, among others, from dystonia improvement and increased metabolic activity measured by PET in the basal ganglia. These data suggest that an appropriate combination of cannabinoids may affect disease progression (Valdeolivas et al. 2017). In another recent report, a different composition of compounds showed symptoms of improvement. In fact, cannabinoids improved dystonia, gait, and fine motor skills in early-onset Huntington's disease patients. In some individuals, the treatment was associated with less hypersalivation and less apathy and irritability (Saft et al. 2018).

Spinocerebellar ataxias are hereditary neurodegenerative disorders some of which share with Huntington's the involvement of proteins with poly-glutamine expansions. From word with both models of spinocerebellar ataxia and post-mortem samples from patients, it is found that the CB_1 receptors are upregulated in neurons and the CB_2 receptors are upregulated in the Purkinje cells and glial cells present in different regions of the cerebellum. It is thought that activating the CB_1 and CB_2 receptors or inhibiting the enzymes that degrade endocannabinoids could result in neuroprotection (Gómez-Ruiz et al. 2019). However, it was noted in a rodent model, the targetted deletion of an endocannabinoid-producing enzyme, monoacylglycerol lipase, affords protection for huntingtin-induced medium spiny neuronal loss and motor impairment. The authors suggest that a reduction in the availability of the product of the reaction, 2-arachidonoylglycerol, may be beneficial and that the enzyme is a potential therapeutic target for the disease (Ruiz-Calvo et al. 2019). As it stands, it is not yet clear where, when, and how decreased endocannabinoid tone is neuroprotective and where, when, and how activation or blockage of cannabinoid receptors result in neuroprotection.

One possible approach for drug development is considering functional selectivity and receptor functionality (Franco et al. 2018). In practice, medicinal chemists are designing "biased" agonist, i.e. molecules that upon binding to a given receptor lead to differential signaling that may affect the viability of cells expressing mutant huntingtin (Laprairie et al. 2016). In a recent review, the authors state: "*Recent studies have found that Gαi/o-biased CB_1 agonists activate extracellular signal-regulated kinase (ERK), increase CB_1* (receptor) *protein levels, and improve viability of cells expressing mutant huntingtin*" (Bagher et al. 2018). Tetrahydrocannabinolic acid, another component of *Cannabis sativa*, affords neuroprotection in two Huntington's disease cell models, one expressing a mutated form of huntingtin (STHdhQ111/Q111 cells) and another expressing a protein with 94 poly-glutamine repeats (mHtt-q94). The neuroprotective action is mediated by agonism at PPARγ (Nadal et al. 2017). It remains to be elucidated whether the compound may also act as a biased agonist at the cannabinoid receptors. The VCE-003.2 synthetic molecule, which was above described in the section devoted to Parkinson's disease, also displays potential to combat Huntington's disease as it afforded progenitor cell survival without losing the capacity to activate PPARγ. Hence, VCE-003.2 improved motor deficits, reduced neuroinflammation, and prevented medium spiny neuronal loss in the rodent models of the disease (Díaz-Alonso et al. 2016).

It is well-known that Huntington's disease has no canonical drug to be used for delay neuronal death. Sativex™ was used in a double-blind, randomized, placebo-controlled clinical trial with patients. Symptoms were not improved but the positive aspect is that Sativex™ was safe and well-tolerated (López-Sendón Moreno et al. 2016). Apart from longer treatment and higher doses, the main issue in neurodegenerative diseases is the difficulty in addressing the measure of neuroprotection in humans (see above).

In conclusion, the potential of cannabinoids for neuroprotection is evident but the challenge is to demonstrate select/develop the most appropriate cannabinoid receptor ligand and to demonstrate its usefulness in clinical assays. As pointed out by (Maccarrone et al. 2017): "*The continued characterization of individual cannabinoids in different diseases (Alzheimer's disease, Parkinson's disease, Huntington's disease, and epilepsy) remains important*".

Funding This work was supported by grants from the Spanish Ministry of Innovation and Competitiveness: Ref. No. BFU2016–64405-R and from the Spanish Ministry of Science, Innovation and Universities: Ref. No. 2019_RTI2018–094204-B-I00; both may include EU FEDER funds, and by the U.S. Alzheimer's Association, grant Ref. No. AARFD-17-503612.

Conflict of Interests Authors declare no conflict of interests.

References

Aso E, Andrés-Benito P, Ferrer I (2016) Delineating the efficacy of a cannabis-based medicine at advanced stages of dementia in a murine model. J Alzheimers Dis 54:903–912. https://doi.org/10.3233/JAD-160533

Attili B, Celen S, Ahamed M, Koole M, Van Den Haute C, Vanduffel W, Bormans G (2019) Preclinical evaluation of [^{18}F]MA3: a CB_2 receptor agonist radiotracer for PET. Br J Pharmacol 176:1481–1491. https://doi.org/10.1111/bph.14564

Bagher AM, Laprairie RB, Kelly MEM, Denovan-Wright EM (2018) Methods to quantify cell signaling and GPCR receptor ligand bias: characterization of drugs that target the Endocannabinoid receptors in Huntington's disease. In: Methods in molecular biology. Humana Press, Clifton, N.J, pp 549–571. https://doi.org/10.1007/978-1-4939-7825-0_25

Balenga NA, Martínez-Pinilla E, Kargl J, Schröder R, Peinhaupt M, Platzer W, Bálint Z, Zamarbide M, Dopeso-Reyes IG, Ricobaraza A et al (2014) Heteromerization of GPR55 and cannabinoid CB2 receptors modulates signalling. Br J Pharmacol 171:1–64. https://doi.org/10.1111/bph.12850

Benito C, Núñez E, Tolón RM, Carrier EJ, Rábano A, Hillard CJ, Romero J (2003) Cannabinoid CB2 receptors and fatty acid amide hydrolase are selectively overexpressed in neuritic plaque-associated glia in

Alzheimer's disease brains. J Neurosci 23:11136–11141

Çakır M, Tekin S, Doğanyiğit Z, Erden Y, Soytürk M, Çiğremiş Y, Sandal S (2019) Cannabinoid type 2 receptor agonist JWH-133, attenuates Okadaic acid induced spatial memory impairment and neurodegeneration in rats. Life Sci 217:25–33. https://doi.org/10.1016/j.lfs.2018.11.058

Calabrese EJ, Baldwin LA (2002) Defining hormesis. Hum Exp Toxicol 21:91–97. https://doi.org/10.1191/0960327102ht217oa

Calabrese EJ, Rubio-Casillas A (2018) Biphasic effects of THC in memory and cognition. Eur J Clin Investig 48: e12920. https://doi.org/10.1111/eci.12920

Callén L, Moreno E, Barroso-Chinea P, Moreno-Delgado-D, Cortés A, Mallol J, Casadó V, Lanciego JL, Franco R, Lluis C et al (2012) Cannabinoid receptors CB1 and CB2 form functional heteromers in brain. J Biol Chem 287:20851–20865. https://doi.org/10.1074/jbc.M111.335273

Carai MAM, Colombo G, Maccioni P, Gessa GL (2006) Efficacy of rimonabant and other cannabinoid CB1 receptor antagonists in reducing food intake and body weight: preclinical and clinical data. CNS Drug Rev 12:91–99. https://doi.org/10.1111/j.1527-3458.2006.00091.x

Celorrio M, Fernández-Suárez D, Rojo-Bustamante E, Echeverry-Alzate V, Ramírez MJ, Hillard CJ, López-Moreno JA, Maldonado R, Oyarzábal J, Franco R et al (2016) Fatty acid amide hydrolase inhibition for the symptomatic relief of Parkinson's disease. Brain Behav Immun 57:94–105. https://doi.org/10.1016/j.bbi.2016.06.010

Celorrio M, Rojo-Bustamante E, Fernández-Suárez D, Sáez E, Estella-Hermoso de Mendoza A, Müller CE, Ramírez MJ, Oyarzábal J, Franco R, Aymerich MS (2017) GPR55: a therapeutic target for Parkinson's disease? Neuropharmacology 125:319–332. https://doi.org/10.1016/j.neuropharm.2017.08.017

Cilia R (2018) Molecular imaging of the cannabinoid system in idiopathic Parkinson's disease. Int Rev Neurobiol 141:305–345. https://doi.org/10.1016/bs.irn.2018.08.004

Díaz-Alonso J, Paraíso-Luna J, Navarrete C, del Río C, Cantarero I, Palomares B, Aguareles J, Fernández-Ruiz J, Bellido ML, Pollastro F et al (2016) VCE-003.2, a novel cannabigerol derivative, enhances neuronal progenitor cell survival and alleviates symptomatology in murine models of Huntington's disease. Sci Rep 6:29789. https://doi.org/10.1038/srep29789

Fernández-Ruiz J, Romero J, Ramos JA (2015) Endocannabinoids and neurodegenerative disorders: Parkinson's disease, Huntington's chorea, Alzheimer's disease, and others. Handb Exp Pharmacol 231:233–259. https://doi.org/10.1007/978-3-319-20825-1_8

Fraguas-Sánchez AI, Torres-Suárez AI (2018) Medical use of cannabinoids. Drugs 78:1665–1703. https://doi.org/10.1007/s40265-018-0996-1

Franco R, Aguinaga D, Jiménez J, Lillo J, Martínez-Pinilla E, Navarro G (2018) Biased receptor functionality versus biased agonism in G-protein-coupled receptors. Biomol Concepts 9:143–154. https://doi.org/10.1515/bmc-2018-0013

Franco R, Cedazo-Minguez A (2014) Successful therapies for Alzheimer's disease: why so many in animal models and none in humans? Front Pharmacol 5:146. https://doi.org/10.3389/fphar.2014.00146

Franco R, Fernández-Suárez D (2015) Alternatively activated microglia and macrophages in the central nervous system. Prog Neurobiol 131:65–86. https://doi.org/10.1016/j.pneurobio.2015.05.003

Franco R, Martínez-Pinilla E, Navarro G (2019) Why have transgenic rodent models failed to successfully mimic Alzheimer's disease. How can we develop effective drugs without them? Expert Opin Drug Discovery 14:327–330. https://doi.org/10.1080/17460441.2019.1581169

Franco R, Navarro G (2018) Adenosine A2A receptor antagonists in neurodegenerative diseases: huge potential and huge challenges. Front Psych 9:1–5. https://doi.org/10.3389/fpsyt.2018.00068

Gajardo-Gómez R, Labra VC, Maturana CJ, Shoji KF, Santibañez CA, Sáez JC, Giaume C, Orellana JA (2017) Cannabinoids prevent the amyloid β-induced activation of astroglial hemichannels: a neuroprotective mechanism. Glia 65:122–137. https://doi.org/10.1002/glia.23080

Gaoni Y, Mechoulam R (1964) Isolation, structure, and partial synthesis of an active constituent of hashish. J Am Chem Soc 86:1646–1647. https://doi.org/10.1021/ja01062a046

García C, Gómez-Cañas M, Burgaz S, Palomares B, Gómez-Gálvez Y, Palomo-Garo C, Campo S, Ferrer-Hernández J, Pavicic C, Navarrete C et al (2018) Benefits of VCE-003.2, a cannabigerol quinone derivative, against inflammation-driven neuronal deterioration in experimental Parkinson's disease: possible involvement of different binding sites at the PPARγ receptor. J Neuroinflammation 15:19. https://doi.org/10.1186/s12974-018-1060-5

García-Gutiérrez MS, Navarrete F, Navarro G, Reyes-Resina I, Franco R, Lanciego JL, Giner S, Manzanares J (2018) Alterations in gene and Protein expression of cannabinoid CB2 and GPR55 receptors in the dorsolateral prefrontal cortex of suicide victims. Neurotherapeutics 15:796–806. https://doi.org/10.1007/s13311-018-0610-y

Gómez-Gálvez Y, Palomo-Garo C, Fernández-Ruiz J, García C (2016) Potential of the cannabinoid CB2 receptor as a pharmacological target against inflammation in Parkinson's disease. Prog Neuro-Psychopharmacol Biol Psychiatry 64:200–208. https://doi.org/10.1016/j.pnpbp.2015.03.017

Gómez-Ruiz M, Rodríguez-Cueto C, Luna-Piñel E, Hernández-Gálvez M, Fernández-Ruiz J (2019) Endocannabinoid system in Spinocerebellar ataxia Type-3 and other autosomal-dominant cerebellar ataxias: potential role in pathogenesis and expected relevance as Neuroprotective targets. Front Mol Neurosci 12:94. https://doi.org/10.3389/fnmol.2019.00094

Goncalves MB, Suetterlin P, Yip P, Molina-Holgado F, Walker DJ, Oudin MJ, Zentar MP, Pollard S, Yáñez-Muñoz RJ, Williams G et al (2008) A diacylglycerol lipase-CB2 cannabinoid pathway regulates adult subventricular zone neurogenesis in an age-dependent manner. Mol Cell Neurosci 38:526–536. https://doi.org/10.1016/j.mcn.2008.05.001

González-Naranjo P, Pérez-Macias N, Pérez C, Roca C, Vaca G, Girón R, Sánchez-Robles E, Martín-Fontelles MI, de Ceballos ML, Martin-Requero A et al (2019) Indazolylketones as new multitarget cannabinoid drugs. Eur J Med Chem 166:90–107. https://doi.org/10.1016/j.ejmech.2019.01.030

Janssen B, Vugts D, Windhorst A, Mach R (2018) PET imaging of microglial activation—beyond targeting TSPO. Molecules 23:607. https://doi.org/10.3390/molecules23030607

Kallinen A, Boyd R, Lane S, Bhalla R, Mardon K, Stimson DHR, Werry EL, Fulton R, Connor M, Kassiou M (2019) Synthesis and in vitro evaluation of fluorine-18 benzimidazole sulfones as CB2 PET-radioligands. Org Biomol Chem 17:5086–5098. https://doi.org/10.1039/c9ob00656g

Kargl J, Balenga N, Parzmair GP, Brown AJ, Heinemann A, Waldhoer M (2012) The cannabinoid receptor CB1 modulates the signaling properties of the lysophosphatidylinositol receptor GPR55. J Biol Chem 287:44234–44248. https://doi.org/10.1074/jbc.M112.364109

Katz I, Katz D, Shoenfeld Y, Porat-Katz BS (2017) Clinical evidence for utilizing cannabinoids in the elderly. Isr Med Assoc J IMAJ 19:71–75

Kaur RJ, Singh S, Sidhu P, Sharma PK (2018) TGN-1412 and BIA-2474 trials with tragic end: lessons learnt to make clinical trials safer. Rev Recent Clin Trials 13:252–256. https://doi.org/10.2174/1574887113666180521093529

Kho DT, Glass M, Graham ES (2017) Is the cannabinoid CB 2 receptor a major regulator of the Neuroinflammatory Axis of the neurovascular unit in humans? Adv Pharmacol 80:367–396. https://doi.org/10.1016/bs.apha.2017.03.009

Laprairie RB, Bagher AM, Kelly MEM, Denovan-Wright EM (2015) Cannabidiol is a negative allosteric modulator of the cannabinoid CB1 receptor. Br J Pharmacol 172:4790–4805. https://doi.org/10.1111/bph.13250

Laprairie RB, Bagher AM, Kelly MEM, Denovan-Wright EM (2016) Biased type 1 cannabinoid receptor signaling influences neuronal viability in a cell culture model of Huntington disease. Mol Pharmacol 89:364–375. https://doi.org/10.1124/mol.115.101980

Laun AS, Shrader SH, Brown KJ, Song Z-H (2019) GPR3, GPR6, and GPR12 as novel molecular targets: their biological functions and interaction with cannabidiol. Acta Pharmacol Sin 40:300–308. https://doi.org/10.1038/s41401-018-0031-9

Laun AS, Song Z-H (2017) GPR3 and GPR6, novel molecular targets for cannabidiol. Biochem Biophys Res Commun 490:17–21. https://doi.org/10.1016/j.bbrc.2017.05.165

Libro R, Diomede F, Scionti D, Piattelli A, Grassi G, Pollastro F, Bramanti P, Mazzon E, Trubiani O (2016) Cannabidiol modulates the expression of Alzheimer's disease-related genes in Mesenchymal stem cells. Int J Mol Sci 18:26. https://doi.org/10.3390/ijms18010026

Lim K, See YM, Lee J (2017) A systematic review of the effectiveness of medical cannabis for psychiatric, movement and neurodegenerative disorders. Clin Psychopharmacol Neurosci 15:301–312. https://doi.org/10.9758/cpn.2017.15.4.301

Liu CS, Chau SA, Ruthirakuhan M, Lanctôt KL, Herrmann N (2015) Cannabinoids for the treatment of agitation and aggression in Alzheimer's disease. CNS Drugs 29:615–623. https://doi.org/10.1007/s40263-015-0270-y

Llorente-Ovejero A, Manuel I, Lombardero L, Giralt MT, Ledent C, Giménez-Llort L, Rodríguez-Puertas R (2018) Endocannabinoid and muscarinic signaling crosstalk in the 3xTg-AD mouse model of Alzheimer's disease. J Alzheimers Dis 64:117–136. https://doi.org/10.3233/JAD-180137

López-Sendón Moreno JL, García Caldentey J, Trigo Cubillo P, Ruiz Romero C, García Ribas G, Alonso Arias MAA, García de Yébenes MJ, Tolón RM, Galve-Roperh I, Sagredo O et al (2016) A double-blind, randomized, cross-over, placebo-controlled, pilot trial with Sativex in Huntington's disease. J Neurol 263:1390–1400. https://doi.org/10.1007/s00415-016-8145-9

MacCallum CA, Russo EB (2018) Practical considerations in medical cannabis administration and dosing. Eur J Intern Med 49:12–19. https://doi.org/10.1016/j.ejim.2018.01.004

Maccarrone M, Maldonado R, Casas M, Henze T, Centonze D (2017) Cannabinoids therapeutic use: what is our current understanding following the introduction of THC, THC:CBD oromucosal spray and others? Expert Rev Clin Pharmacol 10:443–455. https://doi.org/10.1080/17512433.2017.1292849

Madeo G, Schirinzi T, Maltese M, Martella G, Rapino C, Fezza F, Mastrangelo N, Bonsi P, Maccarrone M, Pisani A (2016) Dopamine-dependent CB 1 receptor dysfunction at corticostriatal synapses in homozygous PINK1 knockout mice. Neuropharmacology 101:460–470. https://doi.org/10.1016/j.neuropharm.2015.10.021

Mannucci C, Navarra M, Calapai F, Spagnolo EV, Busardò FP, Cas RD, Ippolito FM, Calapai G (2017) Neurological aspects of medical use of Cannabidiol. CNS Neurol Disord Drug Targets 16:541–553. https://doi.org/10.2174/1871527316666170413114210

Martínez-Pinilla E, Reyes-Resina I, Oñatibia-Astibia A, Zamarbide M, Ricobaraza A, Navarro G, Moreno E, Dopeso-Reyes IGG, Sierra S, Rico AJJ et al (2014) CB1 and GPR55 receptors are co-expressed and form heteromers in rat and monkey striatum. Exp Neurol 261:44–52. https://doi.org/10.1016/j.expneurol.2014.06.017

Martínez-Pinilla E, Varani K, Reyes-Resina I, Angelats E, Vincenzi F, Ferreiro-Vera C, Oyarzabal J, Canela EI, Lanciego JL, Nadal X et al (2017) Binding and signaling studies disclose a potential allosteric site for cannabidiol in cannabinoid CB2receptors. Front Pharmacol 8:744. https://doi.org/10.3389/fphar.2017.00744

McCarthy RC, Lu D-Y, Alkhateeb A, Gardeck AM, Lee C-H, Wessling-Resnick M (2016) Characterization of a novel adult murine immortalized microglial cell line and its activation by amyloid-beta. J Neuroinflammation 13:21. https://doi.org/10.1186/s12974-016-0484-z

Mechoulam R, Parker LA (2013) The Endocannabinoid system and the brain. Annu Rev Psychol 64:21–47. https://doi.org/10.1146/annurev-psych-113011-143739

Mechoulam R, Shani A, Edery H, Grunfeld Y (1970) Chemical basis of hashish activity. Science (New York, NY) 169:611–612. https://doi.org/10.1126/science.169.3945.611

Morales P, Isawi I, Reggio PH (2018) Towards a better understanding of the cannabinoid-related orphan receptors GPR3, GPR6, and GPR12. Drug Metab Rev 50:74–93. https://doi.org/10.1080/03602532.2018.1428616

Nadal X, del Río C, Casano S, Palomares B, Ferreiro-Vera C, Navarrete C, Sánchez-Carnerero C, Cantarero I, Bellido ML, Meyer S et al (2017) Tetrahydrocannabinolic acid is a potent PPARγ agonist with neuroprotective activity. Br J Pharmacol 174:4263–4276. https://doi.org/10.1111/bph.14019

Navarro G, Borroto-Escuela D, Angelats E, Etayo I, Reyes-Resina I, Pulido-Salgado M, Rodríguez-Pérez A, Canela E, Saura J, Lanciego JL et al (2018) Receptor-heteromer mediated regulation of endocannabinoid signaling in activated microglia. Role of CB1 and CB2 receptors and relevance for Alzheimer's disease and levodopa-induced dyskinesia. Brain Behav Immun 67:139–151. https://doi.org/10.1016/j.bbi.2017.08.015

Navarro G, Reyes-Resina I, Rivas-Santisteban R, Sánchez de Medina V, Morales P, Casano S, Ferreiro-Vera C, Lillo A, Aguinaga D, Jagerovic N et al (2018) Cannabidiol skews biased agonism at cannabinoid CB1 and CB2 receptors with smaller effect in CB1-CB2 heteroreceptor complexes. Biochem Pharmacol 157:148–158. https://doi.org/10.1016/j.bcp.2018.08.046

Navarro G, Varani K, Reyes-Resina I, Sánchez de Medina V, Rivas-Santisteban R, Sánchez-Carnerero Callado C, Vincenzi F, Casano S, Ferreiro-Vera C, Canela EI et al (2018) Cannabigerol action at cannabinoid CB1 and CB2 receptors and at CB1-CB2 Heteroreceptor complexes. Front Pharmacol 9:632. https://doi.org/10.3389/fphar.2018.00632

Navarro-Dorado J, Villalba N, Prieto D, Brera B, Martín-Moreno AM, Tejerina T, de Ceballos ML (2016) Vascular dysfunction in a transgenic model of Alzheimer's disease: effects of CB1R and CB2R cannabinoid agonists. Front Neurosci 10:422. https://doi.org/10.3389/fnins.2016.00422

Noel C (2017) Evidence for the use of "medical marijuana" in psychiatric and neurologic disorders. Ment Health Clin 7:29–38. https://doi.org/10.9740/mhc.2017.01.029

Ojha S, Javed H, Azimullah S, Haque ME (2016) β-Caryophyllene, a phytocannabinoid attenuates oxidative stress, neuroinflammation, glial activation, and salvages dopaminergic neurons in a rat model of Parkinson disease. Mol Cell Biochem 418:59–70. https://doi.org/10.1007/s11010-016-2733-y

Palomo-Garo C, Gómez-Gálvez Y, García C, Fernández-Ruiz J (2016) Targeting the cannabinoid CB 2 receptor to attenuate the progression of motor deficits in LRRK2-transgenic mice. Pharmacol Res 110:181–192. https://doi.org/10.1016/j.phrs.2016.04.004

Pascual AC, Gaveglio VL, Giusto NM, Pasquaré SJ (2017) 2-Arachidonoylglycerol metabolism is differently modulated by oligomeric and fibrillar conformations of amyloid beta in synaptic terminals. Neuroscience 362:168–180. https://doi.org/10.1016/j.neuroscience.2017.08.042

Peball M, Werkmann M, Ellmerer P, Stolz R, Valent D, Knaus H-G, Ulmer H, Djamshidian A, Poewe W, Seppi K (2019) Nabilone for non-motor symptoms of Parkinson's disease: a randomized placebo-controlled, double-blind, parallel-group, enriched enrolment randomized withdrawal study (the NMS-nab study). J Neural Transm 126:1061–1072. https://doi.org/10.1007/s00702-019-02021-z

Qureshi AR, Rana AQ, Malik SH, Rizvi SFH, Akhter S, Vannabouathong C, Sarfraz Z, Rana R (2018) Comprehensive examination of therapies for pain in Parkinson's disease: a systematic review and meta-analysis. Neuroepidemiology 51:190–206. https://doi.org/10.1159/000492221

Reyes-Resina I, Navarro G, Aguinaga D, Canela EI, Schoeder CT, Załuski M, Kieć-Kononowicz K, Saura CA, Müller CE, Franco R (2018) Molecular and functional interaction between GPR18 and cannabinoid CB2G-protein-coupled receptors. Relevance in neurodegenerative diseases. Biochem Pharmacol 157:169–179. https://doi.org/10.1016/j.bcp.2018.06.001

Rojo-Bustamante E, Abellanas MA, Clavero P, Thiolat ML, Qin L, Luquin MR, Bezard E, Aymerich MS (2018) The expression of cannabinoid type 1 receptor and 2-arachidonoyl glycerol synthesizing/degrading enzymes is altered in basal ganglia during the active phase of levodopa-induced dyskinesia. Neurobiol Dis 118:64–75. https://doi.org/10.1016/j.nbd.2018.06.019

Ruiz-Calvo A, Bajo-Grañeras R, Maroto IB, Zian D, Grabner GF, García-Taboada E, Resel E, Zechner R, Zimmermann R, Ortega-Gutiérrez S et al (2019) Astroglial monoacylglycerol lipase controls mutant huntingtin-induced damage of striatal neurons. Neuropharmacology 150:134–144. https://doi.org/10.1016/j.neuropharm.2019.03.027

Ruthirakuhan MT, Herrmann N, Gallagher D, Andreazza AC, Kiss A, Verhoeff NPLG, Black SE, Lanctôt KL (2019) Investigating the safety and efficacy of nabilone for the treatment of agitation in patients with moderate-to-severe Alzheimer's disease: study protocol for a cross-over randomized controlled trial. Contemporary Clin Trial Comm 15:100385. https://doi.org/10.1016/j.conctc.2019.100385

Ruthirakuhan M, Lanctôt KL, Vieira D, Herrmann N (2019) Natural and synthetic cannabinoids for agitation and aggression in Alzheimer's disease. J Clin Psychiatry 80:18r12617. https://doi.org/10.4088/JCP.18r12617

Ryberg E, Larsson N, Sjögren S, Hjorth S, Hermansson N-O, Leonova J, Elebring T, Nilsson K, Drmota T, Greasley PJ (2007) The orphan receptor GPR55 is a novel cannabinoid receptor. Br J Pharmacol 152:1092–1101. https://doi.org/10.1038/sj.bjp.0707460

Saft C, von Hein SM, Lücke T, Thiels C, Peball M, Djamshidian A, Heim B, Seppi K (2018) Cannabinoids for treatment of dystonia in Huntington's disease. J Huntington's Disease 7:167–173. https://doi.org/10.3233/JHD-170283

Sam AH, Salem V, Ghatei MA (2011) Rimonabant: from RIO to ban. J Obes 2011:432607. https://doi.org/10.1155/2011/432607

Schubert D, Kepchia D, Liang Z, Dargusch R, Goldberg J, Maher P (2019) Efficacy of cannabinoids in a pre-clinical drug-screening platform for Alzheimer's disease. Mol Neurobiol 56(11):7719–7730. https://doi.org/10.1007/s12035-019-1637-8

Sierra S, Luquin N, Rico AJ, Gómez-Bautista V, Roda E, Dopeso-Reyes IG, Vázquez A, Martínez-Pinilla E, Labandeira-García JL, Franco R et al (2015) Detection of cannabinoid receptors CB1 and CB2 within basal ganglia output neurons in macaques: changes following experimental parkinsonism. Brain Struct Funct 220:2721–2738. https://doi.org/10.1007/s00429-014-0823-8

Simard O, Niavarani SR, Gaudreault V, Boissonneault G (2017) Torsional stress promotes trinucleotidic expansion in spermatids. Mutation Research/Fundamental and Molecular Mechanisms of Mutagenesis 800–802:1–7. https://doi.org/10.1016/j.mrfmmm.2017.04.001

Spinelli F, Capparelli E, Abate C, Colabufo NA, Contino M (2017) Perspectives of cannabinoid type 2 receptor (CB2R) ligands in neurodegenerative disorders: structure–affinity relationship (SAfiR) and structure–activity relationship (SAR) studies. J Med Chem 60:9913–9931. https://doi.org/10.1021/acs.jmedchem.7b00155

Udow SJ, Freitas ME, Fox SH, Lang AE (2018) Exacerbation of psychosis triggered by a synthetic cannabinoid in a 70-year-old woman with Parkinson disease. Can Med Assoc J 190:E50–E52. https://doi.org/10.1503/cmaj.170361

Valdeolivas S, Sagredo O, Delgado M, Pozo MA, Fernández-Ruiz J (2017) Effects of a Sativex-like combination of Phytocannabinoids on disease progression in R6/2 mice, an experimental model of Huntington's disease. Int J Mol Sci 18:684. https://doi.org/10.3390/ijms18040684

Vallée A, Lecarpentier Y, Guillevin R, Vallée J-N (2017) Effects of cannabidiol interactions with Wnt/β-catenin pathway and PPARγ on oxidative stress and neuroinflammation in Alzheimer's disease. Acta Biochim Biophys Sin 49:853–866. https://doi.org/10.1093/abbs/gmx073

van Esbroeck ACM, Janssen APA, Cognetta AB, Ogasawara D, Shpak G, van der Kroeg M, Kantae V, Baggelaar MP, de Vrij FMS, Deng H et al (2017) Activity-based protein profiling reveals off-target proteins of the FAAH inhibitor BIA 10-2474. Science 356:1084–1087. https://doi.org/10.1126/science.aaf7497

Wang L, Liu B-J, Cao Y, Xu W-Q, Sun D-S, Li M-Z, Shi F-X, Li M, Tian Q, Wang J-Z et al (2018) Deletion of Type-2 cannabinoid receptor induces Alzheimer's disease-like tau pathology and memory impairment through AMPK/GSK3β pathway. Mol Neurobiol 55:4731–4744. https://doi.org/10.1007/s12035-017-0676-2

Zeissler M-L, Eastwood J, McCorry K, Hanemann OC, Zajicek JP, Carroll CB (2016) Delta-9-tetrahydrocannabinol protects against MPP^+ toxicity in SH-SY5Y cells by restoring proteins involved in mitochondrial biogenesis. Oncotarget 7:46603–46614. https://doi.org/10.18632/oncotarget.10314

Cannabidiol Therapy for Refractory Epilepsy and Seizure Disorders

7

Victoria Golub and D. Samba Reddy

Abstract

Cannabis-derived cannabinoids have neuroactive properties. Recently, there has been emerging interest in the use of cannabidiol (CBD)-enriched products for treatment of drug-resistant epilepsy. In 2018, the FDA approved the use of CBD-rich Epidiolex for two severe forms of epilepsy in children (Lennox-Gastaut and Dravet syndromes). Experimental research supports the use of CBD in many CNS disorders, though the mechanisms underlying its anticonvulsant and neuroprotective effects remain unclear. CBD has been shown to reduce inflammation, protect against neuronal loss, normalize neurogenesis, and act as an antioxidant. These actions appear to be due to the multimodal mechanism of action of CBD in the brain. This chapter briefly describes the current information on cannabis pharmacology with an emphasis on the clinical utility of CBD in the treatment of refractory epilepsies and other related seizure conditions. Clinical trials are ongoing for other forms of epilepsy and refractory seizures associated with infantile spasms, tuberous sclerosis, and Rett syndrome. Overall, adjunct CBD has been found to be generally safe and effective for treatment-resistant seizures in children with severe early-onset epilepsy. Whether an add-on CBD is efficacious for the long-term treatment of various epilepsy and seizure types in adults being tested in various clinical trials.

Keywords

Cannabis · Cannabidiol · CBD, THC · Epilepsy · Marijuana · Refractory seizure

7.1 Introduction

Epilepsy is the most complex and devastating chronic brain disorder in children and adults. Despite the influx of new anticonvulsant drugs (AEDs) over the past 50 years, 30–40% of epileptic patients still experience refractory seizures that are untreatable by current medications. These patients with treatment-resistant seizures are at a much higher risk for persistent brain damage and other secondary consequences of epileptic seizures that adversely influence quality of life. Polypharmacy to manage such seizures is associated with serious side effects such as sedation, somnolence, and cognitive impairment. This is commonly observed in children with certain types of devastating pediatric epilepsies, such as Lennox-Gastaut, Doose, and Dravet syndromes in which afflicted children have an increased risk of death compared to their peers of the same age (Autry et al. 2010). One explanation for this

V. Golub · D. S. Reddy (✉)
Department of Neuroscience and Experimental Therapeutics, College of Medicine, Texas A&M University Health Science Center, Bryan, TX, USA
e-mail: sambareddy@tamu.edu

E. Murillo-Rodriguez et al. (eds.), *Cannabinoids and Neuropsychiatric Disorders*, Advances in Experimental Medicine and Biology 1264, https://doi.org/10.1007/978-3-030-57369-0_7

high percentage of refractory seizures is that preclinical research and screening for AEDs has been biased toward agents that modulate only a single pathology in epileptogenesis; i.e., focused on either reducing hyperexcitation or increasing inhibition via the modulation of ion channels or neurotransmission. This empirical approach often ignores the multimodal intracellular components that might serve as novel targets for the relief of symptomatic seizures. Therefore, the need for different therapeutic options to manage refractory forms of epilepsy is still a current issue.

Cannabis is a generic term for products of the plant *Cannabis sativa L.* This plant has been used therapeutically for thousands of years (Grotenhermen 2006). Marijuana is a dried mixture of cannabis leaves and flowers. It was well known that the cannabis plant had psychotropic effects, inducing a "high" or euphoric effect. Research over the past few decades led to the identification of cannabinoids, which are categorized into three primary classes: phytocannabinoids, endocannabinoids, and synthetic cannabinoids. Hemp, a strain of *Cannabis sativa L.* (historically grown for fibrous materials found in stalks and seeds), contains minimal amounts of THC and low levels of CBD. An oil extracted from cannabis seeds by cold pressing is called Hemp oil or hempseed oil. It contains only trace amounts of cannabinoids. CBD oil or hemp CBD oil is an extract obtained from the flowering portions of the hemp plant, then dissolved in coconut or sesame oil. It contains no THC and has no psychoactive properties. Cannabis oil is a concentrated cannabis extract, often containing very high THC levels.

Recently, the cannabis-derived phytocannabinoids have shown compelling evidence as a potential therapy for medication-resistant seizures (Friedman and Devinsky 2015; Reddy and Golub 2016; O'Connell et al. 2017). Specifically, the nonpsychoactive phytocannabinoid, cannabidiol (CBD), has been well tolerated and retains both anticonvulsant and anti-inflammatory properties, though a mechanism of action that has yet to be fully clarified (Reddy and Golub 2016; Billakota et al. 2019). Some CBD-based pharmaceuticals (Epidiolex, Realm Oil, and others) have been suggested as potential therapies for refractory epilepsy. In June 2018, the FDA approved Epidiolex for the treatment of seizures associated with two rare and very severe forms of pediatric epilepsy, Lennox-Gastaut syndrome and Dravet syndrome. Epidiolex has been approved in patients aged two and older in these syndromes; ongoing clinical trials are rapidly underway to determine the efficacy of CBD-based treatments in young, adult, and aged epileptic patients with other types of refractory seizures, as well as for many different neurological conditions.

The topic of cannabis in medical regimens has been widely debated in both the academic and political communities, with a special focus of the rationale of cannabis products for children. There are numerous stories and anecdotal findings of desperate parents seeking cannabis products to relieve their children's seizures; however, what scientific evidence exists for cannabis' effectiveness? This book chapter briefly discusses the pharmacology of cannabis and reviews the clinical utility of CBD in the treatment of refractory epilepsies.

7.2 Pharmacology of Cannabinoids

The cannabis plant contains over 100 compounds, collectively referred to as phytocannabinoids (ElSohly and Gul 2014). Among these products, the two best characterized cannabinoids are tetrahydrocannabinol (THC) and cannabidiol (CBD) (Fig. 7.1). Cannabidivarin (CBDV) is a variant of CBD with some neuroactive properties. Tetrahydrocannabidivarin, a structurally similar compound to THC, is actually an antagonist of THC. Δ-Tetrahydrocannabinolic acid (THCA), the most abundant cannabinoid in cannabis bred for recreational use, is a nonpsychoactive or non-euphoric precursor of THC. THCA converts to THC when heated or smoked. Most of the data on CBD and THC compounds demonstrate anticonvulsant properties in major experimental seizure models (Rosenberg et al. 2017; Patra et al. 2019; Huizenga et al. 2019). However, there has also been some evidence and discrepancy of proconvulsive properties of THC in healthy

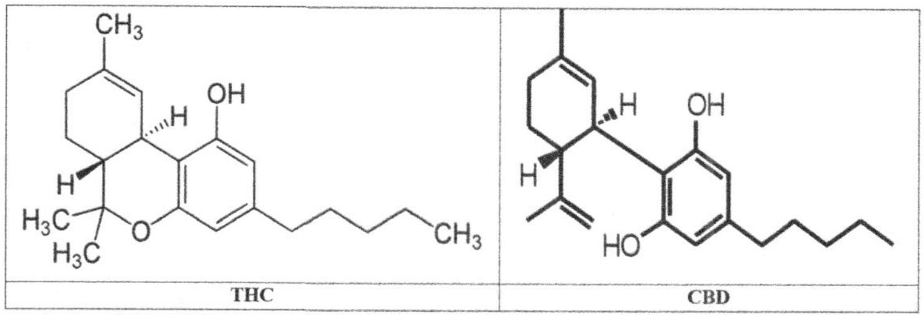

Fig. 7.1 Chemical structures of the two major phytocannabinoids CBD and THC

animals, generally at high doses (Stadnicki et al. 1974; Martin and Consroe 1976). THC is psychoactive and prone to tolerance, and therefore, is a less likely therapeutic option for epilepsy. Conversely, CBD does not exert psychoactive effects. Therefore, CBD has been selected for advanced studies in experimental and clinical trials for epilepsy.

While studying how THC exerting its psychotropic effects, the endocannabinoid system was discovered within the body. The endocannabinoid system can influence many neurophysiological processes, including neuronal excitability, pain and inflammation, feeding and energy regulation, and learning and memory, as well as emotion regulation. Endocannabinoids identified so far include anandamide, 2-arachidonoylglycerol, virodhamine, N-arachidonoyl dopamine, and noladin ether. These physiological functions of endocannabinoids are mediated by cannabinoid receptors (CBRs) that have been found within the central nervous system, reproductive organs, skin, and digestive tract (Grotenhermen 2006).

The primary pharmacological effects of THC and CBD are due to their interaction with the CBRs. There are two subtypes of CBRs, CB_1 and CB_2, which are G-protein-coupled receptors that work primarily to inhibit adenylate cyclase activity through the $G_{i/o}$ pathway (Howlett et al. 1986). Through this mechanism, CB_1 activation can modulate neurotransmitter release in the brain areas the receptors are most highly expressed, i.e., the limbic structures, cerebral cortex, basal ganglia, and select areas of the midbrain and medulla (Tsou et al. 1998). CB_2 receptors are located primarily on immune tissues and specific cell types such as the spleen, lymph nodes, B-cells, macrophages, and microglia (Galiegue et al. 1995). Activation of CB_2 receptors on immune cells or organs may lead to immunosuppressive responses. Furthermore, there is limited expression of CB_2 receptors within the CNS, located only in areas such as the hippocampus and brainstem (Stempel et al. 2016). Unlike activation of CB_1 receptors in these regions, CB_2 activation has been demonstrated to trigger neuronal excitation, which may explain the debate among scientists over THC's antiseizure abilities. However, not all cannabinoids act on the endocannabinoid system, as in the case of CBD. A series of consecutive studies using the maximal electroshock and pilocarpine models of epilepsy observed that the anticonvulsant properties of THC were due to its activity at CB_1 receptors, whereas the antiseizure properties of CBD were CB1 independent and may occur via a different, unknown mechanism (Wallace et al. 2001, 2002, 2003). Pharmacological mechanisms underlying the therapeutic claims for reducing the prevalence of epileptic seizures with CBD-enriched products are currently poorly understood. The following sections discuss the known pharmacokinetics and mechanisms of two major (THC and CBD) cannabinoids.

7.3 Mechanism, Bioavailability, and Metabolism of THC

THC is the most abundant compound found in cannabis and is responsible for the psychoactive effects commonly associated with the recreational smoking of marijuana. It has also been associated with changes in human cognition and perception (Hofmann and Frazier 2013). THC acts as a partial agonist of CB_1 receptors in the endocannabinoid system, which are predominantly found within the CNS, with especially high concentrations in the hippocampus. It is through this binding action that THC exerts it anticonvulsive properties (Consroe et al. 1982). The hippocampus is often a focal point for many forms of epilepsies. Within the hippocampus, CB_1 receptors have been observed in high levels within the CA1–3, and with highest expression in the dentate gyrus (Glass et al. 1997). These receptors have been further characterized onto presynaptic GABAergic neurons of basket cells as well as excitatory neurons (Irving et al. 2000; Kawamura et al. 2006). THC administration can reduce glutamate release in excitatory neurons, as well as inhibit release from cholinergic neurons and GABAergic interneurons (Shen et al. 1996; Gifford et al. 2000; Katona et al. 2000). Furthermore, CB_1 activation has been shown to inhibit voltage-gated Ca^{2+} channels and increase K^+ channel activity to inhibit presynaptic transmission (Shen et al. 1996).

THC also binds to a number of other targets including CB_2 receptors, located primarily on the cells of the immune system; transient receptor potential cation channels TRPA1, TRPM8, and TRPV2; serotonin receptors; G-protein-coupled GPR55 receptor; and the μ- and δ- opioid receptors; as well as some subtypes of Na, K, and Ca channels (Pertwee and Cascio 2014). The mechanism by which THC produces anticonvulsant effects is believed to be through CB_1 receptors (Detyniecki and Hirsch, 2015), but, as previously mentioned, there is some controversy over THC's actual anticonvulsant effects. A recent review of THC's anticonvulsant properties found 34 articles covering animal models of epilepsy. Antiseizure effects were found in 21/34 studies (61.8%), no significant effects were found in 11/34 studies (32.4%), and proconvulsant effects were seen in 1/34 studies (2.9%) (Rosenberg et al. 2015). The mixed results for the anticonvulsant effects of THC are likely due to the complicated pharmacology of the agent as well as the diverse effects of the endocannabinoid system, including the modulation of both excitatory glutamatergic and inhibitory GABAergic neurotransmission.

Both the absorption and bioavailability of THC are highly dependent on the route of administration. Inhaling or smoking THC has the fastest absorption, with peak plasma levels reached after 10 min and a bioavailability of about 25%. Oral consumption of THC provides approximately 6% bioavailability and slower absorption, taking anywhere from 2–6 h to reach peak plasma levels (Gaston and Friedman 2009). Following absorption, THC binds to proteins within the blood to circulate a volume of distribution of 3.4 L/Kg before hepatic metabolism by cytochrome P450 enzymes CYP2C9, CYP2C19, and CYP3A4. Excretion of THC metabolites occurs through both urine and feces (Huestis 2007).

7.4 Mechanism, Bioavailability, and Metabolism OF CBD

Contrary to THC, CBD has a very low affinity for the orthosteric sites of the endocannabinoid receptors. Numerous mechanisms have been proposed to explain the anticonvulsant action of CBD, though the exact mechanisms still remain unknown. CBD shares some targets with currently available antiepileptic drugs (AED) ethosuximide and zonisamide by blocking calcium influxes through T-type voltage-gated calcium channels (Ross et al. 2008). This activity is similar to the actions of levetiracetam, lamotrigine, and eslicarbazepine, which target L-, P/Q, and N- type calcium channels. Recent research has demonstrated that CBD can also target aberrant mutant Nav1.6 sodium channel activity, which is often associated with severe syndromic epilepsies like Dravet syndrome (Patel et al. 2016). Furthermore, Kaplan and colleagues report a substantial reduction in the frequency and duration of seizures in *Scn1a* mutant Dravet mice by enhancing GABAergic neurotransmission and decreasing action potential firing of excitatory neurons (Kaplan et al., 2017). These changes in the excitation/inhibition ratio were both mimicked and occluded by the inhibition of the lipid-activated G-protein coupled receptor GPR55. Their results suggest GPR55 as a target of the CBD's anticonvulsant mechanism. Other potential targets for CBD's anticonvulsant activity include agonism at 5-HT_{1A} serotonin receptors, TRPA1, and TRPV1/2 (Devinsky et al. 2014b, b). Though serotonin receptors seem like an unconventional target for epilepsy, activation of 5-HT_{1A} receptors via flenfluramine administration has demonstrated to be an effective add-on treatment in Dravet syndrome (Ceulemans et al. 2012).

It has been established that CBD binds poorly to the orthosteric sites of the CB_1 and CB_2 endocannabinoid receptors, through which THC exerts its major effects; however, some evidence suggests that CBD can act at CB_1 receptors as a negative allosteric modulator. Recently, Straiker et al. (2018) demonstrated that CBD has no direct effect on excitatory transmission within autaptic hippocampal neurons but does inhibit two forms of endogenous cannabinoid-mediated retrograde synaptic plasticity. CBD reduced depolarization-induced suppression of excitation (DSE), an endogenous 2-arachidonoyl glycerol cannabinoid-mediated effect, as well as metabotropic suppression of excitation without affecting GABA-B receptor function (Straiker et al. 2018). These results expanded on the finding that CBD has a pharmacological profile consistent with potent negative allosteric modulation of CB_1 signaling (Laprairie et al. 2015), though implications for CBD antiseizure properties are still somewhat limited.

The clinical pharmacokinetics and pharmacology profile of CBD is outlined in Table 7.1. Bioavailability and absorption of CBD is also highly dependent of the route of administration, with peak concentrations being reached after 10 min when given intranasally, 2 h when orally consumed, and over 15 h when administered through a transdermal gel (Ohlsson et al. 1986; Deianna et al. 2012). CBD has a low bioavailability of 10% due to high first-pass metabolism in the gut and liver (Devinsky et al. 2014b, b). Both the C_{max} and area under the curve increase when CBD is administered with a high-fat meal or fatty vehicle (i.e., sesame oil- or coconut oil-based suspension). Similar to THC, CBD is highly lipophilic and reaches a volume of distribution of 32 L/kg by traveling through the plasma protein binding (Ohlsson et al. 1986). CBD is metabolized in the liver predominately by the cytochrome P450 enzymes CYP2C19 and CYP3A4 before excretion in feces (Gaston and Friedman 2009). The anticonvulsant effects of the circulating metabolites of CBD, 7-COOH-CBD and 6-OH-CBD, remain undefined.

7.5 Translational Studies of CBD on the Comorbities of Epilepsy

Epilepsy is known first and foremost as a seizure disorder, with patients experiencing a variety of seizure types, i.e., convulsive, absence, atonic, clonic, tonic, and myoclonic. These symptomatic seizures can be induced by genetic, mechanical injury, psychological, or viral causes. In many cases, there is a period between the initial insult and the onset of seizures. During this latency period, several restructuring processes are underway, transforming a normal brain into a hyperexcitable, epileptic one (Clossen and Reddy 2017). These processes have been mirrored in many rodent models, and include neuronal and blood-brain barrier damage, inflammation, alterations in neurogenesis, axonal/dendritic plasticity, and epigenetic changes (Reddy and Kuruba 2013; Younus and Reddy 2017). Furthermore, patients with epilepsy have a higher prevalence for psychiatric disorders such as anxiety and depression. Since epilepsy is associated with such a wide array of neurological disruptions, it is critical that new antiseizure drugs show neuroprotection against some or all these factors. This section describes the current preclinical understanding of the neuroprotective actions of CBD that are related to these epileptogenic processes and comorbidities.

7.5.1 Neuroprotective Actions of CBD

CBD has been shown to have a wide spectrum of actions, suggesting many therapeutic possibilities

Table 7.1 The clinical PK and pharmacological profile of CBD (Epidiolex)

Indications	Treatment of seizures associated with Lennox-Gastaut syndrome or Dravet syndrome in patients 2 years of age and older
Product	Oral solution: 100 mg/ml
Dosage	Starting dosage is 2.5 mg/kg, orally, twice daily (5 mg/kg/day) Maintenance dosage is 5 mg/kg twice daily (10 mg/kg/day) Maximum dosage is 10 mg/kg, twice daily (20 mg/kg/day)
Oral bioavailability	10%
Tmax (5–20 mg dose)	2.5 to 5 h
Volume of distribution (Vd)	~32 L/kg
Metabolism	CYP2C19 and CYP3A4 enzymes, and UGT1A7, UGT1A9, and UGT2B7 isoforms
Half-life ($t_{1/2}$)	56 to 61 h
Clearance (CL)	1111 L/h
Protein binding	>94%
Drug-drug interactions	Potent inhibitor of CYP3A and CYP2C isozymes Strong inducer of CYP3A4 or CYP2C19 isozymes Substrates of CYP1A2 and CYP2B6
Common adverse effects	Somnolence; decreased appetite; diarrhea; fatigue, malaise, asthenia; rash; insomnia, sleep disorder, infections; and suicidal behavior and ideation
Warning and precautions	Hepatocellular injury: It can cause transaminase elevations. Concomitant use of valproate and higher doses of CBD increase the risk of transaminase elevations.
Discontinuation	When discontinuing CBD, the dose should be decreased gradually, because of the risk of increased seizure frequency and status epilepticus.

for the drug. CBD prevented oxidative damage in an NMDA-mediated neurotoxicity rat model (Hampson et al. 1998) and also prevents the exacerbation of reactive oxygen/reactive nitrogen species production (Campos et al. 2017). Reactive oxygen species and reactive nitrogen species (ROS/RNS) accumulation often leads to cell injury and cell death, especially in high glucose-induced mitochondrial production. CBD seems to protect against cell death by enhancing intracellular recycling of cellular components. This autophagic pathway can malfunction under stress, thereby promoting more damaging mechanisms. Hosseinzadeh et al. (2016) suggest the neuroprotective effects of CBD in pilocarpine-induced seizures may be due in part to its activation of hippocampal autophagic machinery to promote cell survivability over cell death. In one study of cerebral ischemia, CBD supplementation improved basal respiration and enhanced mitochondrial function via the pentose-phosphate pathway (Sun et al. 2017). CBD also normalizes caspase-3 expression in rats with brain iron overload and decreased β-amyloid protein deposit-induced neurodegeneration (Da Silva et al. 2014; Esposito et al. 2011).

7.5.2 Anti-inflammatory Actions of CBD

Inflammation within epileptic foci can exacerbate seizures and epileptogenesis. CBD ameliorates expression of pro-inflammatory cytokines, which can aggravate injury (Rajesh et al. 2007). During hypoxic-ischemic brain damage, CBD administration provided neuroprotection by reducing the deleterious effects of glutamate toxicity and inflammatory molecule production such as IL-6, TNF alpha, COX-2, and iNOS (Castillo et al. 2010). These effects were mediated by CBD's activity on CB2 and adenosine receptors. Together, these studies suggest CBD acts as an antioxidant to reduce ROS/RNS, attenuates inflammatory cascades, and significantly reduces neurodegeneration in a variety of injury models.

7.5.3 CBD Facilitation of Neurogenesis

Hippocampal neurogenesis is a major source of plasticity in the brain with a variable rate throughout the lifetime. It has been demonstrated in many rodent models that seizures produce both short- and long-term changes in cell proliferation (Parent et al. 1997; Hattiangady et al. 2004). Prolonged seizure activity leads to a significant increase in neurogenesis in the dentate gyrus cells, which can contribute to acute aberrant network reorganization during epileptogenesis. Cell proliferation returns to baseline approximately 1 month following status epilepticus in rats; however, in cases with extreme neuroinflammation within the temporal regions, reduced neurogenesis after status epilepticus has also been observed (Hattiangady et al. 2004). It is suggested that seizure-generated granule cells have poorer survivability than new-born neurons in a naïve animal, and that a lowered rate of neurogenesis may be correlated with higher initial injury severity (Mohpael et al. 2004).

Neuropsychiatric disorder models provide an opportunity to study the mechanisms of neurogenesis. In rodents, exposure to chronic stressors induces dendritic remodeling and reduced adult hippocampal neurogenesis (Bessa et al. 2009). Reduced neurogenesis is recognized as one of the major mechanisms related to smaller hippocampal volume in patients suffering from mood disorders and schizophrenia (Lucassen et al. 2010). A study in 2010 demonstrated a CBD-rich diet can increase BrdU-labeled newborn neurons in the hippocampus (Wolf et al. 2010); however, the contributory role of neurogenesis in epileptogenesis is still uncertain (Danzer et al. 2019). CBD has also been shown to prevent stress-induced reduction in neurogenesis as well as prevent neurogenic disruption in a genetic mouse model of Alzheimer's disease (Crippa et al. 2018; Esposito et al. 2011). Furthermore, CBD can modulate intracellular pathways related to synaptic remodeling including the Erk1/2 and Akt pathways (Solinas et al. 2013). In an iron overload induced-brain damage model, CBD administration normalized the expression of synaptophysin, a critical protein involved in proper synaptic and vesicular function (Da Silva et al. 2014). These actions of CBD are independent of cannabinoid receptor modulation.

7.5.4 Anxiolytic Actions of CBD

Initial studies investigating the effects of CBD on anxiety-like behaviors yielded somewhat contradictory results. In the early 1980s, Zuardi and Karniol (1983) found 10 mg/kg of CBD attenuated conditioned emotional responses in rats, whereas Silveira Filho and Tufik (1981) found no such effects with 100 mg/kg of CBD. It was later found during a dose-response study that CBD induces anxiolytic effects at low doses, but those effects diminish with higher quantities (Guimaraes et al. 1990). Since then, many studies have demonstrated the anxiolytic effects of CBD in preventing fear expression and reconsolidation in models of generalized anxiety, PTSD, panic disorder, and high-stress (Campos et al. 2012a, 2012b; Resstel et al. 2006; Stern et al. 2015). Additionally, it was also demonstrated that acute or repeated administration of CBD significantly decreased the behavioral and autonomic responses of acute restraint stress (Granjeiro et al. 2011). CBD reduced anxiogenic effects seen in acute stress via predator exposure in rats as well as in chronic unpredictable stress (Campos et al. 2012a, 2012b). The mechanism of the anxiolytic effects of CBD are poorly understood, but is most likely due to its interaction with serotonin 5HT1A receptors. This anxiolytic nature of CBD could also be associated with antidepressant like properties, which have been observed in forced-swim and bulbectomy models in rats (Zanelati et al. 2010; Linge et al. 2016). This is corroborated by a human study that shows that CBD reduces anxiety in social anxiety disorder and that this is related to its effects on activity in limbic and paralimbic brain areas (Crippa et al. 2011).

7.6 Pivotal Clinical Trials That Led to the Approval of Epidiolex

Early case reports and surveys on the use of cannabis for epilepsy have historically been limited and underpowered with inconclusive results. In the late nineteenth century, two prominent British neurologists observed reductions in seizure frequency when they treated their epileptic patients with cannabis (Reynolds 1861; Gowers 1881). Despite these successes, cannabis remained understudied as a possible therapeutic for convulsive disorders until the 1970/80s. A Cochrane review published in 2012 found four controlled studies, which examined the therapeutic potential of CBD for epilepsy (Gloss and Vickrey 2014). This review searched for randomized, controlled clinical trials that showed direct evidence of anticonvulsant effects of CBD in human seizures. They identified only four studies that fit their efficacy criteria, with a total sample of 48 participants. Of these four trials, two found limited improvements on seizure frequency, but all had some methodological flaw including small sample size and inadequate blinding (Mechoulam and Carlini 1978; Cunha et al. 1980; Ames and Cridland 1986; Trembly and Sherman 1990). Dosing ranged from 200 mg - 300 mg per day, and the only side effect reported in any of these studies was somnolence. In general, only short-term tolerability of CBD-enriched therapeutics was demonstrated in these studies.

Two case studies received a lot of media attention when patients became seizure-free after treatment with medical marijuana or extracts of its components. In one case, a 24-year-old man self-treated with 2–5 rolled cigarettes per night containing whole plant marijuana in addition to his prescribed AEDs and experienced a full reduction in refractory seizures (Consroe et al. 1975). The second case report, which convinced many parents with epileptic children to move to Colorado, featured a young girl named Charlotte, who suffered from severe Dravet syndrome. She was started on a sublingual preparation of cannabis extract with a ratio of 16:1 CBD:THC in tandem with her prescribed clobazam. After 3 months of treatment, Charlotte experienced >90% reduction in seizures (Maa and Figi 2014). This strain is now named "Charlotte's Web" as a tribute to the success Charlotte found using this extract. In 2013, an oil-based extract of Charlotte's Web, called Realm Oil, was tested in 13 patients whose diagnoses included Doose, Dravet, and Lennox-Gastaut syndromes. A parent-survey reported that 11/13 patients reported a weekly reduction in seizures, and of these 11, 8 experienced a 98–100% reduction in seizures (Gedde and Maa 2013). An additional caregiver survey yielded positive outcomes of other CBD-enriched products in epilepsy. In one survey, 16/19 responders (84%) reported a reduction in seizure frequency with cannabis therapy (Porter and Jacobson 2013). Response rates appear similar with all products but vary by syndrome (LGS > Dravet). Together with the steady progress in experimental research, these case reports and underwhelming trials paved the way for the neoteric clinical trials and FDA approval of Epidiolex for Lennox-Gastaut and Dravet syndromes in 2018.

Lennox-Gastaut syndrome (LGS) is a rare, severe, and drug-resistant form of epilepsy that begins in early childhood and changes throughout life. LGS is prevalent in ~4% of children with epilepsy. The classic description of LGS focuses on three typical clinical (classic triad) features: (i) multiple seizure types; (ii) cognitive impairment; and (iii) slow spike-wave EEG (Bourgeois et al. 2014). LGS involves a variety of seizure types including tonic and atypical absence; drop seizures occur in at least 50% patients. Many patients with LGS have significant cognitive impairment. A distinct EEG pattern is the third characteristic feature of LGS for most patients. LGS presentation, however, is variable and not all patients exhibit all components of the classic triad at onset. LGS with cognitive and physical impairments elicits a significant impact on patients and caregivers. LGS signs and symptoms may also change over time. Such high-risk seizures may predispose patients to status epilepticus or sudden unexpected death in epilepsy (SUDEP) and head injury (Schmidt and Bourgeois 2000). There is a high risk (14-fold) of mortality in children with LGS. Research is currently being carried out to identify genetic factors in LGS.

The effectiveness of Epidiolex for the symptomatic treatment of seizures associated with LGS was founded on two randomized, double-blind, placebo-controlled trials conducted in 2014–2015 (Thiele et al. 2018; Devinsky et al. 2018a, 2018b). Epidiolex contains >98% CBD and less than <0.15% THC and is prepared in a strawberry-flavored sesame oil-based suspension. These trials included patients aged 2–55 years old (Study 1 $n = 171$; Study 2 $n = 225$) and compared doses of 10 mg/kg/day and 20 mg/kg/day of Epidiolex versus placebo. A period of 4- weeks was used to assess baseline seizures in LGS patients. A minimum of 8 drop seizures, which were inadequately controlled by their current medications during this period, was required to participate in the clinical trial. The most common concomitant AEDs in these studies were clobazam, valproate, lamotrigine, levetiracetam, and rufinamide. A 2-week titration period followed the 4-week baseline.

The primary efficacy measure was percent change in seizure frequency during the 12-week maintenance period. In both studies, Epidiolex-treated patients observed a significantly greater reduction in seizure frequency. In Study 1 (Thiele et al. 2018), a 41% reduction in seizure was observed with 20 mg/kg/day ($p = 0.01$). In Study 2 (Devinsky et al. 2018a, b), a 36% reduction was seen at 10 mg/kg/day ($p < 0.01$) and 38% reduction at 20 mg/kg/day ($p < 0.01$) compared to placebo. Secondary measures included changes in Subject/Caregiver Global Impression of Change (S/CGIC) during the last visit. This scale was conducted using a 7-point comparison impression on the status of the patient's overall condition at the beginning and ending of the clinical trial using phrases such as "much improved", "slightly improved", "no change", "much worse", etc. The S/CGIC scores for Studies 1 and 2 most closely corresponded to "slightly improved" in treated patients vs "no change" in placebo groups. An additional observational study reported 9/23 patients experience >50% decrease in seizures, with a median reduction in seizures of 32% for all patients when treated with 25 mg/kg per day of Epidiolex (cannabidiol) in tandem with prescribed AEDs (Devinsky et al. 2014b, b). A similar report was made in 2015 with 25 patients (Oldham et al., 2015).

Dravet syndrome (DS) is a developmental disorder typified by severe seizures and delayed onset of psychomotor deficits (Dravet et al. 2005). DS has distinct characteristics to confirm diagnosis. Seizures typically develop in the first year of life in infants in children with no apparent developmental disabilities. The initial seizure is often triggered by an illness and may present as a prolonged, febrile and afebrile-generalized seizures and progress to severe and often refractory epileptic encephalopathy; seizures decrease in frequency and severity with sexual maturity (Steel et al. 2017; Gataullina and Dulac 2017). DS has a broad differential diagnosis with a typical and unique quartet characteristics: (i) temperature sensitivity; (ii) prolonged seizures in otherwise normally developing infant; (iii) developmental delays following early normal development; and (iv) myoclonic seizures. DS affects motor and cognitive development. In early childhood, seizures are triggered by hyperthermia, bathing, flashing lights, emotional stress, visual patterns, and overexertion. In adolescence, seizures persist, occurring more often during sleep. In DS, seizures may be worsened by AEDs that target sodium channels. In addition to increasing the risk for SUDEP, many children are frequently taken to the ER with status epilepticus, which can lead to brain damage. The delayed social and cognitive development and movement disorders are additional clinical presentation in adolescent DS patients. In a majority of cases, mutations in the sodium channel gene SCN1A form the genetic basis for DS (Fujiwara 2006; −Steel et al. 2017). SCN1A$^{+/-}$ mice exhibits symptoms reminiscent of human DS; they display both thermally induced and spontaneous seizures, and develop autism-like social deficits (Rubinstein et al. 2015). The loss of function in Nav1.1 channels in SCN1A$^{+/-}$ mice selectively reduces sodium current and excitatory drive in GABAergic interneurons contributing to heightened epileptogenesis.

The effectiveness for Epidiolex for the treatment of seizures associated with DS was demonstrated in one randomized, double-blind, placebo-controlled trial (Devinksy et al. 2017; NCT02091375). Like the LGS trials, this study had a 4-week baseline, 2-week titration, and 12-week maintenance period. The minimum requirements for trial participation were a documented history of DS and at least 4 uncontrolled convulsive seizures while on stable AED medication during the baseline period. Primary efficacy measurement was based on percent change from baseline in seizure frequency, including atonic, tonic, clonic, and tonic-clonic seizures during the treatment period. Patients were aged between 2–18 years and at least 93% (112/120) of participants were taking two or more concomitant AEDs. Most commonly prescribed AEDs during the DS study were clobazam, valproate, stiripentol, levetiracetam, and topiramate. A reduction in seizure frequency was observed after 4-weeks of treatment with 20 mg/kg/day Epidiolex. The median frequency of convulsive seizures decreased from 12.4 per month to 5.9 per month in the treated patients, compared to a decrease from 14.9 to 14.1 seizures per month in the placebo group. Of the treated participants ($n = 61$), 43% had a > 50% reduction in seizures, including 5% who became seizure-free. Among these patients, cannabidiol treatment resulted in a significant reduction in convulsive seizures versus placebo.

Safety and antiseizure effectiveness of an open-label extension trial on the long-term CBD treatment in patients with Dravet syndrome were reported recently (Devinsky et al. 2019). Patients who completed GWPCARE1 Part A (NCT02091206) or Part B, or a second placebo-controlled trial, GWPCARE2 (NCT02224703), were invited to enroll in a long-term open-label extension trial, GWPCARE5 (NCT02224573). A purified CBD oral solution (100 mg/mL) was titrated from 2.5 to 20 mg/kg/day over a 2-week period, along with their existing medications. A total of 264 enrolled in this open-label extension. Common adverse events reported include diarrhea (34.5%), pyrexia (27.3%), decreased appetite (25.4%), and somnolence (24.6%). In patients from GWPCARE1 Part B, the median reduction from baseline in monthly seizure frequency assessed in 12-week periods up to week 48 ranged from 39% to 51% for total seizures. After 48 weeks of treatment, 85% of patients/caregivers reported improvement in the patient's overall condition. Overall, this study confirms that long-term CBD treatment had a reasonable safety profile and led to sustained reductions in seizure frequency in patients with treatment-resistant DS.

Long-term safety and efficacy of CBD in children and adults with treatment resistant LGS or DS from expanded access program were reported recently (Laux et al. 2019). Since 2014, patients with severe treatment-resistant epilepsies (TREs) have been receiving add-on CBD in an ongoing, expanded access program (EAP), which closely reflects clinical practice. Twenty-eight percent of LGS/DS patients withdrew, primarily owing to lack of efficacy (20%). At 12 weeks, add-on CBD reduced monthly seizures by 50% and total seizures by 44%. At 12 weeks, the proportions of patients with ≥50%, ≥75%, and 100% reductions in major motor seizures were 53%, 23%, and 6%; the proportions with corresponding reductions in total seizures were 46%, 26%, and 5%. These results confirm that CBD had an acceptable safety and efficacy during long-term treatment in LGS or DS.

There is emerging evidence on long-term safety and efficacy data for intractable epilepsies beyond LGS and Dravet syndrome, as evident from the open-label expanded access program (Szaflarski et al. 2018). In January 2014, an expanded access program (EAP) was initiated to provide CBD to patients with treatment-resistant epilepsy (TRE). Preliminary efficacy data have been reported previously (Devinsky et al. 2016). They have published an updated paper, reporting pooled results for safety outcomes up to 144 weeks and efficacy endpoints up to 96 weeks in more than 600 patients (Szaflarski et al. 2018). Of the 607 patients treated (median treatment duration, 48 weeks; range 2–146 weeks), 76% of patients remained on treatment. CCBD was associated with 51% and 48% reductions in median monthly convulsive and total seizures, respectively, after 12 weeks of treatment. Reductions in median monthly convulsive and total seizures were similar among visit windows through 96 weeks of treatment. At visits between weeks 12 and 96, inclusive, the ≥50%, ≥75%, and 100% response rates were notable and similar among time points. Overall, these results from this ongoing EAP support previous observational and clinical trial data showing that add-on CBD may be an efficacious long-term treatment option for TRE (Szaflarski et al. 2018b). CBD was generally well tolerated; treatment-emergent adverse events were consistent with those reported previously.

To date, most clinical trials have focused on the assessment of safety and/or efficacy of CBD in combination with prescribed antiepileptic drugs. Studies focusing on genetically based epilepsies such as Dravet, Lennox-Gastaut, and West syndromes have been the most common. However, there are currently over 20 clinical trials ongoing to study the efficacy of CBD for a number of neurological conditions including the seizure-associated disorders of infantile spasms (NCT02953548; NCT02954887), tuberous sclerosis complex (NCT02544750; NCT02544763), Rett syndrome (NCT03848832), and refractory epilepsy in adults (NCT02607904; NCT02564952; NCT02565108; and NCT02286986). These investigations of CBD are providing rigorous information on the pharmacological basis of their clinical use. A higher dose of CBD is associated with a greater chance for seizure improvement, though children may respond better to lower doses of CBD than adults (Hernando et al. 2018). There is promising early indications of CBD's effectiveness as an adjunct AED for children with intractable generalized epilepsy (Cilio et al. 2018; Carney et al. 2017). A transdermal CBD gel is proposed as an adjunctive therapy for the treatment of focal seizures in adults (Sebree et al. 2016; O'Brien et al. 2018), but further evaluation in double blind studies is warranted to confirm such predictions.

7.7 Adverse Effects and Drug Interactions

In controlled and uncontrolled trials in patients with LGS and DS, 689 patients were treated with CBD, including 533 patients treated for more than 6 months, and 391 patients treated for more than 1 year. In an expanded access program and other compassionate use programs, 161 patients with DS and LGS were treated with CBD, including 109 patients treated for more than 6 months, 91 patients treated for more than 1 year, and 50 patients treated for more than 2 years (Devinsky et al. 2014b, b; Devinksy et al. 2017; Szaflarski et al. 2018). In these clinical trials of CBD, the most common adverse reactions that occurred in CBD-treated patients (incidence at least 10% and greater than placebo) were somnolence; decreased appetite; diarrhea; transaminase elevations; fatigue, malaise, and asthenia; rash; insomnia, sleep disorder, and poor quality sleep; and infections. CBD can cause weight loss (Szaflarski et al. 2018b). However, CBD does not produce THC-like behavioral (rewarding) responses and it does not produce physical dependence. There is increasing attention on the importance of the relationship between sleep and epilepsy. CBD therapy has been associated with alterations in sleep architecture (Drake et al. 2018). Further research on the effects of CBD on sleep parameters and related measures is needed.

CBD is metabolized by CYP3A4 and CYP2C19. Therefore, co-administration with a moderate or strong inhibitor of CYP3A4 or CYP2C19 can increase CBD plasma concentrations, which may result in a greater risk of adverse reactions. Conversely, co-administration with a strong CYP3A4 or CYP2C19 inducer can decrease CBD plasma concentrations, which may lower the efficacy of the drug. Co-administration of CBD increases plasma concentrations of drugs that are metabolized by CYP2C19 (e.g., diazepam) and may increase the risk of adverse reactions with these substrates. In addition, co-administration of CBD produces a three-fold increase in plasma concentrations of N-desmethylclobazam, the active metabolite of clobazam (a substrate of CYP2C19. This drug interaction may increase the risk of clobazam-related adverse reactions. In addition, diet had a significant interaction on the pharmacokinetics of CBD. In clinical pharmacokinetic studies, high fat diet was found to increase Cmax and AUC of CBD (Epidiolex) ~4 fold (Taylor et al. 2018). In another pharmacokinetic study, purified CBD capsules were administered orally with and without food to adults with refractory epilepsy. An average of 14-fold and 4-fold increases in Cmax and AUC respectively were found in patients in the fed condition (Birnbaum et al. 2019).

Some studies were performed to investigate drug-drug interactions between CBD and frequently prescribed AEDs. These trials found significant changes in AED serum levels when CBD and clobazam were taken in tandem (Friedman et al. 2014; Geffrey et al. 2015). For those patients, the dose of clobazam was reduced. These studies also boasted a 50–55% reduction in seizures in 11/13 patients, though the remaining 2/13 patients experienced increased seizure activity (Geffrey et al. 2015). The discrepancy in efficacy could be due to varying etiologies of epilepsy, since the mechanism of CBD is still unknown.

As all completed clinical trials have tested CBD with concurrent AED medications, the clinical relevance of drug-drug interactions is extremely important. Both CBD and THC inhibit hepatic enzymes CYP2C19 at low levels, and CYP3A4 at high levels. These enzymes are induced by carbamazepine, topiramate, and phenytoin, and are inhibited by sodium valproate, and are responsible for the metabolism of many AEDs (Gaston and Friedman 2009). An open label CBD study found drug-drug interactions with several AEDs taken by both children and adults, including increases in levels of rufinamide, topiramate, zonisamide, eslicarbazepine, clobazam, and N-desmethylclobazam (Devinsky et al. 2018a, 2018b; Jiang et al. 2013). The only significant interactions observed were with clobazam and N-desmethylclobazam, which increased to levels above the therapeutic range. Clinically significant drug interactions were observed with CBD and clobazam and N-desmethylclobazam, which increased to levels above therapeutic range. Such actions of CBD have raised questions on whether the observed antiseizure effects and side effects of CBD therapy were related to a direct action of CBD, or were due to CBD's indirect effects on elevation of N-desmethylclobazam levels when coadministered with clobazam, a antiseizure medication for DS. In a recent case report from five patients receiving adjunctive treatment with CBD exhibited increases in brivaracetam levels by 95% to 280% (Klotz et al. 2019). Another clinically significant drug interaction between BCD and tacrolimus was reported (Leino et al. 2019). Therefore, it is advised that when taking these medications concurrently with CBD, dosing may be altered to reduce risk of interactions and serious adverse effects.

7.8 Conclusions and Future Perspectives

Among epileptic patients, 30% remain untreated with currently available medications. Recent clinical trials and experimental research have brought new insights to the potential use of cannabinoids for the treatment of epileptic seizures. CBD, the active ingredient in Epidiolex, is a cannabinoid that naturally occurs in the *Cannabis sativa* plant. Preclinical models have demonstrated that CBD not only possesses anticonvulsant properties but may also be able to disrupt maladaptive processes associated with epileptogenesis. CBD has been shown to act as an antioxidant, reduce proinflammatory cytokines, rescue neurogenesis, and ameliorate neurodegeneration. Furthermore, CBD can modulate neuropsychiatric disorders such as anxiety and depression, which are often seen as comorbidities in epileptic patients. Clinical efforts from several well-designed trials of a plant-based CBD (Epidiolex, Greenwich Biosciences, Carlsbad, CA, USA) have demonstrated the antiseizure efficacy in patients >2 years old, especially when taken in tandem with currently prescribed AEDs (Elliott et al. 2019). Patients exhibited a favorable adverse effect profile. However, CBD therapy has associated adverse effects, with somnolence, decreased appetite, and diarrhea being the most common. Moreover, there is some concern for drug-drug interactions, specifically with clobazam, topiramate, rufinamide, phenytoin, clonazepam, and carbamazepine. Similar beneficial outcomes have been reported by an Israeli clinical study with CBD-enriched medical cannabis in a population of children and adolescents (Tzadok et al. 2016). Despite such overwhelming therapeutic claims, the molecular basis of CBD therapy for epilepsy remains unclear. It does not appear to exert its anticonvulsant effects through interaction with cannabinoid receptors. It is critical to know how CBD controls seizures so chemists can design novel synthetic compounds for epilepsy to surpass the hurdles of mixed CBD extracts such as extraction, purification, and standardization. In addition, it remains to be determined if plant-based vs synthetic CBDs are identical in terms of pharmacological outcomes, both in effectiveness and side effects. Recently, a synthetic, non-intoxicating 8,9-dihydrocannabidiol (H2CBD) has been prepared and demonstrated to have effectiveness comparable to CBD both for decreasing the frequency and severity of experimental seizures in rats (Mascal et al. 2019). It is claimed that H2CBD cannot be converted by any reasonable synthetic route into THC, and thus could be a safe, noncontroversial drug for seizure therapy.

Patient responsiveness to cannabis-enriched therapeutics varies substantially, and in some cases has been suggested to even exacerbate seizures. This inconsistency in therapeutic responses could be contributed to qualitative and quantitative chemical variability in medical products, individual differences in the etiology of seizures between patients, or even in the pathological reorganization of epileptic circuits between forms of epilepsy. Therefore, the consensus among neurologists is to first clinically test FDA-approved formulations of cannabis products in specific epilepsy syndromes. However, marketing unapproved hemp or CBD-containing products with uncertain formulations is often seen in some dispensaries and wellness aisles of grocery stores. These should not be considered substitutes or generics for FDA-approved medicines. These false advertisements can keep patients from accessing appropriate and recognized therapies (such as Epidiolex) to treat serious, and in certain cases, fatal diseases. The rigorous FDA-approval process is bypassed for such dispensary products. FDA-approved CBD products are available by prescription in both specialty and retail pharmacies, but not dispensaries. With adequate and well-controlled clinical studies, physicians will have more confidence in the drug's potency and efficacy to support appropriate dosing in a variety of patients. To this end, the FDA approved the first CBD-based medication, Epidiolex, for two specific forms of severe pediatric epilepsy: Lennox-Gastaut and Dravet syndromes. There are ongoing clinical trials to expand upon the efficacy, safety, and dosing for other epilepsies, as well as different age groups. CBD may impact a variety of brain conditions (Pisanti et al. 2017); further research is warranted to profile the full spectrum of CBD pharmacology.

Cannabis and all its cannabinoids are controlled substances and regulated by the Drug Enforcement Administration. While most cannabinoids are classified as Schedule I (banned substances), the regulatory statutes are rapidly changing with recent legislation. As of October 30, 2018, Hemp, defined as a cannabis plant containing $<0.3\%$ THC has been descheduled. Thus, on a federal account, the Agicultural Improvement Act (also called the Farm bill) extended the protections of Hemp research to include plant products including Hemp-derived CBD. This provision recognizes the importance of experimental research to discover medicinal uses and mechanistic pathways of cannabinoids for epileptic disorders and other conditions. Presently, four products that are approved by the FDA are de-classified from this schedule. The clinically available THC and THC analogs are listed in Schedule II/III and the plant-derived CBD (Epidiolex) is listed in Schedule V. Recently, a synthetic CBD (H2CBD) has been prepared and tested in experimental seizure models (Mascal et al. 2019). It is claimed that the synthetic CBD alternative is easier to purify than a plant extract, eliminates the need to use agricultural land for hemp cultivation, and could avoid legal complications with cannabis-related products. The synthetic H2CBD and CBD were found to be equally effective for the reduction of both the frequency and severity of seizures. Unlike the plant-derived CBD, H2CBD cannot be converted by synthetic route into THC, and thus has fewer regulatory hurdles.

Conflict of Interest
None.

References

Ames FR, Cridland S (1986) Anticonvulsant effect of cannabidiol. S Afr Med J 69:14–18

Autry AR, Trevathan E, Van Naarden Braun K, Yeargin-Allsopp M (2010) Increased risk of death among children with Lennox-Gastaut syndrome and infantile spasms. J Child Neurol 25(4):441–447

Bessa JM, Ferreira D, Melo I, Marques F, Cerqueira JJ, Palha JA, Almeida OF, Sousa N (2009) The mood improving actions of antidepressants do not depend on neurogenesis but are associated with neuronal remodeling. Mol Psychiatry 14(8):764–773

Billakota S, Devinsky O, Marsh E (2019) Cannabinoid therapy in epilepsy. Curr Opin Neurol 32(2):220–226

Birnbaum AK, Karanam A, Marino SE, Barkley CM, Remmel RP, Roslawski M, Gramling-Aden M, Leppik IE (2019) Food effect on pharmacokinetics of cannabidiol oral capsules in adult patients with refractory epilepsy. Epilepsia 60(8):1586–1592

Bourgeois BF, Douglass LM, Sankar R (2014) Lennox-Gastaut syndrome: a consensus approach to differential diagnosis. Epilepsia 55(Suppl 4):4–9

Campos AC, Ferreira FR, Guimaraes FS (2012a) Cannabidiol block long-lasting behavioral consequences of predator threat stress: possible involvement of 5HT1A receptors. J Psychiatr Res 46:1501–1510

Campos AC, Fogaca MV, Scarante FF, Joca SRL, Sales AJ, Gomes FV, Sonego AV, Rodrigues NS, Galve-Roperh I, Guimaraes FS (2017) Plastic and neuroprotective mechanisms involved in the therapeutic effects of cannabidiol in psychiatric disorders. Front Pharmacol 8:269

Campos AC, Moreira FA, Gomes FV, Del Bel EA, Guimaraes FS (2012b) Multiple mechanisms involved in the large-spectrum therapeutic potential of cannabidiol in psychiatric disorders. Philos Trans R Soc Lond Ser B Biol Sci 367:3364–3378

Carney PR, Evans V, Febo M, DeMarse T, Johnson CR, Anderson C (2017) An evaluation of effectivenss of cannabidiol as an antiepileptic drug for children with intractable generalized epilepsy. In: Abstract 1.048. American Epilepsy Society, Washington, D.C

Castillo A, Tolon MR, Fernandez-Ruiz J, Romero J, Martinez-Orgado J (2010) The neuroprotective effect of cannabidiol in an in vitro model of newborn hyper-ischemic brain damage in mice is mediated by CB2 and adenosine receptors. Neurobiol Dis 37(2):434–440

Ceulemans B, Boel M, Leyssens K, Van Rossem C, Neels P, Jorens PG et al (2012) Successful use of fenfluramine as an add-on treatment for Dravet syndrome. Epilepsia 53:1131–1139

Cilio MR, Wheless JW, Dlugos D, Parikh N, Miller I (2018) Long-term safety of pharmaceutical cannabidiol oral solution as adjunctive treatment for pediatric patitents with treatment-resistant epilepsy. In: Abstract 3.294. American Epilepsy Society, New Orleans, LA

Clossen BL, Reddy DS (2017) Novel therapeutic approaches for disease-modification of epileptogenesis for curing epilepsy. Biochim Biophys Acta Mol basis Dis 1863(6):1519–1538

Consroe P, Benedito MA, Leite JR, Carlini EA, Mechoulam R (1982) Effects of cannabidiol on behavioral seizures caused by convulsants drugs or current in mice. Eur J Pharmacol 83:293–298

Consroe PF, Wood GC, Buchsbaum H (1975) Anticonvuldant nature of marihuana smoking. JAMA 234:306–307

Crippa JA, Derenusson GN, Ferrari TB, Wichert-Ana L, Duran FL, Martin-Santos R, Simões MV, Bhattacharyya S, Fusar-Poli P, Atakan Z, Santos Filho A, Freitas-Ferrari MC, McGuire PK, Zuardi AW, Busatto GF, Hallak JE (2011) Neural basis of anxiolytic effects of cannabidiol (CBD) in generalized social anxiety disorder: a preliminary report. J Psychopharmacol 25(1):121–130

Crippa JA, Guimaraes FS, Campos AC, Zuardi AW (2018) Translational investigation of the therapeutic potential of cannabidiol: Toward a new age. Front Immunol 9:2009. https://doi.org/10.3389/fimmu.2018.02009

Cunha JM, Carlini EA, Pereira AE, Ramos OL, Pimentel C, Gagliardi R, Sanvito WL, Lander N, Mechoulam R (1980) Chronic administration of cannabidiol to healthy volunteers and epileptic patients. Pharmacol 21:175–185

Da Silva VK, de Freitas BS, da Silva Dornelles A, Nery LR, Falavigna L, Ferreira RD, Bogo MR, Hallak JE, Zuardi AW, Crippa JA, Schroder N (2014) Cannabidiol normalizes caspase 3, synaptophysin, and mitochondrial fission protein DNM1L expression levels in rats with brain iron overload: implications for neuroprotection. Mol Neurobiol 49(1):222–233

Danzer SC (2019) Adult neurogenesis in the development of epilepsy. Epilepsy Curr 19(5):316–320

Deianna S, Watanabe A, Yamasaki Y, Amada N, Arthur M, Fleming S et al (2012) Plasma and brain pharmacokinetic profile of cannabidiol (CBD), cannabidivarine (CBDV), delta-9-tetrahydrocannabivarin (THCV) and cannabigerol (CBG) in rats and mice following oral and intraperitoneal administration and CBD action on obsessive-compulsive behavior. Psychopharmacology 219(3):859–873

Devinksy O, Cross JH, Laux L, Marsh E, Miller I, Nabbout R, Scheffer IE, Thiele EA, Wright S, Cannabidiol in Dravet Syndrome Study Group (2017) Trial of Cannabidiol for drug-resistant seizures in the Dravet Syndrome. N Engl J Med 376(21):2011–2020

Devinsky O, Cilio MR, Cross H, Fernandez-Ruiz J, French J, Hill C, Di Marzo V, Jutras-Aswad D, Notcutt WG, Martinez-Orgado J, Robson PJ, Rohrback BG, Thiele E, Whalley B, Friedman D (2014b) Cannabidiol: Pharmacology and potential therapeutic role in epilepsy and other neuropsychiatric disorders. Epilepsia 55:791–802

Detyniecki K, Hirsch L (2015) Marijuana use in epilepsy: The myth and the reality. Curr Neurol Rep 15, 65

Devinsky O, Nabbout R, Miller I, Laux L, Zolnowska M, Wright S, Roberts C (2019) Long-term cannabidiol treatment in patients with Dravet syndrome: An open-label extension trial. Epilepsia 60(2):294–302

Devinsky O, Marsh E, Friedman D et al (2016) Cannabidiol in patients with treatment-resistant epilepsy: an open-label interventional trial. Lancet Neurol 15:270–278

Devinsky O, Patel AD, Cross JH, Villanueva V, Wirrell EC, Privitera M, Greenwood SM, Roberts C, Checketts D, Van Landingham KE, Zuberi SM, GWCARE3 Study Group (2018b) Effect of cannabidiol on drop seizures in the Lennox-Gastaut syndrome. N Engl J Med 378(20):1888–1897

Devinsky O, Patel AD, Thiele EA et al (2018a) Randomized, dose-ranging safety trial of cannabidiol in Dravet syndrome. Neurology 90:e1204–e1211

Devinsky O, Sullivan J, Friedman D, Thiele E, Marsh E, Laux L, Hedlund J, Tilton N, Bluvstein J, Cilio M (2014a. Seattle, WA) Efficacy and safety of Epidiolex (cannabidiol) in children and young adults with treatment-resistent epilepsy: Initial data from an expanded access program. In: Proceedings of the American Epilepsy Society Annual Meeting. American Epilepsy Society, Chicago, IL

Drake A, Chapman KE, Bear JJ (2018) Alterations in sleep architecture with CBD use. In: Abstract 2.269. American Epilepsy Society, New Orleans LA

Dravet C, Bureau M, Oguni H, Fukuyama Y, Cokar O (2005) Severe myoclonic epilepsy in infancy: Dravet syndrome. Adv Neurol 95:71–102

Elliott J, DeJean D, Clifford T, Coyle D, Potter BK, Skidmore B, Alexander C, Repetski AE, Shukla V, McCoy B, Wells GA (2019) Cannabis-based products for pediatric epilepsy: A systematic review. Epilepsia 60(1):6–19

ElSohly M, Gul W (2014) Constituents of Cannabis sativa. In: Pertwee RG (ed) Handbook of Cannabis. Oxford University Press, Oxford, UK, pp 3–22

Esposito G, Scuderi C, Valenza M, Togna GI, Latina V, De Filippis D, Cipriano M, Carratu MR, Luvone T, Sterado L (2011) Cannabidiol reduces Aβ-induced neuroinflammation and promotes hippocampal neurogenesis through PPARγ involvement. PLoS One 6(12):e28668

Friedman D, Cilio MR, Tilton N, Sullivan J, Hedlund J, Rosenburg E, Bluvstein J, Devinksy O (2014. 5-9; Seattle, WA. Abst. 2.309) The effect of Epidiolex (cannabidiol) on serum levels of concomitant anti-epileptic drugs in children and young adults with treatment-resistant epilepsy in an expanded access program. In: Proceedings of the American Epilepsy Society Annual Meeting. American Epilepsy Society, Chicago, IL

Friedman D, Devinsky O (2015) Cannabinoids in the treatment of epilepsy. N Engl J Med 373:1048–1458

Fujiwara T (2006) Clinical spectrum of mutations in SCN1A gene: severe myoclonic epilepsy in infancy and related epilepsies. Epilepsy Res 70(Suppl 1): S223–S230

Galiegue S, Mary S, Marchand J, Dussossoy D, Carriere D, Carayon P et al (1995) Expression of central and peripheral cannabinoid recepors in human immune tissues and leukocyte subpopulations. Eur J Biochem 232:54–61

Gaston TE, Friedman D (2009) Pharmacology of cannabinoids in the treatment of epilepsy. Epilepsy Behav 15(2):S3–S4. https://doi.org/10.1016/j.yebeh.2009.04.027

Gataullina S, Dulac O (2017) From genotype to phenotype in Dravet disease. Seizure 44:58–64

Gedde MM, and Maa E (2013) Whole cannabis extract of high concentration cannabidiol may calm seizures in highly refractory pediatric epilepsies. In: American Epilepsy Society annual meeting, Seattle, WA. Abst. 3.330

Geffrey AL, Pollack SF, Bruno PL, Thiele EA (2015) Drug-drug interaction between clobazam and cannabidiol in children with refractory epilepsy. Epilepsia 56:1246–1251

Gifford AN, Bruneus M, Gatley SJ, Volkow ND (2000) Cannabinoid receptor-mediated inhibition of acetylcholine release from hippocampal and cortical synaptosomes. British J Pharm 131(1):645–650

Glass M, Dragunow M, Faull RL (1997) Cannabinoid receptors in the human brain: A detailed anatomical and quantitative autoradiographic study in the fetal, neonatal, and adult human brain. Neurosci 77 (2):299–318

Gloss D, Vickrey B (2014) Cannabinoids for epilepsy. Cochrane Database Syst Rev 6(3):CD009270

Gowers W (1881) Epilepsy and other chronic convulsive disorders. Churchill, London

Granjeiro EM, Gomes FV, Guimaraes FS, Correa FM, Resstel LB (2011) Effects of intracisternal administration of cannabidiol on the cardiovascular and behavioral responses to acute restraint stress. Pharmacol Bicochem Behav 99(4):743–748

Grotenhermen F (2006) Cannabinoids and the endocannabinoid system. Can Underwrit 1:10–14

Guimaraes FS, Chiaretti TM, Graeff FG, Zuardi AW (1990) Antianxiety effect of cannabidiol in the elevated plus maze. Psychopharm 100(4):558–559

Hampson AJ, Grimaldi M, Axelrod J, Wink D (1998) Cannabidiol and (−)Delta9-tetrahydrocannabidiol are neuroprotective antioxidants. Proc Natl Acad Sci U S A 95(14):826–8273

Hattiangady B, Rao MS, Shetty AK (2004) Chronic temporal lobe epilepsy is associated with severely declined dentate neurogenesis in the adult hippocampus. Neurobiol Dis 17:473–490

Hernando K, Bebin EM, Gaston TE, Grayson L, Moreadith R, Szaflarski JP. (2018) Higher cannabidiol levels are associated with better seizure response following treatment with a pharmaceutical grade cannabidiol. *Abstract 1.468, American Epilepsy Society, New Orleans, LA*

Hofmann ME, Frazier CJ (2013) Marijuana, endocannabinoids, and epilepsy: Potential and challenges for improved therapeutic intervention. Exper Neurol 244:43–50

Hosseinzadeh M, Nikseresht S, Kohdagholi F, Naderi N, Maghsoudi N (2016) Cannabidiol post-treatment alleviates rat epileptic-related behaviors and activates hippocampal cell autophagy pathway along with antioxidant defense in chronic phase of pilocarpine-induced seizure. J Mol Neurosci 58(4):432–440

Howlett AC, Qualy JM, Khachatrian LL (1986) Involvement of gi in the inhibition of denylate cyclase by cannbimimetic drugs. Mol Pharmacol 29(3):307–313

Huestis MA (2007) Human cannabinoid pharmacokinetics. Chem Biodivers 4(8):1770–1804

Huizenga MN, Sepulveda-Rodriguez A, Forcelli PA (2019) Preclinical safety and efficacy of cannabidivarin for early life seizures. Neuropharmacology 148:189–198

Irving AJ, Coutts AA, Harvey J, Rae MG, Mackie K, Bewick GS, Pertwee RG (2000) Functional expression of cell surface cannabinoid CB1 receoptros on presynaptic inhibitiory terminals in cultured rat hippocampal neurons. Neurosci 98(2):253–262

Jiang R, Yamaori S, Okamoto Y et al (2013) Cannabidiol is a potent inhibition of the catalytic activity of cytochrome P450 2C19. Drug Metab Pharmacokinetic 28:332–338

Kaplan JS, Stella N, Catterall WA, Westenbroek RE (2017) Cannabidiol attenuates seizures and social deficits in mouse model of Dravet syndrome. Proc Natl Acad Sci USA 114(42):11229–11234

Katona I, Sperlagh B, Magloczky A, Santha E, Kofalvi A et al (2000) GABAergic interneurons are the targets of cannabinoid actions in the human hippocampus. Neurosci. 100(4):797–804

Kawamura Y, Fukaya M, Maejima T, Yoshida T, Miura E, Watanabe M et al (2006) The CB1 cannabinoid receptor is the major cannabinoid receptor at excitatory presynaptic sites in the hippocampus and cerebellum. Neurosci 26(11):2991–3001

Klotz KA, Hirsch M, Heers M, Schulze-Bonhage A, Jacobs J (2019) Effects of cannabidiol on brivaracetam plasma levels. Epilepsia 60(7):e74–e77

Laux LC, Bebin EM, Checketts D, Chez M, Flamini R, Marsh ED, Miller I, Nichol K, Park Y, Segal E, Seltzer L, Szaflarski JP, Thiele EA, Weinstock A; CBD EAP study group (2019) Long-term safety and efficacy of cannabidiol in children and adults with treatment resistant Lennox-Gastaut syndrome or Dravet syndrome: Expanded access program results. Epilepsy Res 154:13–20

Laprairie RB, Bagher AM, Kelly ME, Denovan-Wright EM (2015) Cannabidiol is a negative allosteric modulator of the cannabinoid CB1 receptor. Br J Pharmacol 172:4790–4805

Leino AD, Emoto C, Fukuda T, Privitera M, Vinks AA, Alloway RR (2019) Evidence of a clinically significant drug-drug interaction between cannabidiol and tacrolimus. Am J Transplant 19(10):2944–2948

Linge R, Jimenez-Sanchez L, Campa L, Pilar-Cuellar F, Vidal R, Pazos A, Adell A, Diaz A (2016) Cannabidiol induces rapid-acting antidepressant-like effects and enhances cortical 5-HT/glutamate neurotransmission: role of 5-HT1A receptors. Neuropharmacology 103:16–26

Lucassen PJ, Meerlo P, Naylor AS, van Dam AM, Dayer AG, Fuchs E, Oomen CA, Czeh B (2010) Regulation of adult neurogenesis by stress, sleep disruption, exercise and inflammation: Implications for depression and antidepressant action. Eur Neuropsychopharmacol 20 (1):1–17

Maa E, Figi P (2014) The case for medical marijuana in epilepsy. Epilepsia 55:783–786

Martin P, Consroe P (1976) Cannabinoid induced behavioral convulsions in rabbits. Science 194 (4268):965–967

Mascal M, Hafezi N, Wang D, Hu Y, Serra G, Dallas ML, Spencer JPE (2019) Synthetic, non-intoxicating 8,9-dihydrocannabidiol for the mitigation of seizures. Sci Rep 9(1):7778

Mechoulam R, Carlini EA (1978) Toward drugs derived from cannabis. Naturwissenschaften 65:174–179

Mohpael P, Ekdahl CT, Lindvall O (2004) Status epilepticus severity influences the long-term outcome of neurogenesis in the adult dentate gyrus. Neurobiol Dis 15:196–205

O'Brien TJ, Berkovic SF, French J, Messenheimer J, Gutterman D. (2018) Transdermal cannabidiol (CBD) gel for the treatment of focal epilepsy in adults. *Abstract 2.253, American Epilepsy Society, New Orleans, LA*

O'Connell BK, Gloss D, Devinsky O (2017) Cannabinoids in treatment-resistant epilepsy: A review. Epilepsy Behav 70(Pt B):341–348

Ohlsson A, Lindgren JE, Andersson S, Agurell S, Gillespie H, Hollister LE (1986) Single-dose kinetics of deuterium-labelled cannabidiol in man after smoking ad intravenous administration. Biomed Environ Mass Spectrom 13(2):77–83

Oldham M, Sullivan J, Singhal N, Tilton N, and Cilio M (2015) Long-term efficacy and tolerability of add-on cannabidiol for drug-resistant pediatric epilepsies, in: Proceedings of the American Epilepsy Society Annual Meeting; 2015 Dec. 4–8; Philadelphia, PA. Abst. 2.296, American Epilepsy Society, Chicago, IL

Parent JM, Yu TW, Leibowitz RT, Geschwind DH, Sloviter RS, Lowenstein DH (1997) Dentate granule cell neurogenesis is increased by seizures and contributes to aberrant network reorganization in the adult rat hippocampus. J Neurosci 17:3727–3738

Patel RR, Barbosa C, Brustovetsky T, Brustovetsky N (2016) & Cummins, T.R. Aberrant epilepsy-associated mutant Nav1.6 sodium channel activity can be targeted with cannabidiol. Brain 139:2164–2181

Patra PH, Barker-Haliski M, White HS, Whalley BJ, Glyn S, Sandhu H, Jones N, Bazelot M, Williams CM, McNeish AJ (2019) Cannabidiol reduces seizures and associated behavioral comorbidities in a range of animal seizure and epilepsy models. Epilepsia 60 (2):303–314

Pertwee RG, Cascio MG (2014) Known pharmacological actions of delta-9-tetrahydrocannabinol and of four other chemical constituents of cannabis that activate cannabinoid receptors. In: Pertwee RG (ed) Handbook of cannabis. Oxford University Press, Oxford, UK

Pisanti S, Malfitano AM, Ciaglia E, Lamberti A, Ranieri R, Cuomo G, Abate M, Faggiana G, Proto MC, Fiore D, Laezza C, Bifulco M (2017) Cannabidiol: State of the art and new challenges for therapeutic applications. Pharmacol Ther 175:133–150

Porter BE, Jacobson C (2013) Report of a parent survey of cannabidiol-enriched cannabis use in pediatric treatment-resistant epilepsy. Epilepsy Behav 29:574–577

Rajesh M, Mukhopadhyay P, Batkai S, Hasko G, Liaudet L, Drel VR, Obrosova IG, Pacher P (2007) Cannabidiol attenuates high glucose-induced endothelial cell inflammatory response and barrier disruption. Am J Physiol Heart Circ Physiol 293:H610–H619

Reddy DS, Golub V (2016) The pharmacological basis of cannabis therapy for epilepsy. J Pharmacol Exp Therap 357:45–55

Reddy DS, Kuruba R (2013) Experimental models of status epilepticus and neuronal injury for evaluation of therapeutic interventions. Int J Mol Sci 14 (9):18284–18318

Resstel LB, Joca SR, Moreira FA, Correa FM, Guimaraes FS (2006) Effects of cannabidiol and diazepam on behavioral and cardiovascular responses induced by contextual conditioned fear in rats. Behav Brain Res 172:294–298

Reynolds JR (1861) Epilepsy: its symptoms, treatment, and relation to other chronic convulsive diseases. UK. John Churchill, London

Rosenberg EC, Patra PH, Whalley BJ (2017) Therapeutic effects of cannabinoids in animal models of seizures, epilepsy, epileptogenesis, and epilepsy-related neuroprotection. Epilepsy Behav 70(Pt B):319–327

Rosenberg E, Tsien C, Richard W, Whalley BJ, Devinksy O (2015) Cannabinoids and epilepsy. Neurotherapeutics 12(4):747–768

Ross HR, Napier I, Conner M (2008) Inhibition of recombinant human T-type calcium channels by delta9-tetrahydrocannabinol and cannabidiol. J Bio Chem 283:16124–16134

Rubinstein M, Han S, Tai C, Westenbroek RE, Hunker A, Scheuer T, Catterall WA (2015) Dissecting the phenotypes of Dravet syndrome by gene deletion. Brain 138(8):2219–2233

Schmidt D, Bourgeois B (2000) A risk-benefit assessment of therapies for Lennox-Gastaut syndrome. Drug Saf 22(6):467–477

Sebree T, O'Neill C, Messenheimer J, Gutterman D. (2016) Safety and tolerability of ZYN002 (synthetic cannabidiol) transdermal gel in healthy subjects: two phase 1, randomized, double-blind, placebo-controlled studies. *Abstract 2.214, American Epilepsy Society, Houston,* TX

Shen M, Piser TM, Seybold VS, Thayer SA (1996) Cannabinoid receptor agonists inhibit glutamatergic synaptic transmission in rat hippocampal cultures. Neurosci. 16(14):4322–4334

Silveira Filho NG, Tufik S (1981) Comparative effects between cannabidiol and diazepam on neophobia, food intake, and conflict behavior. Res Commun Psychol Psychiatry Behav 6:25–26

Solinas M, Massi P, Cinquina V, Valentia M, Bolognini D, Garboldi M, Monti E, Rubino T, Parolaro D (2013) Cannabidiol, a non-psychoactive cannabinoid compound, inhibits proliferation and invasion in U87-MG and T98G glioma cells through a multitarget effect. PLoS One 8(10):e76918

Stadnicki SW, Schaeppi U, Rosenkrantz H, Braude MC (1974) Detla9-tetrahydrocannabidiol: subcortical spike bursts and motor manifestations in a Fischer rat treated orally for 109 days

Steel D, Symonds JD, Zuberi SM, Brunklaus A (2017) Dravet syndrome and its mimics: beyond SCN1A. Epilepsia 58:1807–1816

Stempel AV, Stumpf A, Zhang HY, Ozdogan T, Pannasch U, Theis AK et al (2016) Cannabinoid type 2 receptors mediate a cell type-specific plasticity in the hippocampus. Neuron 90:795–809

Stern CA, Gazarini L, Vanvossen AC, Zuardi AW, Galve-Roperh I, Guimaraes FS et al (2015) Delta9-tetrahydrocannabinol alone and combined with cannabidiol mitigate fear memory through reconsolidation disruption. Eur Neuropsychopharmacol 25:958–965

Straiker A, Dvorakova M, Zimmowitch A, Mackie K (2018) Cannabidiol inhibits endocannabinoid signaling in autaptic hippocampal neurons. Mol Pharmacol 94(1):743–748

Sun S, Hu F, Wu J, Zhang S (2017) Cannabidiol attenuates OGD/R-induced damage by enhancing mitochondrial bioenergetics and modulating glucose metabolism via pentose-phosphate pathway in hippocampal neurons. Redox Biol 11:577–585

Szaflarski JP, Bebin EM, Comi AM et al (2018a) CBD EAP Study Group. Long-term safety and treatment effects of cannabidiol in children and adults with treatment-resistant epilepsies: expanded access program results. Epilepsia 59(8):1540–1548

Szaflarski JP, Bebin EM, Cutter G, DeWolfe J, Dure LS, Gaston TE, Kankirawatana P, Liu Y, Singh R, Standaert DG, Thomas AE, Ver Hoef LW, Program UC (2018b) Cannabidiol improves frequency and severity of seizures and reduces adverse events in an open-label add-on prospective study. Epilepsy Behav 87:131–136

Thiele EA, Marsh ED, French JA, Mazurkiewicz-Beldzinska M, Benbadis SR, Joshi C, Lyons PD, Taylor A, Roberts C, Sommerville K (2018) GWPCARE4 Study Group. Cannabidiol in patients with seizures associated with Lennox-Gastaut syndrome (GWCARE4): a randomized, double-blind, placebo-controlled phase 3 trial. Lancet 391 (10125):1085–1096

Taylor L, Gidal B, Blakey G, Tayo B, Morrison G (2018) A Phase I, Randomized, Double-Blind, Placebo-Controlled, Single Ascending Dose, Multiple Dose, and Food Effect Trial of the Safety, Tolerability and Pharmacokinetics of Highly Purified Cannabidiol in Healthy Subjects. CNS Drugs 32(11):1053–1067

Trembly B, Sherman M (1990. July 8-11) Double-blind clinical study of cannabidiol as a secondary anticonvulsant, presented at: *marijuana '90 International Conference on Cannabis and Cannabinoids.* International Association for Cannabinoid Medicines, Kolymbari, Crete

Tsou K, Brown S, Sanduo-pena M, Mackie K, Walker J (1998) Immunohistochemical distribution of cannabinoid CB1 receptors in the rat central nervous system. Neuroscience 83:393–411

Tzadok M, Uliel-Siboni S, Linder I, Kramer U, Epstein O, Menascu S, Nissenkorn A, Yosef OB, Hyman E, Granot D, Dor M, Lerman-Sagie T, Ben-Zeev B (2016) CBD-enriched medical cannabis for intractable pediatric epilepsy: the current Israeli experience. Seizure 35:41–44

Wallace MJ, Blair RE, Falenski KW, Martin BR, DeLorenzo RJ (2003) The endogenous cannabinoid system regulates seizure frequency and duration in a model of temporal lobe epilepsy. J Pharmacol Exp Ther 307:129–137

Wallace MJ, Martin BR, DeLorenzo RJ (2002) Evidence for a physiological rle of endocannabinoids in the modulation of seizure threshold and severity. Eur J Pharmacol 452:295–301

Wallace MJ, Wiley JL, Martin BR, DeLorenzo RJ (2001) Assessment of the role of CB1 receptors in cannabinoid anticonvulsant effects. Eur J Pharmacol 428:51–57

Wolf SA, Bick-Sander A, Fabel K, Leal-Galicia P, Tauber S, Ramirez-Rodriquez G, Muller A, Melnik A, Waltinger TP, Ullrich O, Kempermann G (2010) Cannabinoid receptor CB1 mediates baseline and activity-induced survival of new neurons in adult hippocampal neurogenesis. Cell Commun Signal 8:12

Younus I, Reddy DS (2017) Epigenetic interventions for epileptogenesis: a new frontier for curing epilepsy. Pharmacol Ther 177:108–122

Zanelati TV, Biojone C, Moreira FA, Guimaraes FS, Joca SR (2010) Antidepressant-like effects of cannabidiol in mice: possible involvement of 5-HT1A receptors. Br J Pharmcol 159:122–128

Zuardi AW, Karniol IG (1983) Changes in the conditioned emotional response of rats induced by 9-THC, CBD and mixture of the two cannabinoids. Arq Biol Tecnol 26:391–397

Cannabinoid-Based Medicines and Multiple Sclerosis

8

Clementina Manera and Simone Bertini

Abstract

The emerging role of the endocannabinoid system (ECS) in the control of symptoms and disease progression in multiple sclerosis (MS) has been highlighted by recent studies. MS is a chronic, immune-mediated, and demyelinating disorder of the central nervous system with no cure so far. It is widely reported that cannabinoids might be used to control MS symptoms and that they also might exert neuroprotective effects and slow down disease progression. The aim of this chapter is to give an overview of the main endogenous and synthetic cannabinoids used for the symptomatic amelioration of MS and their beneficial outcomes, providing new possible perspectives for the treatment of this disease.

Keywords

Multiple sclerosis · Endocannabinoid system · Cannabinoid receptors · Monoacylglycerol lipase (MAGL) inhibitors · Fatty acid amide hydrolase (FAAH) inhibitors · Arachidonoylethanolamine (AEA) reuptake inhibitors

8.1 Introduction

Multiple sclerosis (MS) is an important neurological disease that affects the central nervous system (CNS). It is the most common neurological disorder in young adults and affects approximately 2.3 million people worldwide (Browne et al. 2014). MS is more common in women than in men (Koch-Henriksen and Sørensen 2010; Koch-Henriksen et al. 2018), with a prevalence ratio of 3:1 (Dunn et al. 2015b; Dunn et al. 2015a). Regarding its etiology, it is now widely accepted that genetic and environmental factors may contribute to the onset and development of the disease (Hafler et al. 2007; Huynh and Casaccia 2013). MS is a chronic inflammatory immune-mediated condition characterized by demyelination of the axons in the CNS. It gradually leads to progressive neurodegeneration that damages CNS myelin, leading to neuronal dysfunction and a broad spectrum of neurological symptoms that depend upon the site where lesions have occurred in the brain and spinal cord. The symptoms of MS include spasticity, sensory alterations, weakness, painful spasms, bladder dysfunction, tremor, ataxia, optic neuritis, fatigue, and dysphagia (Compston 2008).

Molecular mechanisms of MS progression remain unclear. However, the observed hallmarks are considered as a consequence of three synergistically mechanisms: inflammation, demyelination, and axonal damage. Recent evidence indicates that MS is primarily a

This chapter is based in great parts on the Open Access article: Medicines (Basel) 2018, 5(3): 91. doi: 10.3390/medicines5030091.

C. Manera (✉) · S. Bertini
Department of Pharmacy, University of Pisa, Pisa, Italy
e-mail: clementina.manera@unipi.it

E. Murillo-Rodriguez et al. (eds.), *Cannabinoids and Neuropsychiatric Disorders*, Advances in Experimental Medicine and Biology 1264, https://doi.org/10.1007/978-3-030-57369-0_8

neurodegenerative disease that starts in the brain and then develops because of inflammation (Lassmann et al. 2012). This hypothesis has led to two models of MS immune-pathogenesis: the "inside-out" and "outside-in." In the first model, a dysfunction of brain cells causes the immune response that destroys myelin and leads to blood-brain barrier (BBB) breakdown. In the second model, a dysfunction of the periphery leads to BBB damage, myelin disruption, and axonal death (Stys et al. 2012). The subsequent high presence of lymphocytes in the CNS and the activation of innate immune cells (dendritic cell, macrophages, and microglia) play key roles in MS pathogenesis. The activation of autoimmune cells, resident microglia, astrocytes, and macrophages, results in an immunological storm that involves abundant secretion of reactive species, cytokines, chemokines, autoantibody production, and enhanced excitotoxicity. There is a continuing activation of resident microglia and astrocytes producing pro-inflammatory mediators that potentiate the neuroinflammatory response. This results in oligodendrocytes and axonal damage, and ultimately in demyelination, synaptic alteration, and neuronal loss (Compston 2008; Dutta and Trapp 2011; Calabrese et al. 2015; Mahad et al. 2015). In the early phases of MS, the oligodendrocytes generate new myelin, and this remyelination is one of the reasons why symptoms decrease or temporarily disappear in relapsing-remitting MS (RR-MS) (Peferoen et al. 2014), which is the most common form of MS (approximately 85–90% of all cases) (Compston 2008) and it is typified by unpredictable relapses with full recovery or with sequelae. However, the myelin sheaths are not completely rebuilt by oligodendrocytes, and repeated attacks lead to damage in the axons where scar-like plaques build up with subsequent axonal loss (Cambron et al. 2012), associated with the characteristic symptoms of MS (Polman et al. 2011).

Over the last 15 years, a great amount of preclinical studies has demonstrated that compounds targeting the endocannabinoid system (ECS) exert anti-inflammatory properties, neuroprotective and immunomodulatory effects (Chiurchiù et al. 2015b), allowing them to alleviate symptoms and to limit progressive neurodegeneration in animal models of MS (Chiurchiù et al. 2018).

Cannabinoids exert neuroprotective effects acting at multiple molecular sites that are in all key cellular elements for the control of neuronal survival (e.g., neurons, astrocytes, resting and reactive microglia, and oligodendrocytes) and also in key brain structures (e.g., BBB) (Fernández-Ruiz et al. 2015). These effects are due to activation of two G protein-coupled receptors, the type-1 (CB1R) and type-2 (CB2R) cannabinoid receptors.

CB1R is widely expressed within the CNS (cortical neurons and interneurons, oligodendrocytes, and astrocytes) and also in several leukocytes infiltrating the brain (Galve-Roperh et al. 2013). Initially, CB2R has been restricted exclusively to immune cells (macrophages, mast cells, and B and T lymphocytes) and immune organs (spleen, thymus, and lymph nodes) (Howlett et al. 2002). However, some evidence showed the expression of CB2R in microglia of the CNS (Klegeris et al. 2003), and more recently, it has been also reported to be expressed in brainstem neurons and astrocytes upon cellular activation by an insult or inflammation (Chiurchiù et al. 2015b; Atwood and Mackie 2010; Chiurchiù et al. 2015a).

The multiplicity of action of cannabinoids allows reducing the excitotoxicity by acting through neuronal CB1R, as well as the toxic influence of reactive microgliosis by acting through microglial CB2R, or enhancing the trophic and metabolic support to neurons by acting through astroglial CB1R and/or CB2R. In particular, the activation of CB1R provides neuroprotection regulating glutamate homeostasis (Docagne et al. 2007). In fact, it is well known that glutamate is a key mediator in neuronal and oligodendrocyte damage in MS (Musella et al. 2014), and CB1R agonists exert direct neuroprotective effects by limiting glutamate release and the excitotoxic damage characteristic of several neurodegenerative disorders (Fernández-Ruiz et al. 2010).

Furthermore, the protective effects of CB2R activation in microglial cells upon inflammatory-induced CNS damage have been demonstrated in preclinical models of MS (Fernández-Ruiz et al. 2010). Microglia may be in two activated states: M1 and M2. The classical M1 state is characterized by release of pro-inflammatory factors, i.e., interleukins (IL-1beta, IL-18, and IL-6), prostanoids, and inducible nitric oxide synthase (NOS2)-derived NO. On the other hand, the neuroprotective M2 state, known as "alternative activation", is associated with the release of anti-inflammatory factors, such as IL-10, IL-4, and NGF (Orihuela et al. 2016). Microglia has a functional endocannabinoid signaling system, composed of cannabinoid receptors and the complete machinery for the synthesis and degradation of endocannabinoids. The expression of cannabinoid receptors, mainly CB2R, and the production of endocannabinoids have been related to the activation profile of these cells (Mecha et al. 2016).

In preclinical studies, the beneficial effects of cannabinoids have been reported in different animal models of MS that are highly useful for studying different aspects of inflammation, demyelination, remyelination, and neurodegeneration in the CNS (Lassmann and Bradl 2017). Anyway, so far, none of the available models is able to cover the entire spectrum of clinical, immunological, and pathological features of the disease. For this reason, in order to have the complexity of MS fully replicated, multiple models should be used. The animal models more used are: – Experimental Autoimmune Encephalomyelitis (EAE), which is a useful animal model of MS since many of the pathologies observed in the CNS of mice with EAE show strong similarity to those found in the CNS of MS patients (McCarthy et al. 2012); – chronic relapsing experimental allergic encephalomyelitis (CREAE) is an animal model, which also presents relapsing-remitting paralytic episodes and tremor and spasticity of limb muscles during postrelapse remission strongly similar to those of MS (Heremans et al. 1996); – Theiler's murine encephalomyelitis virus (TMEV), which is an animal model characterized by inflammation and demyelination similar to those described in MS patients (Lassmann and Bradl 2017).

8.2 Medicinal Cannabinoids

The significant interest for introducing cannabinoid-based medicines into clinical practice for the treatment of MS substantially derived from anecdotal reports of MS patients that experienced symptomatic relief after recreational use of cannabis (i.e., smoking) (Clark et al. 2004). These pieces of evidence have indeed stimulated scientific research regarding the use of cannabinoids in this therapeutic field. Therefore, the efficacy of various cannabinoid preparations in symptomatic treatment of MS and other neurological disorders has been evaluated in several clinical studies in human patients (Fife et al. 2015). The medicinal grade cannabinoids that are licensed for MS treatment change from country to country; off-label use also varies widely (Fife et al. 2015; Johnson 2013; Gloss and Maa 2015).

The four licensed cannabinoid-based medicines currently available are Marinol®, Cesamet®, Sativex®, and Epidiolex®. The active principle of Marinol® is dronabinol, i.e., synthetic Δ^9-tetrahydrocannabinol (Δ^9-THC); Cesamet® is based on nabilone (a synthetic analog of Δ^9-THC); nabiximols, a standardized ~1:1 (w/w) mix of Δ^9-THC and cannabidiol (CBD), both extracted from *Cannabis sativa*, is the active principle of Sativex®; Epidiolex®, whose active principle is CBD, has been very recently approved for the treatment of seizures associated with Lennox-Gastaut syndrome or Dravet syndrome in patients of 2 years of age and older.

Other preparations based on natural cannabinoids contain essentially various quantitative ratios of Δ^9-THC and CBD, i.e., Bedrocan® (22% Δ^9-THC and < 1% CBD from *Cannabis sativa*), Bedrobinol® (13.5% Δ^9-THC and < 1% CBD from *Cannabis sativa*), Bediol® (6.5% Δ^9-THC and 8% CBD from *Cannabis sativa*), Bedica® (14% Δ^9-THC and < 1% CBD from *Cannabis indica*), Bedrolite® (<1% Δ^9-THC and 9% CBD from *Cannabis sativa*), and Bedropuur® (20%–24% Δ^9-THC and < 1% CBD from *Cannabis indica*).

The structures of the above-mentioned natural and synthetic cannabinoids that have been studied for the treatment of MS are shown in Table 8.1.

In the following paragraphs of this part, an overview of the current findings about dronabinol, nabilone, and nabiximols in the treatment of MS will be provided.

8.2.1 Dronabinol

The synthetic pure isomer (−)-trans-Δ^9-tetrahydrocannabinol (the main THC isomer found in the cannabis plant) is officially called "dronabinol". Its original indication was the treatment of chemotherapy-induced nausea and vomiting (CINV). Subsequently, its use has been extended to anorexia associated with weight loss in patients affected with AIDS. These indications are still retained.

Soft gelatin capsules with a range of three dosages (2.5, 5, and 10 mg) are the current pharmaceutical form of this drug.

Efficacy and safety of dronabinol in treating MS symptoms has been specifically evaluated in 10 clinical studies published between 1981 and 2013 (Killestein et al. 2002; Zajicek et al. 2003; Clifford 1983; Killestein et al. 2003; Svendsen et al. 2004; Zajicek et al. 2005; Petro and Ellenberger 1981; Ungerleider et al. 1987; Freeman et al. 2006; Zajicek et al. 2013); almost all of these studies are reported in 10 reviews published between 2003 and 2016 (Shakespeare et al. 2003; Whiting et al. 2015; Ben Amar 2006; Zhornitsky and Potvin 2012; Koppel et al. 2014; Karst et al. 2010; Jawahar et al. 2013; Mills et al. 2007; Andrzejewski et al. 2016; Wang et al. 2008) that have been included in a systematic review of reviews on the basis of eligibility criteria of methodological quality (AMSTAR Tool: A Measurement Tool to Assess systematic Reviews) (Nielsen et al. 2018). In particular, with the aim of providing an overview of the current findings about dronabinol, eight key clinical outcomes in MS have been considered: disability/disease progression, pain, spasticity, bladder function, ataxia/tremor, sleep, quality of life, and adverse effects (Nielsen et al. 2018). Table 8.1 shows a summary of clinical evidence for dronabinol.

Regarding the pain related to MS, positive results were found (Zajicek et al. 2003; Svendsen et al. 2004; Zajicek et al. 2005; Petro and Ellenberger 1981). In the assessment of ataxia and tremor, substantially no change was evidenced (Killestein et al. 2002; Zajicek et al. 2003; Clifford 1983); the same for disability and disease progression (Killestein et al. 2002; Zajicek et al. 2003; Clifford 1983; Killestein et al. 2003; Zajicek et al. 2013). Indeed, in some cases, negative effects have been detected (Killestein et al. 2002). In a noteworthy CUPID study (Zajicek et al. 2013), a large amount of data concerning treatment with dronabinol has been provided, showing that it has no overall effect on the progression of MS. Mixed findings, although mostly positive, were highlighted regarding the quality of sleep (Zajicek et al. 2003; Zajicek et al. 2005). For the rest of the clinical outcomes (spasticity (Killestein et al. 2002; Zajicek et al. 2003; Zajicek et al. 2005; Petro and Ellenberger 1981; Ungerleider et al. 1987), bladder function (Zajicek et al. 2003; Freeman et al. 2006), and quality of life (Killestein et al. 2002; Zajicek et al. 2003)), mixed findings were also reported. The main adverse effects reported for dronabinol are dizziness, euphoria, dry mouth, fatigue, and drowsiness (Killestein et al. 2002; Zajicek et al. 2003; Clifford 1983; Svendsen et al. 2004; Zajicek et al. 2005; Petro and Ellenberger 1981; Ungerleider et al. 1987; Freeman et al. 2006); however, these effects have been described more frequently as mild to moderate. With the exception of patients with pre-existing cognitive disfunctions, cognitive impairment associated with the use of dronabinol did not seem to be relevant (Killestein et al. 2002; Zajicek et al. 2003; Svendsen et al. 2004; Zajicek et al. 2005; Freeman et al. 2006).

On the basis of the above-mentioned results, the only significant clinical evidence regarding dronabinol relates to its ability to relieve pain associated with MS, while quite inconsistent

Table 8.1 Summary of clinical evidence of dronabinol, nabilone, and nabiximols[a]

	Dronabinol (synthetic Δ^9-THC)	Nabilone (synthetic analog of Δ^9-THC)	Nabiximols (Δ^9-THC:CBD ~ 1:1 $_{(w/w)}$)
Structure(s)	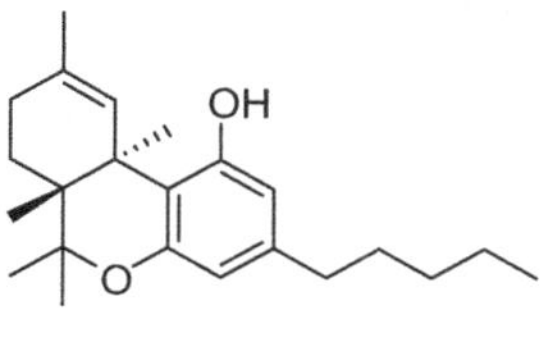	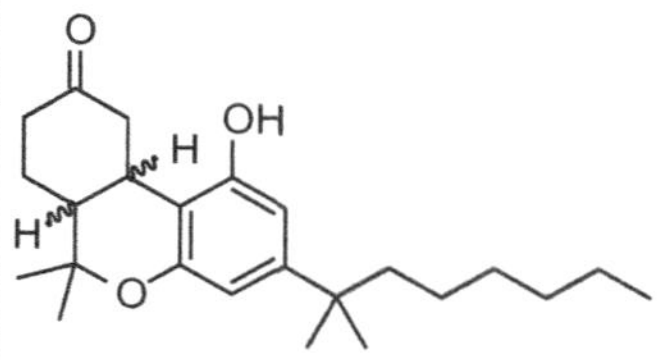	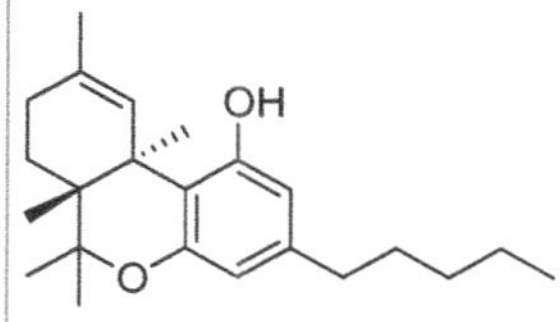 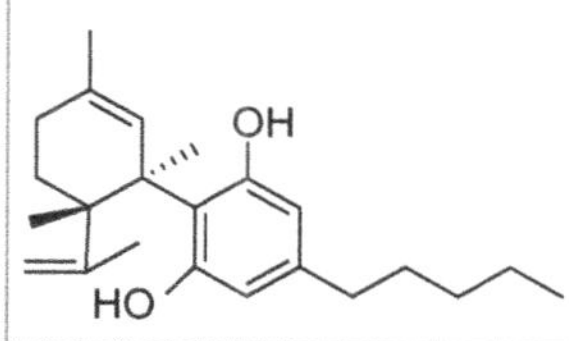
Formulation	Soft gelatin capsules (2.5, 5, and 10 mg)	Capsules (0.25, 0.5, and 1 mg)	Oro-mucosal spray (27 mg of Δ^9-THC and 25 mg of CBD / 1.0 mL)
Disability and disease progression	No evident changes	No studies	No evident changes
Pain	Positive effects	Positive effects	Mixed findings (mostly positive effects)
Spasticity	Mixed findings	Positive effects	Mixed findings (mostly positive effects)
Bladder function	Mixed findings	Positive effects	Mixed findings
Ataxia and tremor	No evident changes	No studies	No evident changes
Sleep	Mixed findings (mostly positive effects)	No studies	Positive effects
Quality of life	Mixed findings	Mixed findings (mostly positive effects)	Mixed findings
Adverse effects	Mild to moderate. Principally dizziness, euphoria, dry mouth, fatigue, and drowsiness	Moderate sedation, dizziness, and moderate weakness in the legs	Mild to moderate. Principally drowsiness, dizziness, headache, fatigue, impaired balance, and disturbance in attention
Num. Of studies	10	3	11
Num. Of reviews	11	5	12
Studies (references)	(Killestein et al. 2002; Zajicek et al. 2003; Clifford 1983; Killestein et al. 2003; Svendsen et al. 2004; Zajicek et al. 2005; Petro and Ellenberger 1981; Ungerleider et al. 1987; Freeman et al. 2006; Zajicek et al. 2013)	(Martyn et al. 1995; Wissel et al. 2006; Turcotte et al. 2015)	(Wade et al. 2004; Collin et al. 2010; Centonze et al. 2009; Rog et al. 2005; Wade et al. 2006; Conte et al. 2009; Rog et al. 2007; Langford et al. 2013; Collin et al. 2007; Leocani et al. 2014; Kavia et al. 2010)

(continued)

Table 8.1 (continued)

	Dronabinol (synthetic Δ^9-THC)	Nabilone (synthetic analog of Δ^9-THC)	Nabiximols (Δ^9-THC:CBD ~ 1:1 $_{(w/w)}$)
Reviews (references)	(Shakespeare et al. 2003; Whiting et al. 2015; Ben Amar 2006; Zhornitsky and Potvin 2012; Koppel et al. 2014; Karst et al. 2010; Jawahar et al. 2013; Mills et al. 2007; Andrzejewski et al. 2016; Wang et al. 2008; Nielsen et al. 2018)	(Whiting et al. 2015; Ben Amar 2006; Koppel et al. 2014; Karst et al. 2010; Nielsen et al. 2018)	(Whiting et al. 2015; Ben Amar 2006; Koppel et al. 2014; Karst et al. 2010; Jawahar et al. 2013; Mills et al. 2007; Andrzejewski et al. 2016; Wang et al. 2008; Nielsen et al. 2018; Martyn et al. 1995; Wissel et al. 2006; Turcotte et al. 2015; Russo and Guy 2006; Novotna et al. 2011; Giacoppo et al. 2017; Wade et al. 2004; Collin et al. 2010; Centonze et al. 2009; Rog et al. 2005; Wade et al. 2006; Conte et al. 2009; Rog et al. 2007; Langford et al. 2013; Collin et al. 2007; Leocani et al. 2014; Kavia et al. 2010; Lakhan and Rowland 2009; Keating 2017)

[a]Adapted from: Medicines (2018) 5:91

conclusions can be made for the other clinical outcomes.

8.2.2 Nabilone

Nabilone is a synthetic dibenzopyran-9-one analog of Δ^9-THC (Table 8.1), available as a racemic mixture of (S,S)-(+)- and (R,R)-(−)-isomers. In 1985, it was originally licensed for the treatment of CINV in patients not responding to conventional antiemetic therapies. The use of nabilone for this therapeutic application has been partially supplanted by the development of serotonin 5-HT_3 receptor antagonists.

The current pharmaceutical form of nabilone consists of capsules in strengths of 0.25, 0.5, and 1 mg. The effectiveness of nabilone in the treatment of neuropathic, chronic and cancer pain, and spasticity related to MS has recently been addressed. However, there has been a minimal amount of research on its use beyond its license, over the last two decades. In fact, it has been specifically evaluated for its effectiveness and safety in treating MS symptoms in only three clinical studies published between 1995 and 2015 (Martyn et al. 1995; Wissel et al. 2006; Turcotte et al. 2015). These studies are reported in four reviews published between 2006 and 2015 (Whiting et al. 2015; Ben Amar 2006; Koppel et al. 2014; Karst et al. 2010) included in a systematic review of reviews on the basis of eligibility criteria of methodological quality (AMSTAR Tool) (Nielsen et al. 2018). As mentioned above, eight MS clinical outcomes have been considered: disability/disease progression, pain, spasticity, bladder function, ataxia/tremor, sleep, quality of life, and adverse effects (Nielsen et al. 2018). A summary of clinical evidence about nabilone is shown in Table 8.1.

Regarding pain (Martyn et al. 1995), spasticity (Martyn et al. 1995; Wissel et al. 2006), and bladder dysfunction (Martyn et al. 1995) related to MS, positive effects due to nabilone were found. Mixed findings (although mostly positive) emerged in the evaluation of quality of life (Martyn et al. 1995; Turcotte et al. 2015): one study reported a significant improvement in objective rating of general health status (Martyn et al. 1995); another study, in which nabilone was evaluated as an adjunctive drug to gabapentin, reported an improvement in patient global impression of change, but no statistically significant difference in "VAS impact" between nabilone and placebo groups (Turcotte et al. 2015) ("VAS impact" refers to influence of pain on patient's daily activities, recorded using a visual analog scale). Currently, there are no studies about the effect of nabilone on sleep quality of MS patients and on disability/disease progression. Moderate sedation, dizziness, and moderate weakness in the legs are the main adverse effects reported for nabilone (Martyn et al. 1995; Wissel et al. 2006).

It can be concluded that concerning three clinical outcomes related to MS, i.e., pain, spasticity, and bladder problems, there is positive evidence for nabilone.

8.2.3 Nabiximols

First approved in Canada in 2005 for the treatment of neuropathic pain associated with MS and suddenly approved (2007) in the same country as adjunctive analgesic treatment of advanced cancer pain, nabiximols is a specific extract from cloned plants of *Cannabis sativa* consisting of an approximate 1:1 fixed ratio of Δ^9-THC and CBD (Russo and Guy 2006).

Several European countries approved the drug in the following years, and today it is available in about 20 countries worldwide for the treatment of MS-related moderate to severe spasticity in patients not responsive to other antispasticity therapies.

Nabiximols was developed in response to widespread anecdotal reports about the usefulness of cannabis for treating various symptoms related to MS. The introduction of nonpsychoactive phytocannabinoid CBD in the drug essentially aims to reduce side effects of Δ^9-THC.

Sativex® is the trade name of the drug based on nabiximols; it is a pharmaceutical product standardized in composition, formulation, and dosage. It is formulated as an oro-mucosal spray containing 27 mg of Δ^9-THC and 25 mg of CBD/1.0 mL, in an aromatized water–ethanol solution. Each spray (or "puff") delivers 0.1 mL of solution, which corresponds to 2.7 mg of Δ^9-THC and 2.5 mg of CBD. Sativex® is available as 5.5 mL spray bottles (maximum 48 sprays) or as 10 mL spray bottles (maximum 90 sprays).

Being the absorption after an oro-mucosal administration slower with respect to inhalation, the high plasma levels that occur when cannabis is smoked or vaporized are avoided. The oro-mucosal administration is more rapid and consistent than the oral administration (Novotna et al. 2011), allowing a more direct access to blood vessels through the mucosa and, as a consequence, a more rapid plateau of plasma concentration, without the problems related to the oral route (Giacoppo et al. 2017). Furthermore, the delivery system in such a formulation is very simple for the patients, allowing them to self-manage a convenient and accurate titration of dosage.

The effectiveness and safety of nabiximols in treating MS symptoms has been widely evaluated in 11 clinical studies published between 2004 and 2014 (Wade et al. 2004; Collin et al. 2010; Centonze et al. 2009; Rog et al. 2005; Wade et al. 2006; Conte et al. 2009; Rog et al. 2007; Langford et al. 2013; Collin et al. 2007; Leocani et al. 2014; Kavia et al. 2010); these studies are reported in 11 reviews published between 2006 and 2017 (Whiting et al. 2015; Ben Amar 2006; Koppel et al. 2014; Karst et al. 2010; Jawahar et al. 2013; Mills et al. 2007; Andrzejewski et al. 2016; Wang et al. 2008; Giacoppo et al. 2017; Lakhan and Rowland 2009; Keating 2017). Most of these reviews have been included in a systematic review of reviews on the basis of eligibility criteria of methodological quality (AMSTAR Tool) (Nielsen et al. 2018). The eight main clinical outcomes related to MS that have been considered are the same reported in the above paragraphs, i.e., disability/disease progression, pain, spasticity, bladder function, ataxia/tremor, sleep, quality of life, and adverse effects (Nielsen et al. 2018). Table 8.1 shows a summary of clinical evidence about nabiximols.

Most of the results support the use of nabiximols for MS-related pain. In fact, a significant reduction in numeric rating scales (NRS) and visual analogic scales (VAS) score was highlighted in the treated group with respect to placebo group in many randomized controlled trials (RCT) (Wade et al. 2004; Collin et al. 2010; Centonze et al. 2009; Rog et al. 2005; Wade et al. 2006; Conte et al. 2009; Rog et al. 2007; Langford et al. 2013). The effectiveness of nabiximols in the treatment of spasticity associated with MS is highlighted in some clinical trials, in particular regarding the patient's subjective evaluation scales (NRS) (Wade et al. 2004; Collin et al. 2010; Collin et al. 2007; Leocani et al. 2014); the data concerning the objective evaluation scales (Ashworth scale (AS) and modified Ashworth scale (MAS)), although in favor of the nabiximols, are not statistically significant in some cases (Collin et al. 2010; Collin et al. 2007). On the other hand, no change in spasticity was found in some studies (Centonze et al. 2009; Conte et al. 2009). The usefulness of nabiximols in ameliorating overall bladder symptoms related to MS is controversial, given the contradictory evidence emerged in diverse studies (Wade et al. 2004; Kavia et al. 2010). Nevertheless, this drug seems to be effective in reducing the number of bladder voids per day (Kavia et al. 2010). While nabiximols has been shown to improve subjective quality of sleep (Wade et al. 2004), no statistically significant positive change in tremor and ataxia associated with MS has been demonstrated to date (Wade et al. 2004; Collin et al. 2010). Mixed findings are in general reported about the quality of life; however, in some cases, a significant average number of MS patients reported an improvement of the global impression of change following the treatment with nabiximols (Wade et al. 2004; Rog et al. 2005; Langford et al. 2013; Collin et al. 2007). The main adverse effects associated with nabiximols are drowsiness, dizziness, headache, fatigue, impaired balance, and disturbance in attention (Wade et al. 2004; Collin et al. 2010; Centonze et al. 2009; Rog et al. 2005; Wade et al. 2006; Conte et al. 2009; Rog et al. 2007; Collin et al. 2007). These effects are referred to as mild to moderate, and generally well tolerated. Clinical studies about nabiximols have not shown any significant change in parameters that can be referred to disability and disease progression; these include the Barthel index of activity of daily living (ADL) and walking time (10 mt) (Wade et al. 2004; Collin et al. 2010).

These extensive clinical evidences indicate overall that pain, spasticity and quality of sleep in MS patients are the indications for which nabiximols could represent a valid therapeutic option, and that in general the incidence of adverse effects (not serious and well tolerated) is quite low.

8.3 Endocannabinoid System Modulators

8.3.1 CB1R and CB2R Ligands

One of the first studies of cannabinoids' effect in animal models of MS was reported by Lyman et al. in 1989 (Lyman et al. 1989), who showed the effects of daily administration of Δ^9-THC, an active component of marijuana with partial CB1R agonist activity and limited effects on CB2R, on EAE progression in rats. Indeed, the development of EAE was ameliorated and the examination of central nervous system tissue revealed a marked reduction of inflammation in the Δ^9-THC-treated animals with respect to control animals (Lyman et al. 1989) indicating that Δ^9-THC was able to inhibit both clinical and histologic EAE. Subsequently, Wirguin et al. (Wirguin et al. 1994) reported the activity of Δ^8-THC (Table 8.2) on EAE. This phytocannabinoid is more stable and less psychotropic than Δ^9-THC and it binds CB1Rs with high affinity. Δ^8-THC was shown to significantly reduce the incidence and severity of neurological manifestations of EAE. This compound was considered a prodrug, indeed the inhibition of the prostanoid production by action of an active metabolite of Δ^8-THC, formed from the first-pass metabolism in the liver, was proposed as potential mechanisms of action. This hypothesis was supported by the evidence that the beneficial influence of Δ^8-THC only occurred on oral administration and not with parenteral injection (Wirguin et al. 1994).

Several studies were developed regarding the role of endogenous and synthetic cannabinoids in

Table 8.2 ECS modulators and their effects shown in different animal models of MS[a]

Structure Name	Origin Activity	Animal model Effects
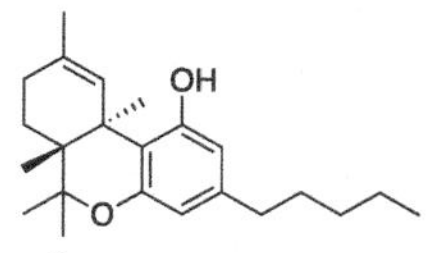 Δ[9]-THC	Phytocannabinoid CB1R partial agonist	In EAE rats: amelioration of EAE progression (Lyman et al. 1989) In CREAE mice: amelioration of tremor and spasticity (Baker et al. 2000)
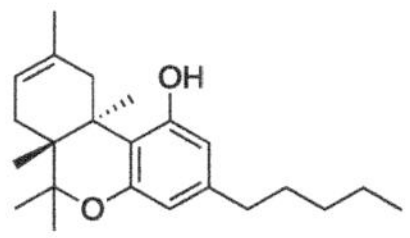 Δ[8]-THC	Phytocannabinoid CB1R ligand	In EAE rats: amelioration of the clinical manifestations of EAE (Wirguin et al. 1994)
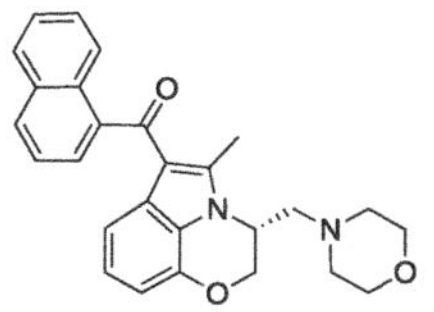 WIN 55,212–2	Synthetic cannabinoid CB2R agonist	In CREAE mice: amelioration of tremor and spasticity (Baker et al. 2000) In TMEV-infected mice: improvement of motor function on established neurological symptomatology; stimulation of the remyelination; reduction of microglial activation and of the number of CD4 + -infiltrated T cells (Arevalo-Martin et al. 2003)
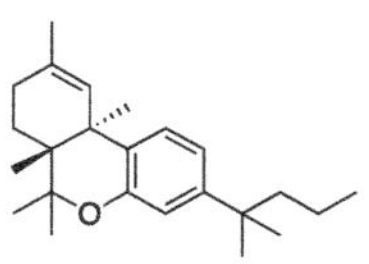 JWH-133	Synthetic cannabinoid CB2R agonist	In CREAE mice: amelioration of tremor and spasticity (Baker et al. 2000) Intrathecal administration in EAE mice: reduction, dose dependently, of both mechanical and cold hypersensitivity without any signs of ataxia or sedation (Fu and Taylor 2015)
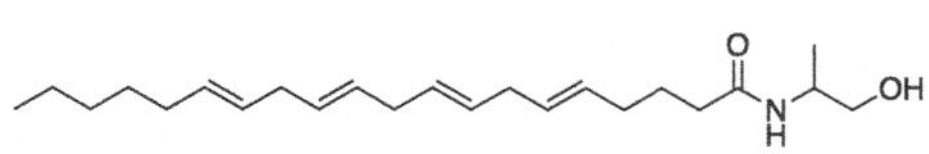 Methanandamide	Endocannabinoid CB1R/CB2R agonist	In CREAE mice: amelioration of tremor and spasticity (Baker et al. 2000)
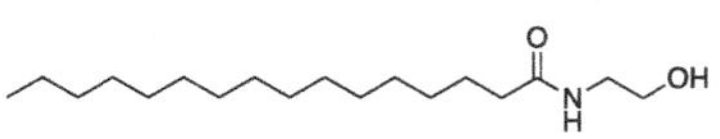 Palmitoylethanolamide (PEA)	Endocannabinoid CB1R/CB2R agonist	In CREAE mice: transient inhibition of spasticity (Baker et al. 2000)
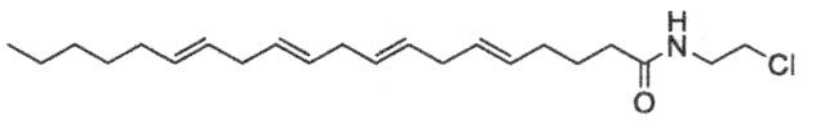 Arachidonyl-2-chloroethylamide (ACEA)	Synthetic cannabinoid CB1R agonist	In TMEV-infected mice: improvement of motor function on established neurological symptomatology; stimulation of the remyelination; and reduction of microglial activation and of the number of CD4 + -infiltrated T cells (Rahimi et al. 2015)
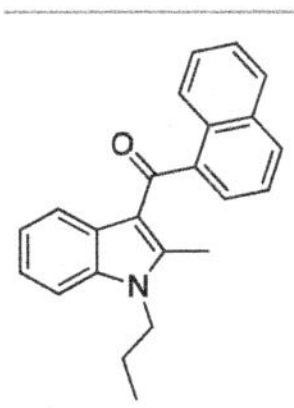 JWH-015	Synthetic cannabinoid CB2R agonist	In TMEV-infected mice: improvement of motor function on established neurological symptomatology; stimulation of the remyelination; and reduction of microglial activation and of the number of CD4 + -infiltrated T cells (Arevalo-Martin et al. 2003)

(continued)

Table 8.2 (continued)

Structure Name	Origin Activity	Animal model Effects
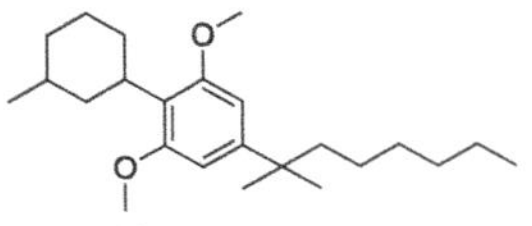 O-1966	Synthetic cannabinoid CB2R agonist	In the chronic EAE model: improved motor function; reduction of rolling and adhesion of endogenous leukocytes to pial microvasculature (Zhang et al. 2009)
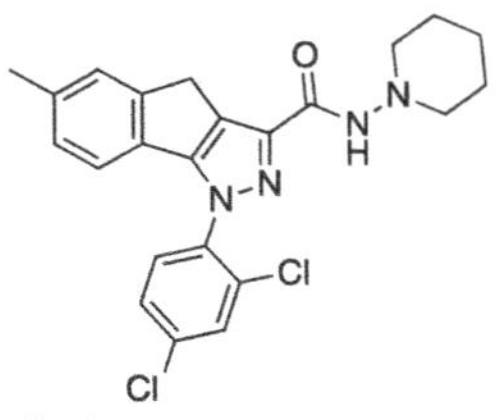 Gp-1a	Synthetic cannabinoid CB2R agonist	In EAE mice: reduction of clinical scores; amelioration of the recovery (Kong et al. 2014)
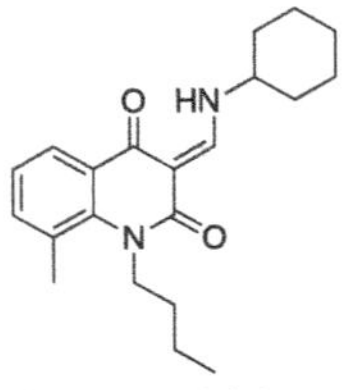 Compound 21	Synthetic cannabinoid CB2R agonist	In EAE mice: - reduction of the clinical scores and symptoms; decrease of leukocyte infiltration in the spinal cord and demyelination in white matter (Han et al. 2015)
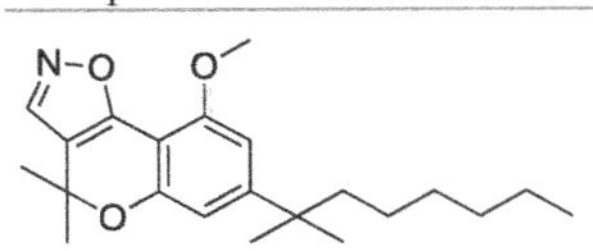 PM-226	Synthetic cannabinoid CB2R agonist	In TMEV-infected mice: dampening of neuroinflammation (Morales et al. 2016; Mecha et al. 2013)
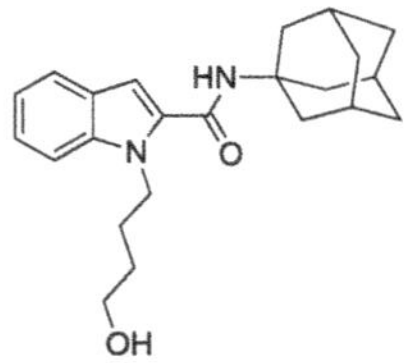 Compound 57	Synthetic cannabinoid CB2R agonist	In EAE mice: alleviation of the clinical symptoms of EAE; protection of the murine central nervous system from immune damage (Shi et al. 2017)
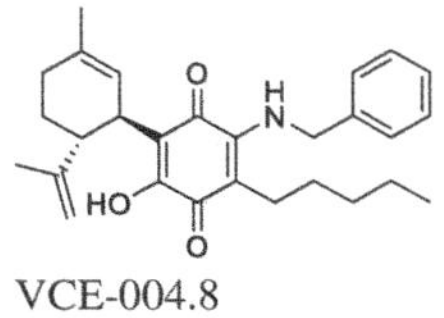 VCE-004.8	Synthetic cannabinoid CB2R agonist	In EAE and TMEV mice: immunomodulatory activity; inhibition of inflammatory chemokines, chemokines receptors, and cytokines; inhibition of the expression of adhesion molecules (VCAM and ICAM-1); and induction of the expression of the hypoxia-inducible factor (HIF) (Navarrete et al. 2018)
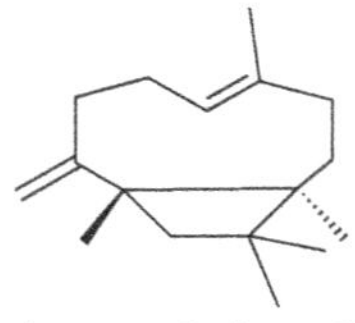 Δ-caryophyllene (BCP)	Phytocannabinoid CB2R agonist	In EAE mice: reduction of mechanical hyperalgesia, inflammation, and pain (Alberti et al. 2017)

[a]Adapted from: Medicines (2018) 5:91

CREAE animal model. Baker et al. evidenced that the CBR agonists R(+)-WIN 55,212–2 (Table 8.2), Δ^9-THC, methanandamide (Table 8.2) and JWH-133 (Table 2) were able to ameliorate both tremor and spasticity in CREAE mice (Baker et al. 2000). In particular, a role of CB1Rs in controlling tremor and of both cannabinoid receptors in the development of spasticity was suggested (Baker et al. 2000). In the same work was reported that the endocannabinoid palmitoylethanolamide (PEA) (Table 8.2) caused a transient inhibition of spasticity (Baker et al. 2000). However, more recently, it was demonstrated that co-administration of PEA with CBD in EAE was not as active as treatment with each compound alone, indicating that these nonpsychoactive cannabinoids could have antagonistic interactions in EAE (Rahimi et al. 2015).

Further research showed that in TMEV-infected mice, WIN 55,212–2 (Table 8.2), arachidonyl-2-chloroethylamide (ACEA), a selective CB1R agonist (Table 8.2), and JWH-015, a weak selective CB2R agonist (Table 8.2), were able to improve motor function, to promote the remyelination, and to reduce microglial activation and the number of CD4+ infiltrated T cells (Arevalo-Martin et al. 2003). Further studies reported that WIN 55,212–2 restored self-tolerance to a myelin self-antigen while ameliorating the disease in a long-term manner. The therapeutic effect of WIN 55,212–2 correlated with a decrease in the activation of $CD4^{+}CD25^{+}Foxp3^{-}$ T cells and an increase in regulatory $CD4^{+}CD25^{+}Foxp3^{+}$ T cells in the CNS, along with alterations in the cytokine and chemokine milieu. These findings demonstrated for the first time that the suppression of autoimmune responses to myelin antigens underlies the therapeutic effect of CBR agonists in the treatment of MS (Arevalo-Martin et al. 2003).

Recent studies were focused on the development and study of CB2R selective agonists as the best therapeutic approach for the treatment of MS, thanks to their lack of central side effects usually associated with a CB1R modulation.

First of all, the resorcinol derivative O-1966 (Table 8.2) was shown to significantly improve motor function in the chronic EAE model, at a concentration of 1 mg kg^{-1} and to reduce rolling and adhesion of endogenous leukocytes (Zhang et al. 2009). Moreover, the 1,4-dihydro-6-methylindeno[1,2-c]pyrazole derivative, Gp-1a (Table 8.2), was demonstrated to be able to reduce clinical scores and ameliorate the recovery in EAE mice presenting a long-term reduction in demyelination and axonal loss. Two different mechanisms were proposed for this compound, indeed at first, it was able to affect Th1/Th17 differentiation in peripheral immune organs and subsequently it affects pathogenic T cell accumulation in the CNS and reduces the expression of chemokine and adhesion molecules in the CNS (Kong et al. 2014).

Furthermore, in 2015, Han et al. reported that a new quinoline-2,4(1H,3H)-dione derivative, compound 21 (Table 8.2), with selective CB2R agonist activity, significantly reduced the clinical scores and symptoms of the EAE mice model, by remarkably decreasing leukocyte infiltration in the spinal cord and demyelination in white matter (Han et al. 2015).

In the same year, Fu et al. (Fu and Taylor 2015) showed that intrathecal administration of JWH-133 (Table 8.2), a selective CB2R agonist, in EAE mice, dose dependently reduced both mechanical and cold hypersensitivity without any signs of ataxia or sedation. The co-administration of JWH-133 with a selective CB2R antagonist dose dependently attenuated the inhibitory effects of JWH-133. These data suggested that the selective targeting of spinal CB2Rs reduced signs of neuropathic pain in EAE mice without any side effects (Fu and Taylor 2015).

During the following year, chromenopyrazole nucleus was identified as the promising scaffold to obtain CBR ligands (Morales et al. 2016). Structural modifications have been studied in order to achieve CB2R selectivity and the structural changes led to the synthesis of chromenoisoxazole derivative PM-226 (Table 8.2) as selective CB2R agonist. This compound dampened neuroinflammation in the TMEV mouse model by reducing microglial activation to levels close to those of the control group (Morales et al.

2016). This decrease in the microglia activation determined a reduction of inflammatory events and an improvement of the neurological status of treated animals (Mecha et al. 2013).

In 2017, Ying Shi et al. reported the identification of new potent and selective indole-based CB2R agonists (Shi et al. 2017) and one of them, compound 57 (Table 8.2), was selected to be studied in a EAE mouse model. This compound significantly showed to be able to alleviate the clinical symptoms and to protect the murine central nervous system from immune damage. Furthermore, histological examination of spinal cords demonstrated a significant reduction of leukocyte infiltration and the extent of demyelination (Shi et al. 2017).

Very recently, Navarrete et al. (Navarrete et al. 2018) provided evidence that VCE-004.8 (Table 8.2), an amino-quinone derivative of CBD, is a promising small molecule with multitarget activity, being a PPAR and CB2R agonist with potent anti-inflammatory activity. VCE-004.8 showed immunomodulatory activity in EAE and TMEV mice models, inhibiting several inflammatory chemokines, chemokines receptors, and cytokines that play a key role in the pathogenesis of MS. In addition, VCE-004.8 inhibited the expression of adhesion molecules such as VCAM and ICAM-1. Remarkable is the finding that VCE-004.8 strongly induced the expression of the hypoxia-inducible factor (HIF), which can have a beneficial role in MS by modulating the immune response and favoring neuroprotection and axonal regeneration (Navarrete et al. 2018).

The sesquiterpene β-caryophyllene (BCP) (Table 8.2) is a CB2R-selective agonist already reported in the literature for its anti-inflammatory and analgesic effects in mouse models of inflammatory and neuropathic pain (Sharma et al. 2016). Very recently, it is reported that BCP is able to attenuate disease progression by reducing mechanical hyperalgesia, inflammation, and pain in the EAE mouse model (Alberti et al. 2017). When BCP was co-administered with a selective CB2R antagonist, the effects were reversed, demonstrating that BCP action was CB2R-mediated (Alberti et al. 2017).

8.3.2 Inhibitors of Metabolic Enzymes of Endocannabinoids

An alternative approach to modulate ECS consists in the blocking of the metabolic enzymes of the two main endocannabinoids (ECs), 2-AG and AEA. This is an interesting therapeutic strategy, as enhancing EC levels is expected to preserve the beneficial effects derived from the direct activation of CBRs but limiting potential side effects mostly associated with direct CB1R agonists. Moreover, in MS patients, there is a significant alteration of the metabolic enzymes, mainly of FAAH and of MAGL (Benito et al. 2007; Chiurchiù et al. 2013). In particular, different studies using TMEV-infected mice showed that the inhibition of FAAH determines an improvement of the motor symptoms, with a reduction of inflammatory response and the downregulation of macrophage and of microglial function (Mestre et al. 2005; Ortega-Gutiérrez et al. 2005). Furthermore, it was demonstrated that chronic and long-term inhibition of FAAH, via genetic ablation, produces clinical remission and ameliorates long-term results in EAE mouse model (Webb et al. 2008).

Other studies showed that 2AG-treatment ameliorated the acute phase of disease with delay of disease onset and reduced disease mortality and long-term clinical disability in EAE models (Lourbopoulos et al. 2011). Moreover, the expression of cannabinoid receptors was increased and it was accompanied by an increase of the M2-macrophages in the perivascular infiltrations. These results indicated that 2-AG treatment may provide direct (via cannabinoid receptors) and immune (via M2 macrophages)-mediated neuroprotection in the EAE model (Lourbopoulos et al. 2011).

As 2-AG is the main endocannabinoid present in the brain, and it is a full agonist of both cannabinoid receptors, many studies were performed on the MAGL inhibitors. However, it was demonstrated that chronic MAGL inhibition and subsequently increase of 2-AG in the brain provoke a functional antagonism of the cerebral ECS, with tolerance to the analgesic effects of acute enzymatic inhibition, cross-tolerance to

CB1R agonists, reduction of expression and function of the CB1Rs, and interruptions in endocannabinoid-dependent synaptic plasticity (Schlosburg et al. 2010).

In contrast, a recent work reported that the chronic administration of JZL184 reduced the neurological consequences of disease progression in EAE mice, reducing the myelin loss and inflammation of spinal cord white matter (Bernal-Chico et al. 2015). Furthermore, it was demonstrated that the repeated administration of JZL184 at a dose of 8 mg kg^{-1} did not provoke changes in CB1R expression in the hippocampus, and there was not tolerance to the anxiolytic and analgesic effects of the MAGL inhibitor (Bernal-Chico et al. 2015).

Recently, Brindisi et al. reported the β-lactam-based compound 4a (Table 8.3) as a very potent hMAGL inhibitor, with high selectivity toward FAAH, other serine hydrolases and cannabinoid receptors (Brindisi et al. 2016). This compound exerted a surprising beneficial effect on the progression of the MS disease in EAE model, due to the CB1R- and CB2R-mediated action. Histological evaluation of myelin demonstrated a significant reduction of the demyelinated area in the EAE mice treated with compound 4a (Brindisi et al. 2016). Finally, oral administration of 4a at 1 mg kg^{-1} dose dependently reversed the lowering of the threshold to cold stimuli (cold plate test) induced by oxaliplatin (OXP), indicating its efficacy in the treatment of neuropathic pain, which clearly depends on the increased levels of 2-AG and the subsequent indirect modulation of cannabinoid receptors (Brindisi et al. 2016).

As above reported, the irreversible MAGL inhibition causes pharmacological tolerance and receptor desensitization. For these reasons, some studies on reversible inhibitors were developed. An interesting example is given by the compound 21 (Table 8.3) synthesized by Hernández-Torres et al. (Hernández-Torres et al. 2014). This compound showed a sub-micromolar IC_{50} value for MAGL inhibition and very good selectivity against FAAH, ABDH6, ABHD12, and both CBRs (Hernández-Torres et al. 2014). In EAE mouse, it demonstrated to significantly increase the levels of 2-AG in spinal cord, improving clinical symptoms and decreasing tissue damage in the spinal cords. Importantly, catalepsy or other motor impairments that are observed after the administration of irreversible MAGL inhibitors, did not occurred.

The negative effects due to the prolonged inhibition of MAGL enzymes do not occur by FAAH inhibition. Indeed, it was demonstrated that the prolonged inhibition of FAAH produced no tolerance or no changes in the expression or function of the CB1R (Pryce et al. 2013). Pryce et al. demonstrated that potent FAAH inhibitors such as CAY10402 (Table 8.3) and CAY10400 (Table 8.3) inhibited spasticity but did not induce any hypothermia, typical of cannabimimetic effects. However, CAY10400 and CAY10402 have poor pharmacokinetics, and therefore, their development is unlikely as therapeutic drugs (Pryce et al. 2013).

A valid alternative is represented by compound URB597 (Table 8.3), which is a potent irreversible FAAH inhibitor with an improved pharmacokinetic profile (Pryce et al. 2013). The administration of URB597 induced spasticity alleviation immediately without an increased effect after four daily doses. However, the use of this inhibitor was not associated with CB1R tolerance. Actually, the study emphasized the benefit because the level of spasticity at the baseline after four administrations was lower than the baseline before treatment.

Nevertheless, the inhibition of the above-reported enzymes, MAGL and FAAH, can also drive to enhanced neurotoxicity due to an increase in the availability of endocannabinoids, which, together with elevated COX-2 activity, may convert endocannabinoids in new oxygenated derivatives, so-called prostamides (derived from AEA) or prostaglandin-glycerylesters (derived from 2-AG), which may be highly toxic for neurons (Alhouayek and Muccioli 2014).

Another effective approach to modulate the ECS is to act on AEA reuptake, and on this basis, many selective inhibitors of cellular reuptake of AEA have been developed. In particular, compounds O-3246 and O-3262 (Table 8.3) were reported to have very high potency as inhibitors

Table 8.3 Inhibitors of metabolic enzymes of ECs and their effects shown in different animal models of MS[a]

Structure Name	Activity	Animal model Effects
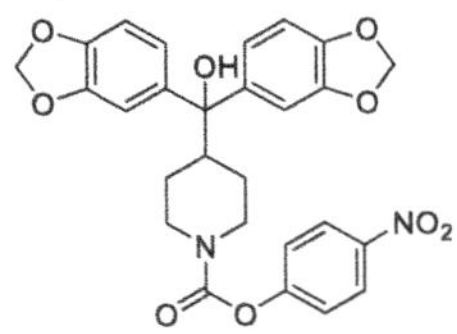 JZL 184	Irreversible MAGL Inhibitor	In EAE mice: reduction of myelin loss; reduction of inflammation on spinal cord white matter (Bernal-Chico et al. 2015)
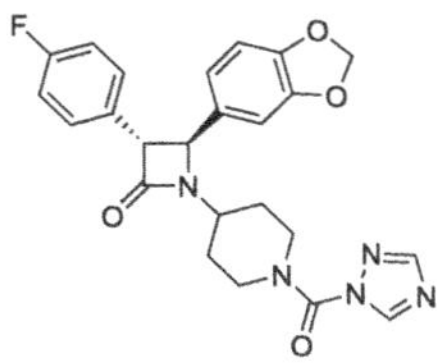 Compound 4	Irreversible MAGL Inhibitor	In EAE mice: analgesic effect (Brindisi et al. 2016)
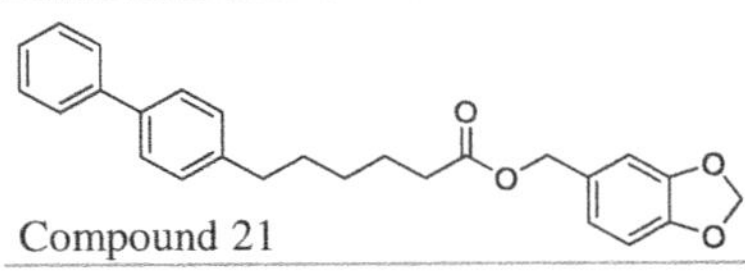 Compound 21	Reversible MAGL Inhibitor	In EAE mice: decrease of tissue damage in the spinal cords (Hernández-Torres et al. 2014)
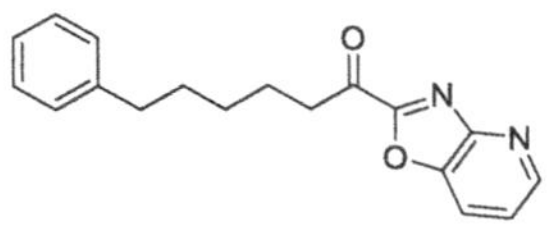 CAY 10402	Irreversible FAAH Inhibitor	In Biozzi ABH mice: inhibition of spasticity (Pryce et al. 2013)
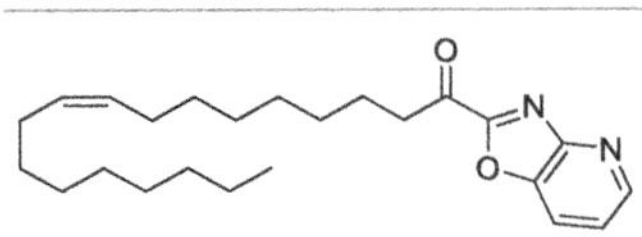 CAY 10400	Irreversible FAAH Inhibitor	In Biozzi ABH mice: inhibition of spasticity (Pryce et al. 2013)
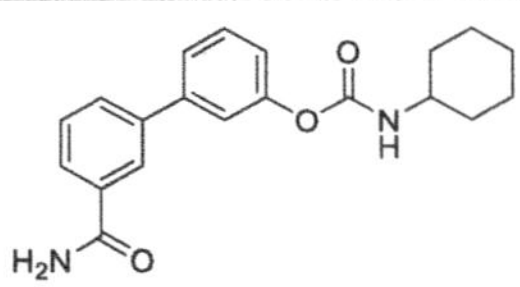 URB597	Irreversible FAAH Inhibitor	In Biozzi ABH mice: inhibition of spasticity (Pryce et al. 2013)
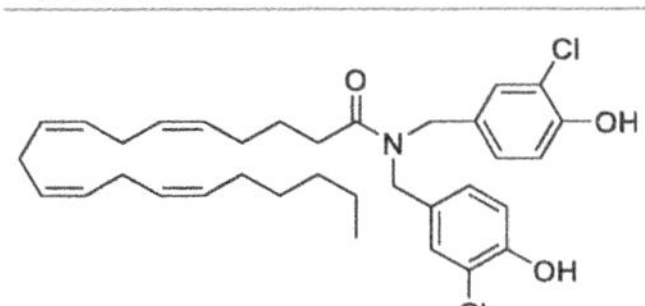 O-3246	AEA reuptake inhibitor	In CREAE mice: inhibition of spasticity (Ligresti et al. 2006)
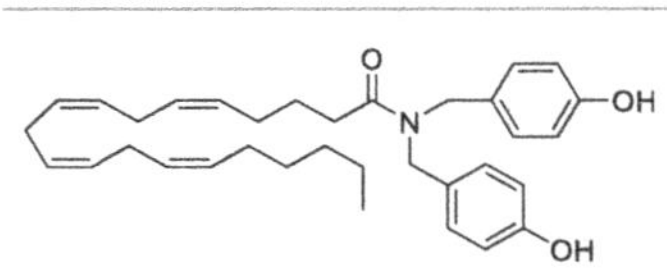 O-3262	AEA reuptake inhibitor	In CREAE mice: inhibition of spasticity (Ligresti et al. 2006)

(continued)

Table 8.3 (continued)

Structure Name	Activity	Animal model Effects
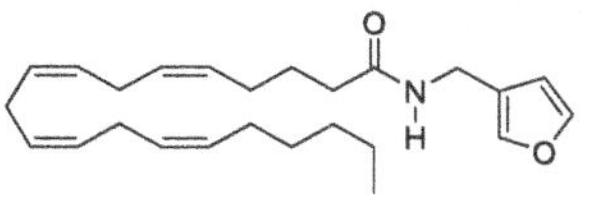 UCM707	AEA reuptake inhibitor	In TMEV-IDD mice: improvement of motor function; reduction of microglial activation; and decrease of cellular infiltrates in the spinal cord (Ortega-Gutiérrez et al. 2005)

[a]Adapted from: Medicines (2018) 5:91

of AEA cellular uptake and a negligible activity as FAAH inhibitors, CB1R and CB2R ligands, and TRPV1 agonists. These compounds have been shown to inhibit spasticity in CREAE mice, confirming the potential utility of selective AEA uptake inhibitors as antispasticity drugs in MS (Ligresti et al. 2006).

Furthermore, it has been found that UCM707 (Table 8.3), a potent and selective inhibitor of the AEA reuptake (Ortega-Gutiérrez et al. 2005), was able to improve the motor function in a TMEV-IDD mouse model, and at the histological level, it reduced microglial activation, diminished major histocompatibility complex class II antigen expression and decreased cellular infiltrates in the spinal cord (Ortega-Gutiérrez et al. 2005). Additionally, in microglial cells, UCM707 decreases the production of the proinflammatory cytokines tumor necrosis factor (TNF)-alpha, interleukin (IL)-1beta, and IL-6; reduces nitric oxide levels and inducible nitric oxide synthase expression; and is able to potentiate the action of a subeffective dose of the endocannabinoid anandamide. These results confirm the role played by the ECS at the level of immunomodulation, and they are in agreement with experiments that describe how the blockade of microglial activation represses the development of the EAE model of MS (Ortega-Gutiérrez et al. 2005).

8.4 Conclusions

Millions of people worldwide are affected by MS, a progressive neurodegenerative disease without any effective cure so far and whose symptoms are still difficult to manage. The modulation of distinct components of ECS (CBRs, degrading enzymes, and AEA transporters) may represent a new and promising therapeutic strategy to control symptoms and disease progression of MS, as demonstrated by recent studies performed in animal models of MS. It has been reported that cannabinoids can relieve symptoms of MS by essentially activating the CB1R. The increase of endocannabinoid levels through the inhibition of the degrading enzymes of AEA and/or 2-AG (FAAH and MAGL, respectively) and of the AEA transporter can lead to the amelioration of spasticity. Moreover, the changes reported for the ECS in different MS models have been associated with adaptive responses for limiting neuronal damage. Specifically, the activation of CB1R regulates glutamate homeostasis and excitotoxic damage, providing neuroprotection. Furthermore, it has been has been shown in preclinical models of MS that CB2R activation has a protective effect in microglial cells upon inflammatory-induced CNS damage. Finally, an enhancing trophic and metabolic support to neurons is mediated by astroglial CB1R and/or CB2R, thus reducing excitotoxicity and leading neuroprotection. On the basis of the above results, it is reasonable to conceive a synergistic action between the simultaneous modulation of more targets of ECS and conventional therapies, producing more beneficial effects. Therefore, the study of multitarget modulators of ECS has been emerging in the last few years, to achieve this goal. These agents offer the possibility of modulating the ECS in a safer and more effective way with respect to a single target modulation, directly and indirectly modulating cannabinoid receptor activity through different mechanisms of action (Chicca et al. 2018). Although there is still a need for extensive preclinical studies, we can hypothesize that multitarget modulators of ECS could be able to control disease progression and symptoms MS, possibly having a great translational potential and representing promising candidates for clinical development.

References

Alberti TB, Barbosa WL, Vieira JL et al (2017) (−)-β-Caryophyllene, a CB2 receptor-selective Phytocannabinoid, suppresses motor paralysis and Neuroinflammation in a murine model of multiple sclerosis. Int J Mol Sci 18:691

Alhouayek M, Muccioli GG (2014) COX-2-derived endocannabinoid metabolites as novel inflammatory mediators trends. Pharmacol Sci 35:284–292

Andrzejewski K, Barbano R, Mink J (2016) Cannabinoids in the treatment of movement disorders: a systematic review of case series and clinical trials. Basal Ganglia 6:73–181

Arevalo-Martin A, Vela JM, Molina-Holgado E et al (2003) Therapeutic action of cannabinoids in a murine model of multiple sclerosis. J Neurosci 23:2511–2516

Atwood BK, Mackie K (2010) CB2: a cannabinoid receptor with an identity crisis. Br J Pharmacol 160:467–479

Baker D, Pryce G, Croxford JL et al (2000) Cannabinoids control spasticity and tremor in a multiple sclerosis model. Nature 404:84–87

Ben Amar M (2006) Cannabinoids in medicine: a review of their therapeutic potential. J Ethnopharmacol 105:1–25

Benito C, Romero JP, Tolón RM et al (2007) Cannabinoid CB1 and CB2 receptors and fatty acid amide hydrolase are specific markers of plaque cell subtypes in human multiple sclerosis. J Neurosci 27:2396–2402

Bernal-Chico A, Canedo M, Manterola A et al (2015) Blockade of monoacylglycerol lipase inhibits oligodendrocyte excitotoxicity and prevents demyelination in vivo. Glia 63:163–176

Brindisi M, Maramai S, Gemma S et al (2016) Development and pharmacological characterization of selective blockers of 2-arachidonoyl glycerol degradation with efficacy in rodent models of multiple sclerosis and pain. J Med Chem 59:612–2632

Browne P, Chandraratna D, Angood C et al (2014) Atlas of multiple sclerosis 2013: a growing global problem with widespread inequity. Neurology 83:1022–1024

Calabrese M, Magliozzi R, Ciccarelli O et al (2015) Exploring the origins of grey matter damage in multiple sclerosis. Nat Rev Neurosci 16:147–158

Cambron M, D'Haeseleer M, Laureys G et al (2012) White-matter astrocytes, axonal energy metabolism, and axonal degeneration in multiple sclerosis. J Cereb Blood Flow Metab 323:413–424

Centonze D, Mori F, Kock G et al (2009) Lack of effect of cannabis-based treatment on clinical and laboratory measures in multiple sclerosis. Neurol Sci 30:531–534

Chicca A, Arena C, Bertini S et al (2018) Polypharmacological profile of 1,2-dihydro-2-oxo-pyridine-3-carboxamides in the endocannabinoid system. Eur J Med Chem 154:155–171

Chiurchiù V, Battistini L, Maccarrone M (2015a) Endocannabinoid signaling in innate and adaptive immunity. Immunology 144:352–364

Chiurchiù V, Cencioni MT, Bisicchia E et al (2013) Distinct modulation of human myeloid and plasmacytoid dendritic cells by anandamide in multiple sclerosis. Ann Neurol 73:626–636

Chiurchiù V, Leuti A, Maccarrone M (2015b) Cannabinoid signaling and neuroinflammatory diseases: a melting pot for the regulation of brain immune responses. J Neuroimmune Pharmacol 10:268–280

Chiurchiù V, van der Stelt M, Centonze D et al (2018) The endocannabinoid system and its therapeutic exploitation in multiple sclerosis: clues for other neuroinflammatory diseases. Prog Neurobiol 160:82–100

Clark AJ, Ware MA, Yazer E et al (2004) Patterns of cannabis use among patients with multiple sclerosis. Neurology 62:2098–2100

Clifford DB (1983) Tetrahydrocannabinol for tremor in multiple sclerosis. Ann Neurol 13:669–671

Collin C, Davies P, Mutiboko IK et al (2007) Randomized controlled trial of cannabis-based medicine in spasticity caused by multiple sclerosis. Eur J Neurol 14:290–296

Collin C, Ehler E, Waberzinek G et al (2010) A double-blind, randomized, placebo-controlled, parallel-group study of Sativex, in subjects with symptoms of spasticity due to multiple sclerosis. Neurol Res 32:451–459

Compston A, Coles A (2008) A Multiple sclerosis. Lancet 372:1502–1517

Conte A, Bettolo CM, Onesti E et al (2009) Cannabinoid-induced effects on the nociceptive system: a neurophysiological study in patients with secondary progressive multiple sclerosis. Eur J Pain 13:472–477

Docagne F, Muñetón V, Clemente D et al (2007) Excitotoxicity in a chronic model of multiple sclerosis: neuroprotective effects of cannabinoids through CB1 and CB2 receptor activation. Mol Cell Neurosci 34:551–561

Dunn SE, Gunde E, Lee H (2015a) Sex-based differences in multiple sclerosis (MS): part II: rising incidence of multiple sclerosis in women and the vulnerability of men to progression of this disease. Curr Top Behav Neurosci 26:57–86

Dunn SE, Lee H, Pavri FR et al (2015b) Sex-based differences in multiple sclerosis (part I): biology of disease incidence. Curr Top Behav Neurosci 26:29–56

Dutta R, Trapp BD (2011) Mechanisms of neuronal dysfunction and degeneration in multiple sclerosis. Prog Neurobiol 93:1–12

Fernández-Ruiz J, García C, Sagredo O, Gómez-Ruiz M et al (2010) The endocannabinoid system as a target for the treatment of neuronal damage. Expert Opin Ther Targets 14:387–404

Fernández-Ruiz J, Romero J, Ramos JA (2015) Endocannabinoids and neurodegenerative disorders: Parkinson's disease, Huntington's chorea, Alzheimer's disease and others. Handb Exp Pharmacol 231:233–259

Fife TD, Moawad H, Moschonas C et al (2015) Clinical perspectives on medical marijuana (cannabis) for neurologic disorders. Neurol Clin Pract 5:44–351

Freeman RM, Adekanmi O, Waterfield MR et al (2006) The effect of cannabis on urge incontinence in patients with multiple sclerosis: a multicenter, randomized placebo-controlled trial (CAMS-LUTS). Int Urogynecol J Pelvic Floor Dysfunct 17:636–641

Fu W, Taylor BK (2015) Activation of cannabinoid CB2 receptors reduces hyperalgesia in an experimental autoimmune encephalomyelitis mouse model of multiple sclerosis. Neurosci Lett 595:1–6

Galve-Roperh I, Chiurchiù V, Díaz-Alonso J et al (2013) Cannabinoid receptor signaling in progenitor/stem cell proliferation and differentiation. Prog Lipid Res 52:633–650

Giacoppo S, Bramanti P, Mazzon E (2017) Sativex in the management of multiple sclerosis-related spasticity: an overview of the last decade of clinical evaluation. Mult Scler Relat Disord 17:22–31

Gloss DS, Maa EH (2015) Medical marijuana. Between a plant and a hard place. Neurol Clin Pract 5:281–284

Hafler DA, Compston A, Sawcer S et al (2007) Risk alleles for multiple sclerosis identified by a genomewide study. N Engl J Med 357:851–862

Han S, Zhang FF, Qian HY et al (2015) Development of Quinoline-2,4(1H,3H)-diones as potent and selective ligands of the cannabinoid type 2 receptor. J Med Chem 58:5751–5769

Heremans H, Dillen C, Groenen M et al (1996) Chronic relapsing experimental autoimmune encephalomyelitis (CREAE) in mice: enhancement by monoclonal antibodies against interferon-gamma. Eur J Immunol 26:2393–2398

Hernández-Torres G, Cipriano M, Hedén E et al (2014) A reversible and selective inhibitor of monoacylglycerol lipase ameliorates multiple sclerosis. Angew Chem Int Ed Engl 53:13765–13770

Howlett AC, Barth F, Bonner TI et al (2002) International union of pharmacology. XXVII. Classification of cannabinoid receptors. Pharmacol Rev 54:161–202

Huynh JL, Casaccia P (2013) Epigenetic mechanisms in multiple sclerosis: implications for pathogenesis and treatment. Lancet Neurol 12:95–206

Jawahar R, Oh U, Yang S et al (2013) A systematic review of pharmacological pain management in multiple sclerosis. Drugs 73:1711–1722

Johnson C (2013) Shared care guideline: Nabilone in the Management of Chronic Neuropathic Pain that has failed to respond to other first and second line treatments. Lincolnshire in Association with United Lincolnshire Hospitals Trust, NHS

Karst M, Wippermann S, Ahrens J (2010) Role of cannabinoids in the treatment of pain and (painful) spasticity. Drugs 70:2409–2438

Kavia RB, De Ridder D, Constantinescu CS et al (2010) Randomized controlled trial of Sativex to treat detrusor overactivity in multiple sclerosis. Mult Scler 16:1349–1359

Keating GM (2017) Delta-9-Tetrahydrocannabinol/Cannabidiol Oromucosal spray (Sativex®): a review in multiple sclerosis-related spasticity. Drugs 77:563–574

Killestein J, Hoogervorst EL, Reif M et al (2002) Safety, tolerability, and efficacy of orally administered cannabinoids in MS. Neurology 58:1404–1407

Killestein J, Hoogervorst EL, Reif M et al (2003) Immunomodulatory effects of orally administered cannabinoids in multiple sclerosis. J Neuroimmunol 137:140–143

Klegeris A, Bissonnette CJ, McGeer PL (2003) Reduction of human monocytic cell neurotoxicity and cytokine secretion by ligands of the cannabinoid-type CB2 receptor. Br J Pharmacol 139:775–786

Koch-Henriksen N, Sørensen PS (2010) The changing demographic pattern of multiple sclerosis epidemiology. Lancet Neurol 9:520–532

Koch-Henriksen N, Thygesen LC, Stenager E et al (2018) Incidence of MS has increased markedly over six decades in Denmark particularly with late onset and in women. Neurology 90:e1954–e1963

Kong W, Li H, Tuma RF et al (2014) Selective CB2 receptor activation ameliorates EAE by reducing Th17 differentiation and immune cell accumulation in the CNS. Cell Immunol 287:1–17

Koppel BS, Brust JCM, Fife T et al (2014) Systematic review: efficacy and safety of medical marijuana in selected neurologic disorders: report of the guideline development Subcommittee of the American Academy of neurology. Neurology 82:1556–1563

Lakhan SE, Rowland M (2009) Whole plant cannabis extracts in the treatment of spasticity in multiple sclerosis: a systematic review. BMC Neurol 9:59

Langford RM, Mares J, Novotna A et al (2013) A double-blind, randomized, placebo-controlled, parallel-group study of THC/CBD oromucosal spray in combination with the existing treatment regimen, in the relief of central neuropathic pain in patients with multiple sclerosis. J Neurol 260:984–997

Lassmann H, Bradl M (2017) Multiple sclerosis: experimental models and reality. Acta Neuropathol 133:223–244

Lassmann H, van Horssen J, Mahad D (2012) Progressive multiple sclerosis: pathology and pathogenesis. Nat Rev Neurol 8:647–656

Leocani L, Nuara A, Houdayer E et al (2014) Effect of THC-CBD oromucosal spray (Sativex) on measures of spasticity in multiple sclerosis: a double-blind, placebo-controlled, crossover study. Mult Scler 20:498

Ligresti A, Cascio MG, Pryce G et al (2006) New potent and selective inhibitors of anandamide reuptake with antispastic activity in a mouse model of multiple sclerosis. Br J Pharmacol 147:83–91

Lourbopoulos A, Grigoriadis N, Lagoudaki R et al (2011) Administration of 2-arachidonoylglycerol ameliorates

both acute and chronic experimental autoimmune encephalomyelitis. Brain Res 1390:126–141
Lyman WD, Sonett JR, Brosnan CF et al (1989) D9-Tetrahydrocannabinol: a novel treatment for experimental autoimmune encephalomyelitis. J Neuroimmunol 23:73–81
Mahad DH, Trapp BD, Lassmann H (2015) Pathological mechanisms in progressive multiple sclerosis. Lancet Neurol 14:183–193
Martyn CN, Illis LS, Thom J (1995) Nabilone in the treatment of multiple sclerosis. Lancet 345:579
McCarthy DP, Richards MH, Miller SD (2012) Mouse models of multiple sclerosis: experimental autoimmune encephalomyelitis and Theiler's virus-induced demyelinating disease. Methods Mol Biol 900:381–401
Mecha M, Carrillo-Salinas FJ, Feliú A et al (2016) Microglia activation states and cannabinoid system: therapeutic implications. Pharmacol Ther 166:40–55
Mecha M, Carrillo-Salinas FJ, Mestre L et al (2013) Viral models of multiple sclerosis: Neurodegeneration and demyelination in mice infected with Theiler's virus. Prog Neurobiol 101–102:46–64
Mestre L, Correa F, Arévalo-Martín A et al (2005) Pharmacological modulation of the endocannabinoid system in a viral model of multiple sclerosis. J Neurochem 92:1327–1339
Mills RJ, Yap L, Young CA (2007) Treatment for ataxia in multiple sclerosis. Cochrane Database Syst Rev 1: CD005029
Morales P, Gomez-Canas M, Navarro G et al (2016) Chromenopyrazole, a versatile cannabinoid scaffold with in vivo activity in a model of multiple sclerosis. J Med Chem 59:6753–6771
Musella A, Sepman H, Mandolesi G et al (2014) Pre- and postsynaptic type-1 cannabinoid receptors control the alterations of glutamate transmission in experimental autoimmune encephalomyelitis. Neuropharmacology 79:567–572
Navarrete C, Carrillo-Salinas F, Palomares B et al (2018) Hypoxia mimetic activity of VCE-004.8, a cannabidiol quinone derivative: implications for multiple sclerosis therapy. J Neuroinflammation 15:64–83
Nielsen S, Germanos R, Weier M et al (2018) The use of cannabis and cannabinoids in treating symptoms of multiple sclerosis: a systematic review of reviews. Curr Neurol Neurosci Rep 18:8
Novotna A, Mares J, Ratcliffe S et al (2011) A randomized, doubleblind, placebo-controlled, parallel-group, enriched-design study of nabiximols (Sativex), as add-on therapy, in subjects with refractory spasticity caused by multiple sclerosis. Eur J Neurol 18:1122–1131
Orihuela R, McPherson CA, Harry GJ (2016) Microglial M1/M2 polarization and metabolic states. Br J Pharmacol 173:649–665
Ortega-Gutiérrez S, Molina-Holgado E, Arévalo-Martín A et al (2005) Activation of the endocannabinoid system as therapeutic approach in a murine model of multiple sclerosis. FASEB J 19:1338–1340
Peferoen L, Kipp M, van der Valk P et al (2014) Oligodendrocyte-microglia cross-talk in the central nervous system. Immunology 141:302–313
Petro DJ, Ellenberger C (1981) Treatment of human spasticity with delta-9-tetrahydrocannabinol. J Clin Pharmacol 21:413S–416S
Polman CH, Reingold SC, Banwell B et al (2011) Diagnostic criteria for multiple sclerosis: 2010 revisions to the McDonald criteria. Ann Neurol 692:292–302
Pryce G, Cabranes A, Fernández-Ruiz J et al (2013) Control of experimental spasticity by targeting the degradation of endocannabinoids using selective- fatty acid amide hydrolase inhibitors. Mult Scler 19:896–1904
Rahimi A, Faizi M, Talebi F et al (2015) Interaction between the protective effects of cannabidiol and palmitoylethanolamide in experimental model of multiple sclerosis in C57BL/6 mice. Neuroscience 290:79–287
Rog DJ, Nurmikko TJ, Friede T et al (2005) Randomized, controlled trial of cannabis-based medicine in central pain in multiple sclerosis. Neurology 65:812–819
Rog DJ, Nurmikko TJ, Young CA (2007) Oromucosal delta9-tetrahydrocannabinol/cannabidiol for neuropathic pain associated with multiple sclerosis: an uncontrolled, open label, 2-year extension trial. Clin Ther 29:2068–2079
Russo E, Guy GW (2006) A tale of two cannabinoids: the therapeutic rationale for combining tetrahydrocannabinol and cannabidiol. Med Hypotheses 66:34–246
Schlosburg JE, Blankman JL, Long JZ et al (2010) Chronic monoacylglycerol lipase blockade causes functional antagonism of the endocannabinoid system. Nat Neurosci 13:1113–1119
Shakespeare DT, Boggild M, Young C (2003) Anti-spasticity agents for multiple sclerosis. Cochrane Database Syst Rev 4:CD001332
Sharma C, Al Kaabi JM, Nurulain SM et al (2016) Polypharmacological properties and therapeutic potential of -caryophyllene: a dietary phytocannabinoid of pharmaceutical promise. Curr Pharm Des 22:3237–3264
Shi Y, Duan YH, Ji YY et al (2017) Amidoalkylindoles as potent and selective cannabinoid type 2 receptor agonists with in vivo efficacy in a mouse model of multiple sclerosis. J Med Chem 60:7067–7083
Stys PK, Zamponi GW, van Minnen J (2012) Will the real multiple sclerosis please stand up? Nat Rev Neurosci 137:507–514
Svendsen KB, Jensen TS, Bach FW (2004) Does the cannabinoid dronabinol reduce central pain in multiple sclerosis? Randomized double blind placebo controlled crossover trial. BMJ 329:253
Turcotte D, Doupe M, Torabi M et al (2015) Nabilone as an adjunctive to gabapentin for multiple sclerosis-induced neuropathic pain: a randomized controlled trial. Pain Med 16:149–159

Ungerleider JT, Andrysiak T, Fairbanks L et al (1987) Δ-9-THC in the treatment of spasticity associated with multiple sclerosis. Adv Alcohol Subst Abuse 7:39–50

Wade DT, Makela P, House H et al (2006) Long-term use of a cannabis-based medicine in the treatment of spasticity and other symptoms in multiple sclerosis. Mult Scler 12:639–645

Wade DT, Makela P, Robson P et al (2004) Do cannabis-based medicinal extracts have general or specific effects on symptoms in multiple sclerosis? A double-blind, randomized, placebo-controlled study on 160 patients. Mult Scler 10:434–441

Wang T, Collet JP, Shapiro S et al (2008) Adverse effects of medical cannabinoids: a systematic review. Can Med Assoc J 178:1669–1678

Webb M, Luo L, Ma JY et al (2008) Genetic deletion of fatty acid amide hydrolase results in improved long-term outcome in chronic autoimmune encephalitis. Neurosci Lett 439:106–110

Whiting PF, Wolff RF, Deshpande S et al (2015) Cannabinoids for medical use: a systematic review and meta-analysis. JAMA 313:2456–2473

Wirguin I, Mechoulam R, Breuer A et al (1994) Suppression of experimental autoimmune encephalomyelitis by cannabinoids. Immunopharmacology 28:209–214

Wissel J, Haydn T, Muller J et al (2006) Low dose treatment with the synthetic cannabinoid nabilone significantly reduces spasticity-related pain: a double-blind placebo-controlled cross-over trial. J Neurol 253:1337–1341

Zajicek J, Ball S, Wright D et al (2013) Effect of dronabinol on progression in progressive multiple sclerosis (CUPID): a randomized, placebo-controlled trial. CUPID investigator group. Lancet Neurol 12:857–865

Zajicek J, Fox P, Sanders H et al (2003) Cannabinoids for treatment of spasticity and other symptoms related to multiple sclerosis (CAMS study): multicentre randomized placebo-controlled trial. Lancet 362:1517–1526

Zajicek JP, Sanders HP, Wright DE et al (2005) Cannabinoids in multiple sclerosis (CAMS) study: safety and efficacy data for 12 months follow up. J Neurol Neurosurg Psychiatry 76:1664–1669

Zhang M, Martin BR, Adler MW et al (2009) Modulation of cannabinoid receptor activation as a Neuroprotective strategy for EAE and stroke. J Neuroimmune Pharmacol 4:249–259

Zhornitsky S, Potvin S (2012) Cannabidiol in humans-the quest for therapeutic targets. Pharmaceuticals 5:529–552

Psychiatric Disorders and Cannabinoid Receptors

9

Neal Joshi and Emmanuel S. Onaivi

Abstract

With the increasing global use of medical and adult recreational use of cannabis and cannabinoids, this chapter provides overview of evidence from animal and human studies on psychiatric disorders and cannabinoid receptors. We review and present evaluation of the relationship between changes in the ECS and psychiatric disorders. Evidence suggests the existence of a relationship between changes in components of the ECS, and some of the symptoms present in psychiatric disorders. Both CB1Rs and CB2Rs are components of the endocannabinoid system with different cellular and tissue localization patterns that are differentially expressed in the CNS and PNS and are emerging targets for the treatment of number psychiatric disorders. As cannabis preparations are widely used for recreation globally, it is predictable that cannabis use disorders (CUDs) will increase and there is currently no available treatment for CUDs. Although major advances have been reported from cannabinoid and ECS research, there are gaps in scientific knowledge on long-term consequences of cannabis use. Adolescent and cannabis use during pregnancy presents further challenges, and more research will uncover the signaling pathways that couple the gut microbiota with the host ECS. Development of cannabis and cannabinoid nanomedicine for nanotherapy will certainly overcome some of the shortcomings and challenges in medicinal and recreational use of cannabis and cannabinoids. Thus, nanotechnology will allow targeted delivery of cannabinoid formulations with the potential to elevate their use to scientifically validated nanotherapeutic applications as the field of cannabis nanoscience matures.

Keywords

Cannabinoid receptors · Endocannabinoids · Endocannabinoid system · Cannabis · Cannabinoids · Psychiatric disorders

Abbreviations

2-AG	2-Arachidonyl glycerol
ABHD6 ABHD12	alpha beta hydrolase domain proteins
AD	Alzheimer's disease
ADHD	Attention hyperactivity disorders
ASDs	Autism spectrum disorders
BT	Brain tumors
CBRs	Cannabinoid receptors

N. Joshi
Rowan University School of Osteopathic Medicine, Stratford, NJ, USA

E. S. Onaivi (✉)
Department of Biology, William Paterson University, Wayne, NJ, USA
e-mail: Onaivie@wpunj.edu

E. Murillo-Rodriguez et al. (eds.), *Cannabinoids and Neuropsychiatric Disorders*, Advances in Experimental Medicine and Biology 1264, https://doi.org/10.1007/978-3-030-57369-0_9

CNR	cannabinoid receptor gene
CNS	central nervous system
CP	cerebral palsy
CUDs	Cannabis use disorders
DSI/DSE	Depolarization-induced suppression of inhibition / excitation
eCB	endocannabinoids
ECS	endocannabinoid system
ENS	enteric nervous system
ERK1/2	Extracellular signal-regulated kinase 1/2
ET	essential tremor
FAAH	Fatty acid amide hydrolase
GABA	Gamma-Aminobutyric acid
GPCR	G protein-coupled receptor
HD	Huntingtin disease
IC	insular cortex
JNK	c-jun N-terminal kinase
LTD	long-term depression
MAGL	Monoacylglycerol lipase
MAOIs	monoamine oxidase inhibitors
MS	Multiple sclerosis
NAc	nucleus accumbens
PBMCs	peripheral blood mononuclear cells
p-CREB	phosphor-cAMP response element-binding protein
PD	Parkinson's disease
PET	positron emission tomography
PI3K/Akt	phosphatidylinositol-3-kinases / protein kinase B
PNS	Peripheral nervous system
PPARs	peroxisome proliferator-activated receptors
SNPs	single nucleotide polymorphisms
SNRIs	selective norepinephrine reuptake inhibitors
SSRIs	selective serotonin reuptake inhibitors
TBI	Traumatic brain injury
TCAs	tricyclic antidepressants
THC	tetrahydrocannabinol
TRPV1	transient receptor potential vanilloid type 1
TS	Tics and Tourette's syndrome

9.1 Introduction

The shifting landscape on cannabis medicalization, legalization, and recreational use is due in part to advances in cannabis and cannabinoid molecular genetics, and the discovery of the endocannabinoid system (ECS) in human body and brain (Onaivi et al. 2012). Review of the accumulating scientific evidence from cannabis plant constituents, animal, and human studies reveals a previously unknown but ubiquitous and complex cannabinoid and ECS that is involved in almost all aspects of mammalian physiology and pathology (Joshi and Onaivi 2019). In this chapter, we review the growing awareness of the pharmacotherapeutic potential and limitation of targeting cannabinoid receptors (CBRs) and other components of the ECS in psychiatry. The components of the ECS are emerging as multifaceted therapeutic targets for cannabis constituents in a number of psychiatric and neurological disorders. The accruing evidence for the therapeutic efficacy of phytocannabinoids is promising for a number of psychiatric disorders. Although cannabis has been used for millennia, it is not benign, and there are side effects associated with use and exposure during pregnancy and in adolescents with psychiatric vulnerability. Furthermore, the type 1 cannabinoid receptors (CB1Rs) are now regarded as one of the most abundant G protein-coupled receptors (GPCRs) in the mammalian brain and have been extensively studied for their role in the biology of depression, pain, behavior, anxiety, neurodegenerative diseases, nausea, and substance abuse disorders. However, the neuronal function of type 2 cannabinoid receptors (CB2Rs) has been less investigated for central nervous system (CNS) function, because they were thought to be predominantly found in immune cells in the periphery and were called peripheral CB2Rs. With the increasing global use of cannabis and the risk of cannabis use disorders (CUDs), there are still lingering doubts, controversy, and debate about the functional neuronal localization and role of CB2Rs (Ghose 2009). The drawback from some previous studies has been the unsuccessful attempts by some groups to generate mice with selective deletion of CB2Rs from neurons and the generation of CB2R-GFP and CB2R-EGFP with peripheral CB2R promoter-driven transgenic reporter mouse line or other flaws in designs that detected microglial but not neuronal expression of CB2Rs (Ghose 2009; Lopez et al. 2018; Schmöle et al. 2015). Indeed, our studies provided the first evidence for neuronal CNS effects of CB2Rs, and its possible role in drug addiction, eating disorders, psychosis, depression, and autism spectrum disorders (ASDs), (Onaivi et al. 2006a; Onaivi et al. 2015; Onaivi et al. 2013; Onaivi et al. 2008; Ishiguro et al. 2007; Ishiguro et al. 2010a; Ishiguro et al. 2010b; Onaivi et al. 2011). Cell type-specific mechanisms of CB2Rs are unclear because the existing CB2R gene knockout mice are constitutive gene knock, with partial/incomplete deletion of CB2Rs, and are not suitable for tissue- and cell type-specific studies at molecular, pharmacological, and behavioral levels. Therefore, using Cre/Lox P technology, we created *Cnr2*-floxed mice to produce CB2R cKO, DAT-*Cnr2,* and Cx3cr1-*Cnr2* mice with deletion of CB2Rs in dopamine neurons and microglia, respectively (Liu et al. 2017). Characterization of these mice provides further evidence for the involvement of CB2Rs in models of psychiatric function and disorders (Liu et al. 2017; Onaivi et al. 2018; Canseco-Alba et al. 2018a; Canseco-Alba et al. 2019; Canseco-Alba et al. 2018b).

With increasing global decriminalization and legalization of adult cannabis use, there are growing concerns for CUDs (Acheson and Fantegrossi 2019; Melis et al. 2017), especially as cannabis and cannabinoids including cannabidiol (Premoli et al. 2019) are thought to be safe in comparison with the effects of alcohol, tobacco products, and opioids. In light of the opioid epidemic in USA, there is distinct CNS localization of mu-opioid receptors—the target of opioids and CBRs—the target of cannabis and cannabinoids. CBRs both CB1Rs and CB2Rs are distributed in different areas of the brain; they are not in the pons and medulla oblongata, areas controlling breathing and respiration. This is why there are no cannabis overdosing resulting in respiratory depression or cardiovascular failure that is associated with the current opioid epidemic. This is because opioid receptors are abundant in the pons and medulla oblongata—brain areas involved in the control of respiration, and overdosing on opioids are associated with respiratory depression from opioid addiction. Thus, there is an increasing focus on the therapeutic effects (Premoli et al. 2019), of cannabis and cannabinoids, and targeting CBRs and components of the ECS in psychiatric disorders as reviewed in this chapter.

9.2 Endocannabinoid Signaling in Psychiatric Disorders

Disruption of the endocannabinoid system is associated with psychopathologies involved in psychiatric disturbances and neurological disorders (Onaivi et al. 2015). The ECS consists of CBRs that are activated by cannabinoids, endocannabinoids (eCBs), and their metabolic enzymes (Onaivi et al. 2012; Joshi and Onaivi 2019; Zou and Kumar 2019). Cannabinoids modulate signal transduction pathways associated with GPCRs, ionotropic, and nuclear receptors to exert their biological and therapeutic effects in psychiatry (Zou and Kumar 2019). G protein-coupled receptor, GPCR-CBR signaling activities in neuronal, glia cells, and ion channels have been demonstrated in vitro and in vivo models. A number of these studies provide CBR signaling network and pathways associated with mitogen-activated protein kinase (MAKP) signaling pathways including extracellular signal-regulated kinase 1/2 (ERK1/2), c-jun N-terminal kinase (JNK), p38, β-arrestins, and phosphatidylinositol-3-kinases (PI3K) / protein kinase B (Akt) pathways. For example, CBR-mediated β-arrestin 1 and 2 translocation is a key mediator of GPCR desensitization and internalization that is species, subtype of CBRs, and agonist dependent (Zou and Kumar 2019; Ibsen et al. 2019). eCBs exert modulatory action on retrograde signaling and act as retrograde messengers at many synapses in the CNS. Some investigators have suggested GPR55 as a putative CB3R as it triggers distinct signaling pathways in response to inflammatory mediators. However, we and others have suggested that TRPV1 or VR1 be classified as CB3R, as anandamide is a CBR partial agonist, but a full agonist at the transient receptor potential cation channel subfamily V member 1, (TRPV1) also called vanilloid receptor 1 (VR1) (Joshi and Onaivi 2019). It turns out that 2-AG is associated with the effects of eCB-mediated retrograde signaling by cannabinoids as the level of 2-arachidonyl glycerol (2-AG) is about 1000 times more than anandamide (AEA) (Zou and Kumar 2019). Therefore, 2-AG serves as retrograde messenger at various types of synapses throughout the brain and is a high efficacy ligand for the translocation of β-arrestins (Ibsen et al. 2019). However, AEA has also been shown to contribute to eCB-mediated synaptic transmission with evidence supporting a tonic role of AEA (Figs. 9.1 and 9.2). The modulatory action of eCB-induced retrograde signaling on GABA-ergic and glutamatergic systems indicates that the main excitatory and inhibitory systems are in part under the influence of the ECS. The discovery that eCBs are principal mediators of retrograde synaptic communication demonstrates the pivotal role that eCBs play as retrograde messengers in GABA-ergic and glutamatergic synapses. One of the major advantages for the physiological actions of CBRs may be associated with the quick response needed for the sequestration of G-proteins and the retrograde signaling of eCBs on presynaptic CB1Rs to inhibit neurotransmitter release. Such retrograde action of eCBs on the inhibition of neurotransmitter release like GABA, glutamate, serotonin, dopamine is modified by CBRs and other components of the ECS that are molecular targets in psychiatric disorders. Depolarization-induced suppression of inhibition (DSI) / excitation (DSE) and in long-term depression (LTD) at both excitatory and inhibitory synapses provided evidence supporting retrograde eCB signaling system in the brain (Castillo et al. 2012; Ohno-Shosaku and Kano 2014).

Thus, eCBs are critically involved in the suppression of synaptic transmission and eCB-mediated communication between neurons and microglia. This is one of the current focuses of study with the identification of CB2R neuro-immune cross talk following conditional deletion of CB2Rs in microglia and dopamine neurons (Liu et al. 2017; Onaivi et al. 2018; Canseco-Alba et al. 2018a; Canseco-Alba et al. 2019; Canseco-Alba et al. 2018b). Therefore, existing evidence suggests that alterations in endocannabinoid signaling are present in a range of psychiatric disorders. Targeting components of ECS components provides therapeutic potential of cannabinoid medicines as CBRs and other components of the ECS that are involved in diverse neural, immune, function, and dysfunction (Robson 2014), in psychiatric disorders.

9.3 ECS Modulation of Neuro-Immune–Microbiome Cross Talk in Psychiatric Disorders

ECS components participate in numerous physiological processes that include immune and metabolic regulatory functions contributing to the maintenance of an organism's homeostasis (Onaivi et al. 2012; Magid et al. 2019; Lin et al. 2013). Increased eCB levels and signaling have been linked with inflammation in animal models and human inflammatory disturbances and CBRs providing protective effects in the inflamed gut (Lin et al. 2013). New and improved knowledge has revealed gut-brain neural communication (Kaelberer et al. 2018), and the gut microbiome has been found to signal in part through the ECS network (Di Marzo 2018). Interactions between gut microorganisms and the ECS highlight a role in the gut microbiota-ECS axis (Lin et al. 2013; Kaelberer et al. 2018; Di Marzo 2018; Cani et al. 2016). Thus, the gut microbiome and endocannabinoidome relationship seems to be bidirectional and their alterations have been linked with dysbiosis, increased microglial neuroinflammation in mental disorders, such as psychosis, depression, anxiety, and neurological disturbances (Di Marzo 2018; Cani et al. 2016; DiPatrizio 2016). CBRs and associated orphan and putative cannabinoid GPCRs are expressed at high levels in the immune and/or central

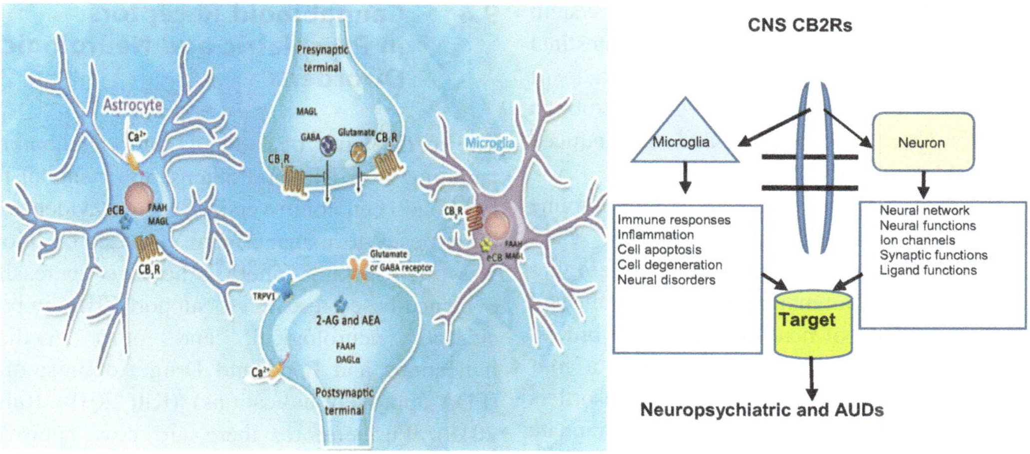

Fig. 9.1 Schematic of ECS neuro-immune signaling in neuropsychiatric disorders. In our studies, we used both IBA1 and CD11b antibodies, as IBA1 is a good marker that does not cross-react with neurons and astrocytes, while CD11b is a good marker for changes in microglia morphology

nervous systems (CNS) and regulate a number of neurophysiological processes, including key events involved in neuroinflammation. As such, these receptors have been identified as emerging therapeutic targets for a number of brain disorders in which neuroinflammation is a key feature, including multiple sclerosis (MS) and Alzheimer's disease (AD). There is increasing attention on the role of the ECS components in mediating immune function with a focus on the immune processes that contribute to neuroinflammatory conditions. The gut

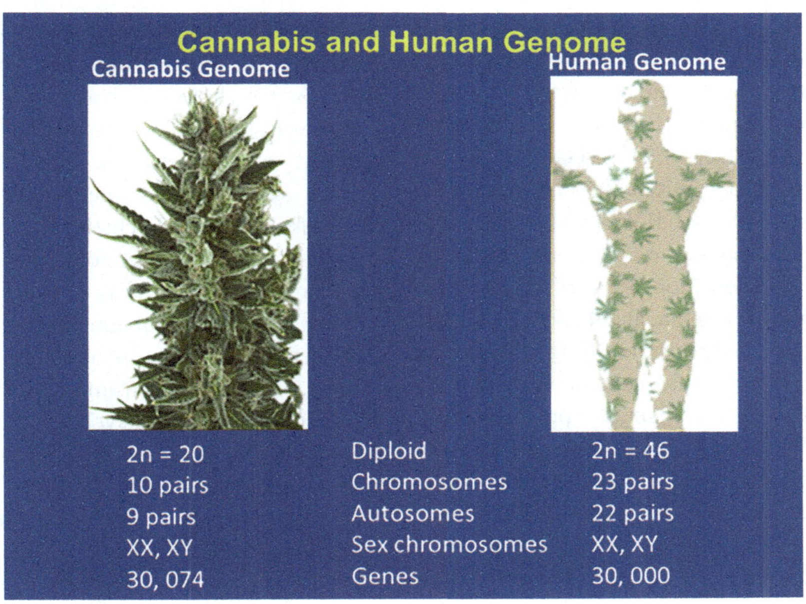

Fig. 9.2 Cannabis and Human genomes. The decoding of the cannabis and human genomes with 10 and 23 chromosomal pairs with 9 and 22 autosomes, respectively, have similar sex chromosomes. There are 30, 074 genes in the decoded cannabis genome and 30, 000 human genes

microbiota plays a critical role in immune system function and regulation of gastrointestinal activities (Kaelberer et al. 2018). Blockade of the CB1R was reported to alter gut microbiota and attenuates inflammation and diet-induced obesity (Mehrpouya-Bahrami et al. 2017).

Microglia cells are the brain's innate immune cells and primary contributors of CNS neuroinflammatory responses. Microglia-mediated neuroinflammation has been implicated in the progression of neuropsychiatric disorders. Neuro-immune signaling is emerging as a contributor to a number of neuropsychiatric disorders and stress increase in brain levels of known innate immune signaling molecules (Crews et al. 2017). Neuroinflammation is also emerging as a key component in the effects of CB2Rs, expressed in macrophages, microglia, and neurons (Fig. 9.1) that are also key regulators of the immune response (Onaivi et al. 2012). eCB-mediated communication between neuron and neuroglia shows that microglial cells and astrocytes are able to produce eCBs (Zou and Kumar 2019). After conducting the initial pioneering work that culminated in the discovery of functional neuronal CB2Rs (Onaivi et al. 2006b), we have now created and generated mice with cell type-specific deletion of CB2Rs to continue to move the field forward. This will allow us to investigate the cell type-specific functional roles of CB2Rs using DAT-*Cnr2* and Cx3Cr1-*Cnr2* cKO mice to determine the neuro-immune basis of CB2R activity in alcohol preference and consumption. With accumulating evidence of an interaction between the immune system, the gut microbiota, and the ECS, we have initiated studies to determine the role of neuro-immune-microbiome endocannabinoid axis in mouse CNS models to identify pro- and anti-inflammatory effects of CBRs. Others have successfully generated Syn-*Cnr2* cKO mice in which synaptic deletion of CB2Rs was shown to mediate a cell type-specific plasticity in the hippocampus (Stempel et al. 2016), and from our recent reports (Liu et al. 2017; Canseco-Alba et al. 2019; Onaivi et al. 2006b).

9.4 Cannabinoid Receptors in Psychiatric and Neurological Disorders

The medical use of cannabis and cannabinoids targeting the CBRs and other components of the ECS had been controversial because evidence of the efficacy to manage many disease conditions was often lacking. There is however increasing evidence-based research to support CBRs in psychiatric, neurological, and other medical indications and Food and Drug Administration (FDA-approved indications) (Kill 2019; Rubin 2018). Furthermore, there are now approved FDA indications supported by high-quality emerging evidence, and current claims and uses for which there is inadequate evidence (Kill 2019). The conclusions of National Academies of Sciences, Engineering and Medicine on the current state of evidence on the health effects of cannabis and cannabinoids provide support for the legitimate study, regulation, and prescription of therapeutic cannabinoids (National Academies of Sciences, Engineering, and Medicine 2017). Not surprisingly, based on positive randomized clinical trials, dronabinol and nabilone have FDA approval for chemotherapy-induced nausea and vomiting and appetite stimulation in conditions that cause weight loss such as in HIV-AIDS. FDA also recently approved cannabidiol for the management of Dravet syndrome and Lennox-Gastaut syndrome that are pediatric epilepsies (Devinsky et al. 2017; Thielle et al. 2018). Thus, disruption of CBRs and endocannabinoid signaling is implicated in an array of psychopathologies ranging from autoimmune diseases associated with neuropsychiatric mental disturbances like anxiety, Autism spectrum disorders (ASDs), depression, insomnia, psychosis, addiction, and neurological disorders such as Alzheimer's, epilepsy, multiple sclerosis, Parkinsonism, stroke, traumatic brain injury. However, acute or chronic activation of CBRs can produce neurologic adverse effects including impaired learning, memory, attention, and motor coordination, while chronic use can lead to cannabis use disorders (CUDs) in vulnerable individuals (Kill

2019). It is also important to consider the effects of cannabis use in those with mental illness and individuals predisposed to developing addictive disorders (Lowe et al. 2019).

9.4.1 Role of CBRs in Anxiety-Related Disorders

Anxiety disorders are a common global mental illness characterized by feelings of fear and anxiogenesis, and the ECS system is involved in the bidirectional regulation of neural anxiety circuits and behavior (Yin et al. 2018). Anecdotally, different subjective effects have been reported in individual who have smoked marijuana, and in some instances, the opposite have been reported. This is not surprising as numerous constituents of the cannabis plant, not only induce biphasic dose-response profile, with low and high doses producing anxiolysis and anxiogenesis, but also have antagonistic effects (Yin et al. 2018; Onaivi et al. 1990). Therefore, systematic research investigating the interactions between CBR activity and which cannabis constituent or mixtures mediates anxiolysis remains an area of study with the changing legal status of medicinal and legal use of cannabis. However, advances and progress in ECS and cannabinoid research using transgenic mouse models and proxy methods of accessing states are providing the underlying mechanisms and brain circuits associated with expression of the different CBR subtypes and ECS machinery involved in the mediation of anxiety. For example, CB1Rs are densely expressed pre-synaptically and involved in retrograde signally associated with the inhibition of neurotransmitter releases, whereas our recent studies with mice with selected deletion of CB2Rs from dopamine neurons showed a reduced anxiety-like behavior in the plus maze and two-compartment black and white models of anxiety (Liu et al. 2017). Impaired 2-AG signaling in hippocampal glutamatergic neurons has been reported to affect anxiety-like behavior (Guggenhuber et al. 2015). CBRs have been identified in the insular cortex (IC) of rodents and humans. IC hyperactivity and its connectivity to amygdala have been linked with affective and anxiety disorders (Andrade et al. 2019). Studies in humans have investigated THC and cannabidiol (CBD) and have reported opposing effects, with higher doses of THC producing anxiogenic and CBD producing anxiolysis and counteracting the effects of THC (Andrade et al. 2019; Papagianni and Stevenson 2019). Anxiety- and other trauma-related disorders are common psychiatric disturbances with inadequate therapeutic options. These anxiety disorders include generalized anxiety, panic, social anxiety, phobias, and separation anxiety, with post-traumatic stress disorder (PTSD) and obsessive-compulsive disorder. The current treatment approaches with benzodiazepines and selective serotonin reuptake inhibitors (SSRIs) have limitations and side effects. Preclinical and ongoing clinical studies suggest that anxiety and related disorders are associated with decreased eCB tone and that CBRs in the brain are involved in the anxiolytic effects of cannabinoids (Papagianni and Stevenson 2019; Sloan et al. 2018; Korem et al. 2016). Therefore, more research is needed to elucidate the constituents of cannabis and what ECS components can be targeted for the therapeutic potential for the treatment of anxiety disorders. Limited positron emission tomography (PET) studies using available radiotracers in humans have revealed dysregulation of CB1Rs. The development of radiotracers for other components of the ECS may reveal novel targets in a number of psychiatric and neurological disorders in anxiety (Papagianni and Stevenson 2019; Sloan et al. 2018). There has also been recent focus on CBD for a number of neuropsychiatric disorders with some success in pediatric epilepsies, and many trials have begun for the treatment of other neuropsychiatric diseases. CBD appears to be a multi-target drug, and the molecular mechanism (s) of action of CBD in many of the psychiatric disorders are of major research efforts (Premoli et al. 2019). In vivo imaging is a powerful way to quantify CB1R radioligand binding in patients. A study looked at the CB1R radioligand, [^{11}C]OMAR, in patients with PTSD using PET. In the PTSD group, [^{11}C]OMAR volume of distribution was increased compared to healthy controls and trauma-exposed controls. Interestingly, but also consistent with the fact that women are more prone to anxiety disorders, this finding was more pronounced in women. Peripheral anandamide concentrations were lower in the PTSD group compared to healthy and trauma controls, which may be involved in why there was receptor upregulation in the PTSD group. Peripheral cortisol levels in the PTSD group and trauma control groups were lower compared to the healthy control. These findings may pave way for understanding and determining if medical marijuana is efficacious for PTSD. In some states, medical marijuana is approved for PTSD and some studies indicate symptomatic improvement from this treatment. Such data would support the notion that marijuana can provide symptomatic relief due to decreased peripheral anandamide concentrations along with CB1R upregulation.

9.4.2 Role of CBRs in Depression

Neuroanatomical distribution of CBRs and components of the ECS in neural circuits associated with processing and regulation of human emotion have been linked to the formation and development of depression (Zhou et al. 2017; Arjmand et al. 2019; Ishiguro et al. 1836; Ibarra-Lecue et al. 2018). Evidence from preclinical and clinical studies indicates an impairment of the ECS pathway in animal models and in patients with depression (Poleszak et al. 2018). Pathway analysis following a bipolar disorder genome-wide association study identifies and implicates ECS gene sets, indicating that bipolar disorder is part of a spectrum of highly correlated psychiatric and mood disorders. Further evidence from patients with bipolar disorders indicates a putative role of CB2Rs (Zhou et al. 2017; Arjmand et al. 2019; Ishiguro et al. 1836; Ibarra-Lecue et al. 2018). Although most studies have focused on CB1Rs, our studies have implicated the involvement of CB2Rs in rodent models of CNS disorders and in human subjects with psychiatric disorders like substance abuse, depression, and psychosis (Onaivi et al. 2008; Ishiguro et al. 2007; Ishiguro et al. 2010a). There are other findings and reports showing alteration in eCB levels, metabolizing enzymes, and CBRs in patients with psychiatric disorders. Table 9.1 shows polymorphisms in the CB1R and CB2R genes that are associated with psychiatric disorders. CBRs and components of the ECS influence the activity and function of neural circuits associated with hypothalamic-pituitary-adrenal (HPA) axis, involved in affective behaviors. The use of CB1R antagonist as anti-obesity medication was withdrawn due to its anxiety-, depression-, and suicide-inducing effects in some patients. This revealed a role of CB1Rs in depression (Ibarra-Lecue et al. 2018; Onaivi 2010). Polymorphisms in eCB metabolizing enzymes along with alterations in eCB levels have also been implicated in depression. Furthermore, the ECS has been shown to modulate multiple monoaminergic systems, which can also influence mood and cognition. The raphe nuclei, locus coeruleus, and ventral tegmental area are targets of many classes of pharmacological agents such as selective serotonin reuptake inhibitors (SSRIs), selective norepinephrine reuptake inhibitors (SNRIs), tricyclic antidepressants (TCAs), and monoamine oxidase inhibitors (MAOIs). All of these areas and many of their inputs and outputs are modulated by the ECS. Therefore, it is no surprise that the ECS may have potential therapeutic effects in depression and other mood disorders. CB1R density and their downstream signaling were increased in the dorsolateral prefrontal cortex (DLPFC) in post-mortem subjects with depression-related suicides. The increased levels of CB1R-mediated signaling using a [^{35}S]GTPγS binding assay in suicide subjects were comparable to controls (Hungund et al. 2004). These results may point to a potential therapeutic treatment in depression and suicidal patients. Although many studies have looked at CB1Rs, exciting studies are now emerging looking at central CB2Rs in depression and anxiety. A recent study investigated a Cnr2 knockout mouse using assays for depressive behavior, anxiety, and motor activity. Using an immune stressor, poly I:C, heterozygous *Cnr2*-KO mice experienced significantly decreased amount of time spent in the open arms of the zero maze

Table 9.1 Endocannabinoid system and CNR gene polymorphisms in psychiatric disorders

CNR Genes Polymorphism (Onaivi et al. 2013)	Linkage or Association
CB2R	Associate with mouse model of impulsivity behavior
CNR2 (Q63R) but not (H316Y)	Associated with alcoholism and depression
CNR2 (rs41311993)	Associated with bipolar disorder
CNR2 (FAAH C385A)	Associated with childhood trauma, anxiety, and depression
CNR2 (R63Q)	Associated with childhood trauma, anxiety, and depression
CNR1/ FAAH gene	Associated impulsivity and marijuana use
CNR1 rs806375, rs806371, rs806368	Associated with drug addiction
CNR1 rs806380, rs806368, rs754387	Associated with cannabis dependence
1359 G/A *CNR1* variant	Associated with alcohol dependence
1359 G/A *CNR1* variant	Not associated with Tourette syndrome
1359 G/A *CNR1* variant	Not associated with alcohol withdrawal tremens
CNR1, FAAH, DRD2 gene	Associated with comorbidity of alcoholism & antisocial
CNR1 SNPs	No association with anorexia nervosa
CNR1 (AAT)n repeats	Associated with restricting and binging/purging anorexia nervosa
CNR1 (AAT)n repeats	Associated with depression in Parkinson's disease
CNR1 SNPs	Associated to striatal responses to facial expression
(AAT)n repeats	Association with attention deficit hyperactivity disorder (ADHD) in alcoholics
CNR1 SNP haplotype	Risk factor for ADHD and post-traumatic stress disorder (PTSD)
1359 G/A *CNR1* variant	Associated with schizophrenia
(AAT)n repeats	Not associated with schizophrenia and mood disorders
(AAT)n repeats	Associated with schizophrenia
(AAT)n repeats	Associated with hebephrenic schizophrenia
CNR1 rs6454674	Associated with schizophrenia severity
CNR1 variants	Associated with depression and anxiety
CNR1 variants and (AAT)n repeats	Associated with impulsivity
1359 G/A CNR1 tag SNP	Associated with antipsychotic response but not schizophrenia
CNR1 SNPs	No association with cognitive impairment in MS
*CNR1*rs12720071	Associated with cognitive performance in schizophrenia
CNR1 rs1049353	Associated with reduction in caudate volume in psychosis
CNR1 rs2023239	Associated with a protective effect against lifetime MDD

[a]The above includes gene polymorphisms catalog indicating inconsistencies in variation of the ECB system CB1R and CB2R genes in some neuropsychiatric disturbances. Further studies with more number of participants from different ethnic backgrounds are required, with consideration of epigenetic and exposomic factors (Onaivi et al. 2013)

compared to saline-treated controls, indicating an increase in anxiety-like behavior. Furthermore, anxiety-like behavior was further bolstered in the rotated pulley model, which was induced by poly I:C, showed that the *Cnr2*-KO mice had a higher number of pulley rotations compared to controls (Ishiguro et al. 2018). ECS involvement in depression presents components of the ECS as therapeutic targets for the development antidepressants (Poleszak et al. 2018).

9.4.2.1 Role of CBRs in the Antidepressant Action of Ketamine

The recent approval by the FDA for a nasal ketamine spray (esketamine; Spravato) for treatment-resistant depression has made the inquiry into its mechanism even more pressing. A recent study probed into the possibility of the ECS and its interaction with ketamine in relation to depression in mice. First, they injected 5 and 10 mg/kg of ketamine, which reduced immobility time in the forced swim test (FST), compared to control, which shows an antidepressant effect. A CB1R agonist, ACPA, and an antagonist, AM251, were also used. Administration of ACPA showed decreased immobility time compared to control in the FST, indicating an antidepressant effect. Interestingly, when administering combined non-effective doses of AM251 or a CB2 inverse agonist, AM630 with non-effective doses of ketamine resulted in antidepressant effects compared to control (Khakpai et al. 2019). The link with the ECS might partially explain the mechanism behind ketamine's antidepressive effects. Upon activation of the CBRs along with the retrograde signaling associated with the modulation of other neurotransmitters, the hypothalamic-pituitary-adrenal axis is associated with providing therapeutic benefits (Onaivi et al. 2015; Onaivi et al. 2013). This is the bases for the modulation of the ECS as therapeutic targets, or as adjunctive treatment in generalized depression. While the role of the ECS in depression is incompletely understood, a number of anecdotal reports and individual variation in antidepressant and depression-like symptoms after consumption of cannabis have been reported (Onaivi et al. 2015; Onaivi et al. 2013).

9.4.3 Role of CBRs in Schizophrenia

Schizophrenia also known as psychosis is a devastating psychiatric syndrome affecting 1% of the world population without a cure, and antipsychotic medications are not effective in all patients. Despite the efforts to elucidate the causes of schizophrenia, the etiopathogenesis remains elusive (Ibarra-Lecue et al. 2018). In vulnerable individuals, the use of cannabis has been known to precipitate psychotic episodes. Alterations in the ECS components in the brains of patients with schizophrenia have been reported (Castillo et al. 2012), and CB1Rs have been implicated in many psychiatric disorders and recent advances have sparked more interest into the ECS role in schizophrenia. As discussed above, ECS system components are expressed in the immune and/or central nervous systems (CNS) and regulate a number of neurophysiological processes, including key events involved in neuroinflammation. ECS interaction with neuroglia changes has been implicated in the pathobiology of schizophrenia. Results on the changes and alterations in CBRs from imaging and postmortem studies of brains from schizophrenia patients compared to controls have yielded inconsistent data with either increase or decrease in CB1R and gene expression in brain areas involved in schizophrenia (Ibarra-Lecue et al. 2018). Many factors have been considered for the inconsistent results when studying CBRs and other components of the ECS in brains of schizophrenics, including age, sex, and the subtypes of schizophrenia. The anterior cingulate gyrus in patients with schizophrenia, bipolar disorder, and major depression were evaluated in postmortem brains, and CB1R immunohistochemical stain was used. In the major depression group, the intake of SSRIs reduced the density of cortical CB1R immunoreactive neurons compared to control. In bipolar disorder, patients taking first-generation antipsychotics had reduced glial CB1R immunoreactive density (Koethe et al. 2007). The results from CB1R changes in different schizophrenic brain regions, provided an indication for the involvement of CB1Rs in this pathology (Ibarra-Lecue et al. 2018). The role of CB2Rs in schizophrenia has not been well investigated when compared to CB1Rs. This was because many investigators were not able to detect the presence of neuronal CB2Rs in healthy brains, and therefore, their role in neuropsychiatric disorders has been much less well characterized. We and others, and many recent studies have reported the discovery and functional characterization of neuronal brain CB2Rs. Indeed, our studies provided the first evidence for neuronal CNS effects of CB2Rs and its possible role in schizophrenia and neuropsychiatric disorders (Onaivi et al. 2012; Onaivi et al. 2015; Ishiguro et al. 2010a; Ishiguro et al. 1836). It must be noted and acknowledged as discussed below that the role of CB2Rs in CNS disturbances involving neurodegenerative diseases associated with neuroinflammation and neuropathic pain has been extensively reported (Hill et al. 2012; Russo 2018; Basavarajappa et al. 2017). CB2Rs in glial cells are modulators during inflammatory conditions. Therefore, modulation of eCB signaling in glia cells may provide pharmacological targets to treat and prevent neuroinflammatory response (Fig. 9.1, and Fig. 9.2), white matter deficits, and pathological mechanisms seen in schizophrenia (de Aleida and Martins-de-Souza 2018).

In our ongoing studies, many features of CBR gene structures, SNPs, copy number variations (CNVs), CPG island, microRNA regulation, and the impact in neuropsychiatry and where possible in rodent models are evaluated. Accumulating evidence suggests the importance of CNVs in the etiology of neuropsychiatric disorders. The clinical consequences of CNV in the coding and non-coding *CNR* gene sequences associated with human phenotypes and disorders are mostly unknown and under investigation. With advances in genomic technologies and the analysis and identification of *CNR* gene CNVs may uncover the relationship (if any), between *CNR* gene CNVs to phenotype and disease. While *CNR1* and *CNR2* SNPs have been associated with a number of neuropsychiatric disorders (see Table 9.1), it is unclear to what extent *CNR* gene CNVs are involved in psychiatric disorders. Therefore, more studies are needed to determine the role and contribution of *CNR* gene CNV to conditions of endocannabinoid dysregulation in psychological and psychiatric disorders. However, a number of focused studies on the polymorphisms in components of ECS have shown that *CNR1* and limited studies for *CNR2* genes (Table 9.1) contribute to the pathogenesis of specific subtypes of schizophrenia (Onaivi et al. 2013; Ishiguro et al. 2010a; Ibarra-Lecue et al. 2018). We investigated genetic associations between *CNR2* gene polymorphisms and schizophrenia in a selected population, and the results from our studies identified that polymorphism in *CNR2* gene (Table 9.1) with low CB2R expression and function indicates an increased risk of schizophrenia (Ishiguro et al. 2010a) when combined with other risk factors. Other studies have evaluated eCB levels and eCB enzymes in schizophrenic patients with reports of alterations in 2-AG metabolizing enzyme and in eCB levels in patients with schizophrenia. These findings warrant further investigation into ECS changes and functional implications of ECS in schizophrenia subtypes. Identification of the roles of ECS components may be valuable in designing new medication targets in the treatment of schizophrenia.

9.4.4 Role of CBRs Addictive Disorders

Addictive disorders are now not only limited to drugs of abuse, with the emergence of digital technology, but also to cell phone, various gaming platforms, and food and gambling. Generally, therefore addictive disorders are chronic relapsing disorders characterized by impulsivity and compulsive disorders that are known to be present in psychiatric (De Luca and Fattore 2015; Onaivi 2008; Malloy-Diniz et al. 2007) and addictive disorders. There is accumulating evidence indicating a central role for this previously unknown but ubiquitous ECS in the regulation of the rewarding effects of abused substances. Thus, an endocannabinoid hypothesis of drug reward and addiction was postulated (Onaivi 2008). eCBs mediate retrograde signaling in neuronal tissues and are involved in the regulation of synaptic transmission to suppress neurotransmitter release by the presynaptic CBRs. This powerful modulatory action on synaptic transmission has significant functional implications and interactions with the effects of abused substances (De Luca and Fattore 2015; Onaivi 2008). In humans, recent studies have revealed diverse responses by the ECS to long-term exposure to several drugs of abuse, and cannabis, ethanol, opioids, nicotine, and cocaine were found to alter the ECS regardless of their diverse pharmacological mechanism of action. Our data, along with those from other investigators, provide strong new evidence for a role for ECS modulation in the effects of drugs of abuse, and specifically for involvement of CBRs in the neural basis of addiction (De Luca and Fattore 2015; Onaivi 2008; Malloy-Diniz et al. 2007). We suggested that cannabinoids and eCBs appear to be involved in adding to the rewarding effects of addictive substances, including, nicotine, opiates, alcohol, cocaine, and BDZs. The results suggest that the ECS may be an important natural regulatory mechanism for drug reward and a target for the treatment of addictive disorders. In animal models, the ECS appears to be also involved in the ability of drugs and drug-associated cues to reinstate drug-seeking behavior in animal models of relapse (De Luca and Fattore 2015). As shown in Table 9.1, polymorphisms in the CB1R and CB2R genes have been associated with substance dependence and drug-related behaviors (De Luca and Fattore 2015; Onaivi 2008). Consequently, the ECS is involved in reward mechanisms that facilitate the hedonic value of natural and drug rewards (De Luca and Fattore 2015; Onaivi 2008). This system participates in the primary rewarding effects of cannabinoids, nicotine, alcohol, and opioids and in the common mechanisms, underlying drug addiction and relapse to drug-seeking behavior. In turn, many drugs of abuse, including cannabinoids, opioids, and alcohol, and nicotine, can alter differently the levels of eCBs in selected brain regions (De Luca and Fattore 2015). Therefore, substantial data now point to a role for the ECS in triggering and/or preventing reinstatement of drug-seeking behavior. It appears that the effects of perturbation of the ECS by drugs of abuse can be ameliorated by restoring the perturbed system using cannabinoid ligands. It is not surprising that CB1R antagonists were briefly approved as an anti-obesity medication in Europe and its potential promise in the reduction in drug use, in smoking cessation, and reduction in alcohol consumption. Nevertheless, due to serious side effects of depression and suicide, rimonabant was withdrawn from use (Onaivi 2010). The promiscuous action and distribution of CBRs in most relevant biological systems provide the EPCS with limitless signaling capabilities for cross talk within, and possibly between, receptor families, which may explain the myriad behavioral effects associated with smoking marijuana. The ECS therefore appears to play a central role in regulating the neural substrate underlying many aspects of drug addiction, including craving and relapse. The findings that the ECS is involved in the reinstatement model provided evidence of the ECS in the neural machinery-underlying relapse. In summary, there is a lot more to learn and research to be done to better understand the nature and neurobiology of the eCB physiological role in addictive- and feeding-related disorders.

9.4.5 Role of CBRs in Other Psychiatric Disorders: Autism Spectrum Disorders (ASDs) and Attention Deficit Hyperactivity Disorders (ADHD)

Several studies highlight a key involvement of ECS in ASD pathophysiology (Krebs et al. 2019; Crume et al. 2018; Gunn et al. 2016; Chakrabarti et al. 2015; Maccarrone et al. 2010; Zhang and Alger 2010; Zamberletti et al. 2017; Zou et al. 2019; Brigida et al. 2017; Crespi et al. 2010). In addition to autism, the ECS is also involved in several other psychiatric disorders (ADHD, anxiety, major depression, bipolar disorder, and schizophrenia). ECS is a key regulator of metabolic and cellular pathways involved in autism, such as food intake, energy metabolism, and immune system control. ASD is also characterized by immune system dysregulation. The mRNA and protein for CB2R and ECS enzymes were significantly dysregulated, further indicating the involvement of the ECS in ASD-associated immunological disruptions. Alterations in eCB signaling in neurodevelopmental disorders may be associated with exposure to cannabis and cannabinoids in utero or during adolescence (Krebs et al. 2019; Crume et al. 2018; Gunn et al. 2016). The use of cannabis during pregnancy may increase adverse outcomes for women and their neonates. As the use of cannabis continues to gain social acceptance, pregnant women and their medical providers could benefit from health education on potential adverse effects of use of cannabis during pregnancy. Babies exposed to cannabis while in the womb may suffer significant and permanent changes in neuroplasticity that could alter brain maturation and cause long-lasting changes that persist in the adult brain (Krebs et al. 2019; Crume et al. 2018; Gunn et al. 2016). This is because exogenous cannabinoids interfere with eCB regulation of brain maturation during adolescence. The role of ECS in ASD is a relatively understudied topic. Some studies and hypotheses may lead us to think the ECS plays a role in the multiple aspects of ASD such as social reward responsivity, circadian rhythm, anxiety-related symptoms, and neuronal development (Chakrabarti et al. 2015). The existing evidence shows potential ECS involvement in fragile X syndrome, in which 10–30% of patients are also diagnosed with ASD. In the highly characterized fragile X mental retardation (Fmr1) knockout mice, some studies showed that ECS-mediated responses of GABAergic synapses are increased in the dorsal striatum and hippocampus in the Fmr1-KO mouse model. In the hippocampus, this effect was indirect via group I mGluR activation which can upregulate the levels of eCBs (Maccarrone et al. 2010; Zhang and Alger 2010). Other studies report the presence of alterations in the ECS as well as the effects of its pharmacological manipulations in animal models of ASD-like behaviors (Onaivi et al. 2018; Zamberletti et al. 2017; Zou et al. 2019). In our studies, we used the BTBR T + tf/J mice that have been shown to exhibit autism-like behavioral phenotypes. The data indicated the BTBR mice have an abnormal regulation of dopamine functioning with an upregulated CBR2A gene expression in naïve BTBR mouse model of ASD. These results along with our findings indicating an increased risk of schizophrenia in patients with low CB2R function (Ishiguro et al. 2010a), which is in agreement with the hypothesis that autism and schizophrenia represent diametric conditions (Crespi et al. 2010). Moreover, more research needs to be done to understand the nature of the neurochemical changes recorded in our preliminary study in the hippocampus, striatum, and frontal cortex, where the levels of dopamine and serotonin and their metabolites were differentially altered in the BTBR and C57BL/6 J mice. Thus, our data provide a basis for further studies in evaluating the role of ECS and monoaminergic systems in the etiology of ASDs. Collectively, the findings to date indicate that the ECS plays a key role in the pathophysiology of ASD and may provide new insights into potential interventions and could represent a novel target and strategy for ASD pharmacotherapy.

9.5 CBRs and Comorbidity between Psychiatric Disorders and Neurological Disturbances

Several lines of evidence suggest a primary function and involvement of components of the ECS in the degenerative process (Basavarajappa et al. 2017). Psychiatric disorders are common in many neurological disorders, including epilepsy, migraine, Alzheimer's disease (AD), Parkinson's disease (PD), Huntingtin disease (HD), Multiple sclerosis (MS), Brain tumors (BT), essential tremor (ET), and cerebral palsy (CP), Tics and Tourette's syndrome (TS), epilepsy, migraine, and stroke (Hesdorffer 2016). Traumatic brain injury (TBI) is a common injury characterized by a change in brain function after an external blow to the head and is highly comorbid with psychiatric disorders. TBI is associated with substance abuse, problem gambling, psychological distress, risk-taking, and impulsivity and especially antisocial both violent and non-violent (Vaughn et al. 2019; Turner et al. 2019). Our current knowledge on the emerging role of the ubiquitous ECS in neuro-immune-microbiome cross talk provides targets to study comorbidity between psychiatric disorders and neurological illness. However, these comorbidities increase disease burden and may complicate the treatment of the combined disorders. Initial studies of the comorbidity of psychiatric and neurological disorders were cross-sectional, and time order of the associations was impossible to elucidate. Work that is more recent has clarified time associations between psychiatric disorders and neurological disorders, particularly in epilepsy and stroke where epidemiological evidence suggests that there is a bidirectional relationship. Although these relationships are understood in many neurological disorders, routine screening for psychiatric disorders in neurological disorders is infrequent. The brief summary below capitulates the contribution of the ECS to the development of neurodegenerative generative and other neurological disorders in a number of animal models and human studies. With the ubiquitous distribution of all components of the ECS in most cells and tissues in man and mouse, changes in the ECS machinery have been reported in brain areas associated with the symptoms of AD. Studies indicating alterations of specific component (s) in ECS cellular, and molecular machinery in brain circuits underlying the symptoms in HD, PD, MS, TBI, BT, ET, CP, TS, epilepsy, migraine, and stroke have been reported (Hill et al. 2012; Russo 2018; Basavarajappa et al. 2017). For example, there is evidence to suggest that synthetic and natural cannabinoids may help patients with Alzheimer's disease-related aggression and agitation. Since AD is a disease of the elderly, many medications are limited to control certain symptoms of AD because of their side effect profile and have not been overly beneficial. Cannabinoids have been of great interest in many neurodegenerative diseases since they pose relatively less risk for abuse and adverse effects. In a study that investigated CB1R and social interaction and aggression, they showed that grouped housed CB1R KO mice spent more time in threat behavior compared to grouped wild type mice. Along with this notion, the grouped CB1R KO had an increased time engaged in attack behavior with increased time in each attack compared to WT mice (Rodriguez-Arias et al. 2013). There is evidence to suggest that synthetic and natural cannabinoids may help patients with Alzheimer's disease-related aggression and agitation (Liu et al. 2015). Since AD is a disease of the elderly, many medications are limited to control certain symptoms of AD because of their side effect profile and have not been overly beneficial. Cannabinoids have been of great interest in many neurodegenerative diseases since they pose relatively less risk for abuse and adverse effects.

There is growing awareness of the involvement of gut microbiome and inflammation in a number of psychiatric and neurological disorders. The ECS system interaction with the microbiome and neuroglia cells in the brain is providing new therapeutic targets but also in understanding underlying mechanism (s) associated with psychiatric and neurological disorders. With the increasing use of cannabis and cannabinoids, an ECS pharmacogenomic test on individualized basis could be used for diagnosis and whether or not medical cannabis can be of therapeutic benefit. Much more mechanistic studies using in vitro and in vivo techniques, need to be done to improve the detection and treatment of patients affected by neurological and psychiatric disorders when cannabis and cannabinoids are indicated. Cannabinoids and CBRs have the ability to control both anti-inflammatory, anti-oxidant, neuroprotective, and neuromodulatory functions. In addition, the ECS modulates other biochemical pathways that could complement their effects on other receptors, ion channels, and enzymes. Thus, the use of cannabinoids provides interesting, unique, and potential therapeutic investigations in psychiatric and neurological disorders. Therefore, an overwhelming number of studies now document CB2R expression in neuronal, endothelial, and glial cells. Mounting evidence also shows that CB2Rs and its gene variants may play possible roles in psychiatric and neurological disorders with associated inflammatory reactions.

9.6 CBR Polymorphisms and Epigenetic Mechanisms in Psychiatric Disorders

Table 9.1 presents a brief summary of genetic polymorphisms of cannabinoid receptor genes. Genetic polymorphisms of the endocannabinoid system, including CB1R, CB2R, and FAAH genes, have been linked or associated with a number of psychiatric and neurological disorders (Onaivi et al. 2015; Onaivi et al. 2013; Onaivi et al. 2008; Ishiguro et al. 2007; Ishiguro et al. 2010a). Studies show that single nucleotide polymorphisms (SNPs) of *CNR1* and FAAH may contribute to drug addiction and other neuropsychiatric disorders. CBR gene variants may provide a deeper insight and novel targets for the effects of cannabinoids in drug addiction and other neuropsychiatric disorders. Variations in genes encoding cannabinoid receptors that are involved in drug addiction, and obesity, for example, provide therapeutic targets in endocannabinoid insufficiency syndrome (Onaivi 2010). Previously, we demonstrated the characteristic features in CB1R gene, and such features and variations in CB2R gene have not been fully characterized (Zhang et al. 2004). Many studies have reported that variations in CB1R genes are linked to drug addiction vulnerability in different ethnic groups (Zhang et al. 2004), and some of these variations are associated with mental and neurological disorders (Onaivi et al. 2015; Onaivi et al. 2013; Zhang et al. 2004). Rare genetic variants in CNR1 and DAGLA genes in neurological phenotypes were reported. In this study, variations in *CNR1, CNR2, DAGLA, FAAH,* and *MGLL* genes were studied for any associations with neurological phenotypes. They concluded from their study that mainly *CNR1* and *DAGLA* were associated with neurological phenotypes, but not the other aforementioned genes. Of these genes, *CNR1* were associated with migraines, memory disorders, and sleep disorders along with or without anxiety (Smith et al. 2017). These results and other similar studies provide insight into potential ECS treatments into many neurological and psychiatric disorders. Therefore, the study of the CBR genomic structure, and its polymorphic nature, subtype specificity and their variants, and associated regulatory elements that confer vulnerabilities to a number of neuropsychiatric disturbances, may provide a deeper insight into the underlining mechanisms. Thus, understanding the ECS in the human body and brain will contribute to elucidating this natural regulatory mechanism and provide potential therapeutic targets in health and disease. A preliminary study of ECS regulation showing distinct alterations of CNR1 promoter DNA methylation in patients with schizophrenia was reported. This study examined DNA methylation in peripheral blood mononuclear cells (PBMCs) in patients with schizophrenia, bipolar disorder, and major depressive disorder. The goal of the study was to measure the alterations of the promoter site of the CNR1 gene. The only significant changes were found in schizophrenia, but not any of the previously mentioned disorders. In human PBMCs, there was an upregulation of CNR1 and decreased CpG methylation in schizophrenic patients compared to control. These results were confirmed in an animal model of schizophrenia, which was done by administering prenatal methylazoxymethanol (MAM) acetate that showed an increase in CNR1 expression in the PFC. There was also reduced DNA CpG site methylation in the promoter itself (D'Addario et al. 2017). In rodent studies, prenatal exposure to cannabis triggers epigenetic changes with possible transgenerational immunological consequences (D'Addario et al. 2017).

Epigenetic regulation of ECS components under both physiological and pathological conditions as well as the epigenetic changes induced by eCB signaling is an emerging potential target of ECS 'epigenetic therapy' (D'Addario et al. 2017; Parira et al. 2017; Zumbrun et al. 2015). Initial focus is to understand how CB1Rs evoke epigenetic mechanisms, either by directly interacting with the epigenetic machinery or by indirectly. In studies of histone modifications due to cannabinoid signaling, THC-modulated multiple histone modification sites like H3K4me3, H3K9me3, H3K27me3, and H3K36me3 in differentiating mouse lymph node cells showing histone modifications are associated with THC-mediated alterations in antigen-specific T-cell responses (Parira et al. 2017). Alterations in DNA methylation status to the effects of cannabinoids indicated that parental exposure to THC altered DNA methylation status of genes related to synaptic plasticity in rat nucleus accumbens (Parira et al. 2017). Another study done in mice showed that THC administration increased methylation at the promoter region of DNA methyltransferases 3A and 3B in myeloid-derived immune suppressor cells and correspondingly reduced expression of the same DNA methyltransferases (Parira et al. 2017). Furthermore, methylation at the promoter regions of Arg1 and STAT3 was decreased by THC, which led to further increases in levels of Arg1 and STAT3 expression. Arg1, which can metabolize

L-arginine, suppresses T-cell function while increasing activation and function of these immunosuppressive cells. Cannabinoid signaling has also been shown to be associated with modulation in certain microRNA-based epigenetic mechanisms. In other studies, data from animal models suggest that in utero exposure to cannabinoids results in profound T-cell dysfunction and a greatly reduced immune response to viral antigens. Furthermore, evidence from animal studies indicates that the immunosuppressive effects of cannabinoids can be mediated through epigenetic mechanisms such as altered microRNA, DNA methylation, and histone modification profiles. Such studies support the hypothesis that that parental or prenatal exposure to cannabis can trigger epigenetic changes that could have significant immunological consequences for offspring as well as long-term transgenerational effects. Epigenetic effects of cannabinoids reveal the ability of cannabinoids to modify neuronal and immune cell functionality either via histone modifications like H3 lysine methylations or by altering DNA methylation (D'Addario et al. 2017; Parira et al. 2017; Zumbrun et al. 2015). A better understanding of the epigenetic regulation of eCB signaling as well as the eCB regulation of epigenetic mechanisms will be of great value for the possible design of more specific eCB epigenetic drugs. Therefore, more studies on the potential maternal and paternal transgenerational and/or epigenetic effects of cannabinoid abuse are needed with global changing landscape in recreational and medical use of cannabis and cannabinoids.

9.7 Emerging Trends in Targeting CBRs in Psychiatric and Neurological Disorders

Advances and new discoveries due to evidence-based in vivo and in vitro studies on molecular and cellular mechanisms of CBRs have implications in psychiatric and neurological disorders. These recent advances in understanding the biological actions of cannabis products are expanding the therapeutic indications and opportunities. Of course, there are continuing trials and tribulations in clinical trials of cannabinoid pharmacotherapy, leading to the withdrawal of CB1R antagonist in obesity and clinical trial tragedy of the FAAH inhibitor that was stopped. The involvement of the ECS in most biological systems shows their vital importance in homeostasis, and disruption in functioning of the ECS leads to catastrophic adverse effects, with suicides and brain-dead patients during clinical trials and use in obesity, respectively. One explanation for the lethal side effects is CBR ligand bias, which is beginning to be accessed (Laprairie et al. 2017a). Ligand-biased binding to its receptor shifts the equilibrium of receptor-dependent signaling toward other possible pathways. Apparently, the limited success of the development of cannabinoid-based therapeutics may be associated with ligand bias (Laprairie et al. 2017a). Progress in understanding of CBR structure and quantification of ligand bias, dimerization, and allosteric modulation of CBRs will optimize drug design, and selection of cannabinoids to match patient indication (Laprairie et al. 2017a; Callén et al. 2012; Alaverdashvili and Lapraire 2018; Tham et al. 2019; Laprairie et al. 2017b; Laprairie et al. 2019). Most cannabinoid drug development and targeting of orthosteric site on CB1R at which eCBs and THC bind. Adverse psychotropic effects limit the clinical utility of CB1R orthosteric agonists (Tham et al. 2019; Laprairie et al. 2017b; Laprairie et al. 2019). This has prompted the search and development of CB1R positive allosteric modulators (PAMs) that enhance orthosteric ligand binding, with improved reduced psychotropic side effects when used in psychiatric and neurological disorders (Tham et al. 2019; Laprairie et al. 2017b; Laprairie et al. 2019). Therefore, allosteric modulation of CB1R holds great therapeutic potential. This is because allosteric modulators do not possess intrinsic efficacy, but instead augment PAM or diminish negative allosteric modulation (NAM), the receptor's response to endogenous ligand. Consequently, CB1R allosteric modulators have an effect ceiling, which allows for the tempering of CB1R signaling without the desensitization, tolerance, dependence, and psychoactivity associated with orthosteric compounds. Several challenges exist for the development of CB1R allosteric modulators, such as receptor subtype specificity, translation to in vivo systems, and mixed allosteric/agonist/inverse agonist activity. Despite these challenges, elucidation of crystal structures of CB1R and CB2R and compound design based on structure-activity relationships will advance the field. This recent work on the determination of the crystal structures of CB1R and CB2R (Fig. 9.3) reveals a yin-yang relationship and functional profile of CB2R antagonism versus CB1R agonism (Hua et al. 2016; Li et al. 2019). The formation of functional CB1R and CB2R heteromers in neuronal cells in the brain indicates that activation of either receptor leads to negative modulation of the partner receptor via heteromers (Callén et al. 2012). Dimerization of CBRs has therapeutic implication and impact on CNS function and it was suggested that CBR1-CB2R heteromers must be taken into account when designing therapeutic approaches toward alterations involving the ECS (Callén et al. 2012). The direct activation of CBRs results in several beneficial effects; therefore, several CBRs ligands have been synthesized and tested in vitro and in vivo, with disappointing advancement for clinical development due mainly to side effects on the CNS. However, other approaches are being developed for allosteric modulators that might offer a novel therapeutic approach to achieve potential therapeutic benefits avoiding inherent side effects of orthosteric ligands.

The evidence-based scientific knowledge supports novel approaches to cannabinoid-based medication and cannabis use in psychiatric disorders and medicine. With the rapidly expanding nanomedicine formulation of nanoparticulate cannabinoid products presents new opportunities and approaches for cannabis use in health and disease. Development of cannabis and cannabinoid nanomedicine for nanotherapy will certainly overcome some of the shortcomings and challenges in medicinal and recreational use of cannabis and cannabinoids. As cannabinoids are a class of lipophilic compounds, the use of different surfactants and delivery systems to improve cannabinoid solubility and enhance bioavailability must be considered. To overcome these limitations, nanoparticle formulations may offer an attractive alternative. In practice, this means that the cannabinoids will be 'packed' in endogenous nanoparticles, offering the opportunity to effectively deliver the cannabis and cannabinoid formulations to the diseased sites, by applying nanotechnological advances to nanomedicine. The application of nanotechnology to medicine involves employing nanoparticles not only to enhance the action of drugs, but also expected to improve diagnosis and therapy of diseases and reduce health care cost (Ngwa et al. 2017). Thus, nanotechnology will allow targeted delivery of cannabinoid formulations with the potential to elevate their use to scientifically validated nanotherapeutic applications as the field of cannabis nanoscience matures. In last decade, the success of new and creative nanosynthetic tools has fashioned new opportunities in drug design, which allows creation of drugs, prodrugs, or diagnostic gears at nanosize regime (Patra et al. 2018). Advances in nanoparticle formulations of cannabinoids as therapeutic molecules for different routes of delivery into peripheral, CNS, and wearable skin patches are being developed for nanomedicine outcomes. A number of ongoing studies are building nanoparticles for nanodelivery of cannabis, cannabinoids, and endocannabinoid system components as nanotherapeutics (Singh et al. 2018). Now it is possible to predictively synthesize nanosized objects with well-defined surface chemistry and predefined size and morphology, which permit certain degree of control over therapeutic outcomes and minimizes side effects of the drugs. Encapsulation strategies for cannabis products and edibles will require greater physical stability, protection against oxidation, and flavor masking using liposomes, micelles, polyplexus, polymersomes, and silica nanoparticles as the industry matures and develops. While research in nanoengineering area is still not sufficiently mature, but it is conceivable that the surface of nano-objects can be tailored to control the solubility of the drugs, ensuring effective circulation

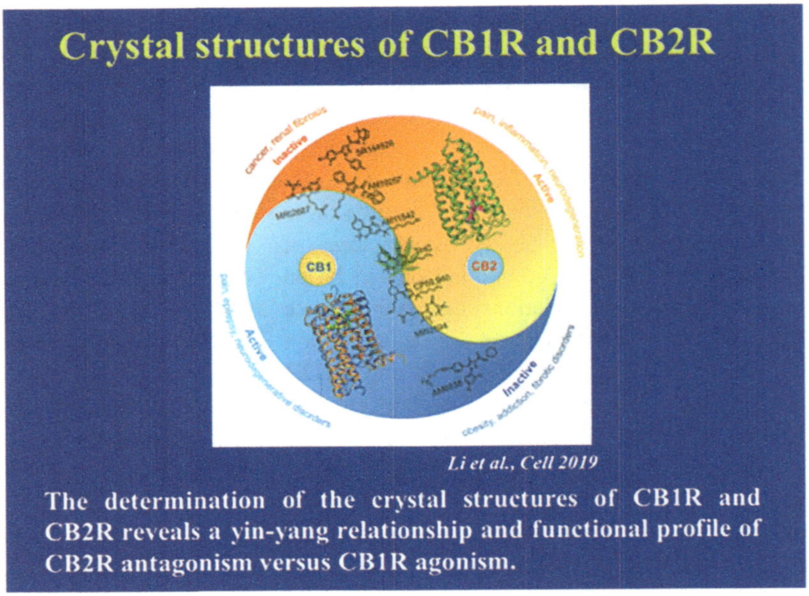

Fig. 9.3 The determination of the crystal structures of CB1R and CB2R reveals a yin-yang relationship and functional profile of CB2R antagonism versus CB1R agonism (Hua et al. 2016; Li et al. 2019)

time, and limiting the biodistribution. In addition, this strategy allows one to control drug release, which in turn reduces and diminishes immunogenicity. One of the very promising strategies has been to conjugate the nano-object surface with microenvironment-specific and/or receptor-specific biomacromolecules such as peptides, proteins, aptamers. A very elegant review of such studies provides a comprehensive picture of accomplishments and bottlenecks in the research in this area (Spicer et al. 2018).

9.8 Current State of Evidence on the Health Effects of Cannabis and Cannabinoids

The conclusions of the National Academies of Sciences, Engineering and Medicine (National Academies of Sciences, Engineering, and Medicine 2017) provide support for the legitimate study, regulation, and prescription of therapeutic cannabinoids. Evidence supports reform to allow the legitimate study, regulation, and prescription of therapeutic cannabinoids (National Academies of Sciences, Engineering, and Medicine 2017). The following conclusions have invigorated the debate over cannabis use and present opportunities for quality improvement and approaches to bridge the gaps between safe cannabis products, science, and medicine:

a). Conclusive and substantial evidence that cannabis and cannabinoids are effective in nausea and vomiting and in multiple sclerosis spasticity, b). Moderate evidence that cannabis or cannabinoids are effective in sleep apnea, fibromyalgia, and some chronic pain, and acute cognitive impairment, c). Limited evidence that cannabis or cannabinoids are effective in Tourette syndrome, anxiety, dementia, glaucoma, HIV/AIDS, and appetite, and d). There is substantial evidence of a statistical association between cannabis use and increased risk of motor vehicle crashes, lower birth weight of the offspring exposed to cannabis and cannabinoids in-utero, and psychosis in vulnerable individuals.

9.9 Concluding Remarks

Existing evidence suggests that alterations in eCB signaling are present in a range of psychiatric disorders. Targeting components of ECS components provides therapeutic potential of cannabinoid medicines as CBRs and other components of the ECS are involved in diverse neural, immune, function, and dysfunction, in psychiatric disorders. The ECS evidently gives novel ideas and options in the field of antidepressant treatment; however, further studies are needed to determine which group of patients could benefit from this type of therapy. The ECS is also involved in the pathogenesis and treatment of depression, though its role in this psychiatric disorder has not been fully understood. Both the pro- and antidepressant activities have been reported after cannabis consumption, and a number of preclinical studies have demonstrated that both agonist and antagonist of the endocannabinoid receptors act similarly to antidepressants. Identification of the roles of ECS components may be valuable in designing new medications targets in the treatment of schizophrenia. The ECS may be an important natural regulatory mechanism for drug reward and a target for the treatment of addictive disorders. Several studies highlight a key involvement of ECS in ASD pathophysiology. In addition to autism, the ECS is also involved in several other psychiatric disorders (ADHD, anxiety, major depression, bipolar disorder, and schizophrenia). Our findings indicating an increased risk of schizophrenia in patients with low CB2R function, is in agreement with the hypothesis that autism and schizophrenia represent diametric conditions. Collectively, the findings to date indicate that the ECS plays a key role in the pathophysiology of ASD and may provide new insights into potential interventions and could represent a novel target and strategy for ASD pharmacotherapy. Our current knowledge on the emerging role of the ubiquitous ECS in neuro-immune-microbiome cross talk provides targets to study comorbidity between psychiatric disorders and neurological illness. There is also growing awareness of the involvement of gut microbiome and inflammation in a number of psychiatric and neurological disorders. The ECS system interaction with the microbiome and neuroglia cells in the brain is providing new therapeutic targets but also in understanding underlying mechanism (s) associated with psychiatric and neurological disorders. Mounting evidence shows that CB2Rs and its gene variants may play possible roles in psychiatric and neurological disorders with associated inflammatory reactions. With the increasing use of cannabis and cannabinoids, an ECS pharmacogenomic test on individualized basis could be used for diagnosis and whether or not medical cannabis can be of therapeutic benefit. A better understanding of the epigenetic regulation of eCB signaling as well as the eCB regulation of epigenetic mechanisms will be of great value for the possible design of more specific eCB epigenetic drugs. Therefore, more studies on the potential maternal and paternal transgenerational and/or epigenetic effects of cannabinoid abuse are needed with the global changing landscape in recreational and medical use of cannabis and cannabinoids. Development of cannabis and cannabinoid nanomedicine for nanotherapy will certainly overcome some of the shortcomings and challenges in medicinal and recreational use of cannabis and cannabinoids. As cannabinoids are a class of lipophilic compounds, the use of different surfactants and delivery systems to improve cannabinoid solubility and enhance bioavailability must be considered. To overcome these limitations, nanoparticle formulations may offer an attractive alternative. In practice, this means that the cannabinoids will be 'packed' in endogenous nanoparticles, offering the opportunity to effectively deliver the cannabis and cannabinoid formulations to the diseased sites, by applying nanotechnological advances to nanomedicine. The application of nanotechnology to medicine involves employing nanoparticles not only to enhance the action of drugs, but also expected to improve diagnosis and therapy of diseases and reduce health care cost. Thus, nanotechnology will allow targeted delivery of cannabinoid formulations with the potential to elevate their use to scientifically validated nanotherapeutic applications as the field of cannabis nanoscience matures.

Acknowledgments This work was partly supported by William Paterson University and Assigned Release Time to ESO who was funded by NIH grant DA032890 and is currently a Guest researcher at the NIDA intramural research program.

References

Acheson A, Fantegrossi E (2019) Introduction to special issue: therapeutic and abuse-related effects of cannabis and cannabinoids. Exp Clin Psychoparmaco 27:299–230

Alaverdashvili M, Lapraire RB (2018) The future of type 1 cannabinoid receptor allosteric ligands. Drug Met Rev 50(1):14–25. https://doi.org/10.1080/03602532.2018.1428341

Andrade AK, Renda B, Murray JE (2019) Cannabinoids, interoception, and anxiety. Pharmacol Biochem Behav 180:60–73

Arjmand S, Behzadi M, Kohlmeier KA, Mazhari S (2019) Bipolar disorder and the endocannabinoid system. Acta Neuropsychiatrica 31:193–201

Basavarajappa BS, Shivakumar M, Joshi V, Subbanna S (2017) Endocannabinoid system in neurodegenerative disorders. J Neurochem 142:624–648

Brigida AL, Schultz S, Cascone M, Antonuchi N, Siniscalco D. Endocannabinod signal Dysregulation in autism Spectrum disorders: a correlation link between inflammatory state and Neuro-immune alterations. Int J Mol Sci. 18(7). pii: E1425. doi: https://doi.org/10.3390/ijms18071425. 2017

Callén L, Moreno E, Barroso-Chinea P, Moreno-Delgado-D, Cortés A, Mallol J et al (2012) Cannabinoid receptors CB1 and CB2 form functional heteromers in brain. J Biol Chem 287:20851–20865

Cani PD, Plovier H, Hul MV, Geurts L, Delzenne NM, Druart C et al (2016) Endocannabinoids- at the crossroads between the gut microbiota and host metabolism. Nat Rev Endocrinology 12:133–146

Canseco-Alba A, Hammouda M, Schanz N, Liu QR, Onaivi ES (2018a) Neurodevelopmental and behavioral alterations after conditional deletion of type 2 cannabinoid receptors in dopamine neurons. Soc Neurosc Abst, San Diego

Canseco-Alba A, Schanz N, Ishiguro H, Liu QR, Onaivi ES. Behavioral Evaluation of Seeking and Preference of Alcohol in Mice Subjected to Stress. Bio Protoc. 8 (20). pii: e3061. doi: https://doi.org/10.21769/BioProtoc.3061. 2018b

Canseco-Alba A, Schanz N, Sanabria B, Zhao J, Lin Z, Liu QR et al (2019) Behavioral effects of psychostimulants in mutant mice with cell-type specific deletion of CB2 cannabinoid receptors in dopamine neurons. Behav Brain Res 360:286–297. https://doi.org/10.1016/j.bbr.2018.11.043

Castillo PE, Younts TJ, Chavez AE, Hashimotodani Y (2012) Endocannabinoid signaling and synaptic function. Neuron 76:70–81

Chakrabarti B, Persico A, Battista N, Maccarrone M (2015) Endocannabinoid signaling in autism. Neurotherapeutics 12:837–847

Crespi B, Stead P, Elliot M (2010) Evolution in health and medicine Sackler colloquium: comparative genomics of autism and schizophrenia. Proc Natl Acad Sci 107 (Suppl 1):1736–1741. https://doi.org/10.1073/pnas.0906080106

Crews FT, Lawrimore CJ, Walter TJ, Coleman LG Jr (2017) The role of neuroimmune signaling in alcoholism. Neuropharmacology 122:56–73

Crume TL, Juhl AL, Brooks-Russel A, Hall KE, Wymore E, Borgelt LM (2018) Cannabis use during the perinatal period in a state with legalized recreational and medical marijuana: the association between maternal characteristics, breastfeeding patterns, and neonatal outcomes. J Pediatr 197:90–96

D'Addario C, Micale V, Di Bartolomeo M, Stark T, Pucci M, Sulcova A et al (2017) A preliminary study of endocannabinoid system regulation in psychosis: distinct alterations of CNR1 promoter DNA methylation in patients with schizophrenia. Schizophr Res 188:132–140

de Aleida V, Martins-de-Souza D (2018) Cannabinoids and glial cells: possible mechanism to understand schizophrenia. Eur Arch Psych Clin Neurosci 268 (7):727–737. https://doi.org/10.1007/s00406-018-0874-6

De Luca MA, Fattore L (2015) Cannabinoids and drug addiction. In: Fattore L (ed) Cannabinoids in neurologic and mental disease. Elsevier, pp 289–314

Devinsky O, Cross JH, Laux L, Marsh E, Miller I, Nabbout R et al (2017) Cannabidiol in Dravet syndrome study group. Trial of cannabidiol for drug-resistant seizures in Dravet syndrome. N Engl J Med 37:2011–2020. https://doi.org/10.1056/NEJMoa1611618

Di Marzo V (2018) New approaches and challenges to targeting the endocannabinoid system. Nat Rev Drug Discov 17:623–639

DiPatrizio NV (2016) Endocannabinoids in the gut. Cannabis Cannabinoid Res 1:1. https://doi.org/10.1089/can.2016.0001

Ghose H. (2009) Cannabinoid controversy. http://www.the-scientist.com/news/display/55969/)

Guggenhuber S, Romo-Parra H, Bindila L, Leschik J, Lomazzo E, Remmers F et al (2015) Impaired 2-AG signaling in hippocampal neurons: aggravation of anxiety-like behavior and unaltered seizure susceptibility. Int J Neuropsychopharmacol. https://doi.org/10.1093/ijn/pyv091

Gunn JK, Rosales CB, Center KE, Nunez A, Gibson SJ, Christ C et al (2016) Prenatal exposure to cannabis and maternal and child health outcomes: a systematic

review and meta-analysis. BML Open 6(4):e009986. https://doi.org/10.1136/bmjopen-2015-009986
Hesdorffer DC (2016) Comorbidity between neurological illness and psychiatric disorders. CNS Spectr 21 (3):230–238. https://doi.org/10.1017/S1092852915000929
Hill A, Williams CM, Whalley BJ, Stephens GJ (2012) Phytocannabinoids as novel therapeutic agents in CNS. Pharmacol Therapeutics 133:79–97
Hua T, Vemuri K, Pu M, Qu L, Han GW, Wu Y et al (2016) Crystal structure of human cannabinoid receptor 1. Cell 167:750–762
Hungund BL, Vinod KY, Kassir SA, Basavarajappa BS, Yalamanchili R, Cooper TB et al (2004) Upregulation of CB 1 receptors and agonist-stimulated [35 S] GTPγS binding in the prefrontal cortex of depressed suicide victims. Mol Psychiatry 9(2):184
Ibarra-Lecue I, Pilar-Cuellar F, Muguruza C, Florensa-Zanuy E, Diaz A, Uriguen L et al (2018) The endocannabinoid system in mental disorders: evidence from human brain studies. Biochem Pharmacol 151:97–107
Ibsen MS, Finlay DB, Patel M, Javitch JA, Glass M, Grimsey NL (2019) Cannabinoid CB1 and CB2 receptor-mediated arrestin translocation: species, subtype, and agonist-dependence. Front Pharmacol 10:350. https://doi.org/10.3389/fphar.2019.00350
Ishiguro H, Carpio O, Horiuchi Y, Shu A, Higuchi S, Schanz N, Benno R et al (2010b) A nonsynonymous polymorphism in cannabinoid CB2 receptor gene is associated with eating disorders in humans and food intake is modified in mice by its ligands. Synapse 64:92–96
Ishiguro H, Horiuchi Y, Ishikawa M, Koga M, Imai K, Suzuki Y et al (2010a) Brain cannabinoid CB2 receptor in schizophrenia. Biol Psychiatry 67:974–982
Ishiguro H, Horiuchi Y, Tabata K, Liu QR, Arinami T, Onaivi ES (2018) Cannabinoid CB2 receptor gene and environmental interaction in the development of psychiatric disorders. Molecules 23. https://doi.org/10.3390/molecules23081836
Ishiguro H, Iwasaki S, Teasenfitz L, Higuchi S, Horiuchi Y, Saito T et al (2007) Involvement of cannabinoid CB2 receptor in alcohol preference in mice and alcoholism in humans. Pharmacogenomics J 7:380–385
Josh N, Onaivi ES (2019) Endocannabinoid system components: overview and tissue distribution. Adv Exp Med Biol 1162:1–12
Kaelberer MM, Buchanan KL, Klein ME, Barth BB, Montoya MM, Shen X (2018) A gut-brain neural circuit for nutrient sensory transduction. Science 361: eaat5236. https://doi.org/10.1126/science.aat5236
Khakpai F, Ebrahimi-Ghiri M, Alijanpour S, Zarrindast MR (2019) Ketamine-induced antidepressant like effects in mice: a possible involvement of cannabinoid system. Biomed Pharmacother 112:108717
Kill KP (2019) Medical use of cannabis in 2019. JAMA. https://doi.org/10.1001/Jama.2019.11868
Koethe D, Llenos IC, Dulay JR, Hoyer C, Torrey EF, Leweke FM et al (2007) Expression of CB 1 cannabinoid receptor in the anterior cingulate cortex in schizophrenia, bipolar disorder, and major depression. J Neural Transm 114(8):1055
Korem N, Zer-Aviv TM, Ganon-Elazar E, Abush H, Akirav I (2016) Targeting the endocannabinoid system to treat anxiety-related disorders. J Basic Clin Physiol Pharmacol 27:193–202
Krebs MO, Kebir O, Jay TM (2019) Exposure to cannabinoids can lead to persistent cognitive and psychiatric disorders. Eur J Pain 23:1225–1233
Laprairie RB, Bagher AM, Denovan-Wright EM (2017a) Cannabinoid receptor ligand bias: implications in the central nervous system. Curr Opin Pharmacol 32:32–43
Laprairie RB, Bagher AM, Rourke JL, Zrein A, Cairns EA, Kelly MEM et al (2019) Positive allosteric modulation of type 1 cannabinoid receptor reduces the signs and symptoms of Huntington's disease in the R6/2 mouse model. Neuropharmacology 151:1–12
Laprairie RB, Kulkarni PM, Deschamps JR, Kelly MEM, Janero DR, Cascio MG et al (2017b) Enantiospecific allosteric modulation of cannabinoid 1 receptor. ACS Chem Neurosci 8:1188–1203
Li X, Hua T, Vemuri K, Ho JH, Wu Y, Wu L et al (2019) Crystal structure of the human cannabinoid receptor CB2. Cell 176:459–467
Lin X-H, Wang Y-Q, Wang H-C, Ren X-Q, Li Y-Y (2013) Role of endogenous cannabinoid system in the gut. Acta Physiologica Sinica 65:451–460
Liu QR, Canseco-Alba A, Zhang HY, Taglaiferro P, Chung M, Dennis E et al (2017) Cannabinoid type 2 receptors in dopamine neurons inhibits psychomotor behaviors, alters anxiety, depression and alcohol preference. Sci Rep 7:17410
Liu CS, Chau SA, Ruthirakuhan M, Lanctôt KL, Herrmann N (2015) Cannabinoids for the treatment of agitation and aggression in Alzheimer's disease. CNS Drugs 29:615–623
Lopez A, Aparicio N, Pazos MR, Grande MT, Barreda-Manso MA, Benito-Cuesta I et al (2018) Cannabinoid CB2 receptors in the mouse brain: relevance for Alzheimer's disease. Neuroinflammation 15:158
Lowe DJE, Sasiadek JD, Coles AS, George TP (2019) Cannabis and mental illness: a review. Europ Arch Psychiat and Clin Neurosci 269:107–120
Maccarrone M, Rossi S, Bari M, De Chiara V, Rapino C, Musella A et al (2010) Abnormal mGlu 5 receptor/endocannabinoid coupling in mice lacking FMRP and BC1 RNA. Neuropsychopharmacology 35:1500–1509
Magid L, Heymann S, Elgali M, Avram L, Cohen Y, Liraz-Zaltsman S et al (2019) Role of CB2 receptor in recovery of mice after traumatic brain injury. J Neurotrauma 36:1836–1846
Malloy-Diniz L, Fuentes D, Leite WB, Correa H, Bechara A (2007) Impulsive behavior in adults with attention deficit/hyperactivity disorder: characterization of

attentional, motor and cognitive impulsiveness. J Int Neuropsychol Soc 13:693–698

Mehrpouya-Bahrami P, Chitrala KN, Ganewatta MS, Tang C, Murphy EA, Enos RT et al (2017) Blockade of CB1 cannabinoid receptors alters gut microbiota and attenuates inflammation and diet-induced obesity. Sci Rep 7:15645. https://doi.org/10.1038/s41598-017-15154-6

Melis M, Frau R, Kalivas PW, Spencer S, Chioma V, Zamberletti E et al (2017) New visitas on cannabis use disorder. Neuropharmacology 124:62–72

National Academies of Sciences, Engineering, and Medicine (2017) The health effects of cannabis and cannabinoids: the current state of evidence and recommendations for research. National Academies Press, Washington, DC. https://doi.org/10.17226/24625

Ngwa W, Kumar R, Moreau M, Dabney R, Herman A (2017) Nanoparticle drones to target lung cancer with radiosensitizers and cannabinoids. Front Oncol 7:208. https://doi.org/10.3389/fonc.2017.00208

Ohno-Shosaku T, Kano M (2014) Endocannabinoid-mediated retrograde modulation of synaptic transmission. Curr Opin Neurobiol 29:1–8

Onaivi ES (2008) An endocannabinoid hypothesis of drug reward and drug addiction. Ann N Y Acad Sci 1139:412–421

Onaivi ES (2010) Endocannabinoid system, pharmacogenomics and response to therapy. Pharmacogenomics 11(7):907–910

Onaivi ES, Benno R, Halpern T, Mehanovic M, Schanz N, Sanders C et al (2011) Consequences of cannabinoid and monoaminergic system disruption in a mouse model of autism spectrum disorders. Curr Neuropharmacol 9:209–214. https://doi.org/10.2174/157015911795017047

Onaivi ES, Canseco-Alba B. D. Sanabria BD, Zhang HY, Eita M, Rohani T et al. Identification of cannabinoid CB2 receptor neuro-immune crosstalk following conditional deletion of type 2 cannabinoid receptors in microglia and dopamine neurons. Soc Neurosc Abst. San Diego. 2018

Onaivi ES, Green MR, Martin BR (1990) Pharmacological characterization of cannabinoids in the elevated plus maze. The J Pharmaco Expl Ther 252:1002–1009

Onaivi ES, Ishiguro H, Gong JP, Patel S, Meozzi PA, Myers L et al (2008) Brain neuronal CB2 cannabinoid receptors in drug abuse and depression: from mice to human subjects. PLoS One 3(2):e1640

Onaivi ES, Ishiguro H, Gong JP, Patel S, Perchuk A, Meozzi PA (2006b) Discovery of the presence and functional expression of cannabinoid CB2 receptors in brain. Ann N Y Acad Sci 1074:514–536

Onaivi ES, Ishiguro H, Gu S, Liu QR (2012) CNS effects of CB2 cannabinoid receptors: beyond neuro-immuno-cannabinoid activity. J Psychopharmacol 26:92–103

Onaivi ES, Ishiguro H, Liu QR (2015) Future perspectives: cannabinoid CB2 receptor ligands and their therapeutic potential in mental diseases. In: Fattore L (ed) Cannabinoids in neurologic and mental disease. Elsevier Inc, pp 425–444

Onaivi ES, Ishiguro H, Sgro S, Leonard CM (2013) Cannabinoid receptor gene variations in drug addiction and neuropsychiatric disorders. J Drug Alcohol Res. https://doi.org/10.4303/jdar/235714

Onaivi ES, Sugiura T, DiMarzo V (2006a) Eds. Endocannabinoids: the brain and body's marijuana and beyond. CRC Press, Taylor and Francis Group, Boca Raton

Papagianni EP, Stevenson CW (2019) Cannabidiol regulation of fear and anxiety: an update. Curr Psychiatric Reports 21:38. https://doi.org/10.1007/s11920-019-1026-z

Parira T, Laverde A, Agudelo M (2017) Epigenetic interactions between alcohol and cannabinergic effects: focus on histone modification and DNA methylation. J Alcohol Depend 5:259. https://doi.org/10.4172/2329-6488.1000259

Patra JK, Das G, Fraceto LF, Campos EVR, Rodriguez-Torres MDP, Acosta-Torres LS et al (2018) Nano based drug delivery systems: recent developments and future prospects. J Nanobiotechnol 16:1–33

Poleszak E, Wosko S, Salwinska K, Szopa A, Wrobel A, Serefko A (2018) Cannabinoids in depressive disorders. Life Sci 213:18–24

Premoli M, Aria F, Bonini SA, Maccarinelli G, Gianoncelli A, Pina SD (2019) Cannabidiol: recent advances and new insights for neuropsychiatric disorders. Life Sci 224:120–127

Robson PJ (2014) Therapeutic potential of cannabinoid medicines. Drug Test Analysis 6:24–30

Rodriguez-Arias M, Navarrete F, Daza-Losada M, Navarro D, Aguilar MA, Berbel P et al (2013) CB1 cannabinoid receptor-mediated aggressive behavior. Neuropharmacology 75:172–180

Rubin R (2018) The path to the first FDA-approved cannabis-derived treatment and what comes next. JAMA 320:1227–1229. https://doi.org/10.1001/jama.2018.11914

Russo EB (2018) Cannabis therapeutics and the future of neurology. Front Integr Neurosci 12:51. https://doi.org/10.3389/fnint.2018.00051

Schmöle AC, Lundt R, Gennequin B, Schrage H, Beins E, Krämer A et al (2015) Expression analysis of CB2-GFP BAC transgenic mice. PLoS One 10(9):e0138986

Singh P, Pandit S, Garnaes J, Tunic S, Mokkapati VR, Sultan A et al (2018) Green synthesis of gold and silver nanoparticles from cannabis sativa (industrial hemp) and their capacity for biofilm inhibition. Int J Nanomedicine 13:3571–3591

Sloan ME, Grant CW, Gowin JL, Ramchandani VJ, Le Foll B (2018) Endocannabinoid signaling in psychiatric disorders: a review of positron emission tomography studies. Act Pharmacologica Sinca 40:342–350

Smith DR, Stanley CM, Foss T, Boles RG, McKernan K (2017) Rare genetic variants in the endocannabinoid system genes CNR1 and DAGLA are associated with neurological phenotypes in humans. PLoS One 12(11): e0187926

Spicer CD, Jumeaux C, Gupta B, Stevens MM (2018) Peptide and protein nanoparticle conjugates versatile platforms for biomedical applications. Chem Soc Rev 47:3574–3620

Stempel AV, Zhang HY, Ozdogan T, Pannasch U, Theis AK (2016) Cannabinoid type 2 receptors mediate a cell type-specific plasticity in the hippocampus. Neuron 90:795–809

Tham M, Yilmaz O, Alaverdasvili M, Kelly MEM, Denovan-Wright EM, Laprairie RB (2019) Allosteric and orthesteric pharmacology of cannabidiol and cannabidiol-dimethylheptyl at the type 1 and type cannabinoid receptors. Br J Pharmacol 176:1455–1469

Thielle EA, Marsh ED, French JA, Mazurkiewicz-Beldzinska M, Benbadis SR, Joshi C et al (2018) Cannabidiol in patients with seizures associated with Lennos-Gastaut syndrome (GWPCARE4): a randomized, double-blind, placebo-controlled phase 3 trial. Lancet 391:1085–1096. https://doi.org/10.1016/S0140-6736(18)30136-3

Turner NE, McDonald AJ, Lalomiteanu AR, Mann RE, McCready J, Millstone D et al (2019) Moderate to severe gambling problems and traumatic brain injury: a population-based study. Psychiatry Res 272:692–697. https://doi.org/10.1016/j.psychres.2018.12.170

Vaughn MG, Sala-Wright CP, John R, Holzer KJ, Qian Z, Veeh C (2019) Traumatic brain injury and psychiatric co-morbidity in the United States. Psychiatry Q 90 (1):151–158. https://doi.org/10.1007/s11126-018-9617-0

Yin A, Wang F, Zhang X (2018) Integrating endocannabinoid signaling in the regulation of anxiety and depression. Act Pharmacologica Sinca. 40:336–341

Zamberletti E, Gabaglio M, Parolaro D. The endocannabinoid system and autism spectrum disorders: insights from animal models. Int J Mol Sci, 18(9). Pii: E1916. Doi: https://doi.org/10.3390/ijms18091916. 2017

Zhang L, Alger BE (2010) Enhanced endocannabinoid signaling elevates neuronal excitability in fragile X syndrome. J Neurosci 30:5724–5729

Zhang PW, Ishiguro H, Ohtsuki T, Hess J, Carillo F, Walther D et al (2004) Human cannabinoid receptor 1:5′ exons, candidate regulatory regions, polymorphisms, haplotypes and association with polysubstance abuse. Mol Psychiatry 9:916–931

Zhou D, Li Y, Tian T, Quan W, Wang L, Shao Q et al (2017) Role of the endocannabinoid system in the formation and development of depression. Pharmazie 72:435–439

Zou S, Kumar U (2019) Cannabinoid receptors and endocannabinoid system: signaling and function in the central nervous system. Int J Mol Sci 19:833. https://doi.org/10.3390/ijms19030833

Zou M, Li D, Wu L, Sun C (2019) Role of the endocannabinoid system in neurological disorders. Int J Dev Neurosci 76:95–102

Zumbrun EE, Sido JM, Nagarkatti PS, Nagagarkatti PS (2015) Epigenetic regulation of immunological alterations following prenatal exposure to marijuana cannabinoids and its long-term consequences in offspring. J Neuroimmune Pharmacol 10:245–254

The manufacturer's authorised representative in the EU is Springer
Nature Customer Service Centre GmbH, Europaplatz 3, 69115 Heidelberg,
Germany. If you have any concerns regarding our products, please
contact ProductSafety@springernature.com

Printed and bound by CPI Group (UK) Ltd, Croydon, CR0 4YY
28/09/2026
02224073-0016